# TRENDS IN LEVELS AND EFFECTS OF PERSISTENT TOXIC SUBSTANCES IN THE GREAT LAKES

# TRENDS IN LEVELS AND EFFECTS OF PERSISTENT TOXIC SUBSTANCES IN THE GREAT LAKES

Articles from the Workshop on Environmental Results, hosted in Windsor, Ontario, by the Great Lakes Science Advisory Board of the International Joint Commission, September 12 and 13, 1996

*Edited by*


MICHAEL GILBERTSON
*International Joint Commission,*
*Windsor, Ontario, Canada N9A 6T3*

GLEN A. FOX
*Canadian Wildlife Service, Environment Canada,*
*Ottawa, Ontario, Canada K1A 0H3*

WILLIAM W. BOWERMAN
*Gale Gleason Environmental Institute,*
*Lake Superior State University,*
*Sault Sainte Marie, Michigan 49783, USA*



Reprinted from Environmental Monitoring Assessment,
Volume 53, No. 1, 1998

International Joint Commission
Commission mixte internationale

A C.I.P. Catalogue record for this book is available from the Library of Congress.

ISBN 978-94-010-6227-5 ISBN 978-94-011-5290-7 (eBook)
DOI 10.1007/978-94-011-5290-7

*Printed on acid-free paper*

## TABLE OF CONTENTS

**TRENDS IN LEVELS AND EFFECTS OF PERSISTENT TOXIX SUBSTANCES IN THE GREAT LAKES**

## PREFACE

The International Joint Commission was set up under the Boundary Waters Treaty of 1909. The purpose of the Treaty was to prevent disputes regarding the use of boundary waters between the United States and the Dominion of Canada. In 1972, after the International Joint Commission had reported on the pollution of the Lower Great Lakes, the United States and Canadian governments negotiated a Great Lakes Water Quality Agreement with provisions for the International Joint Commission to assist in the implementation of the Agreement. In 1978, the Great Lakes Water Quality Agreement was renegotiated and the Parties to the Agreement included a new policy that *'the discharge of any or all persistent toxic substances be virtually eliminated'*. In 1995, the International Joint Commission asked the Great Lakes Science Advisory Board to undertake an evaluation of progress in reducing or eliminating the effects of persistent toxic substances in wildlife and humans. In September 1996, the Board held a Workshop on Environmental Results, in Windsor, Ontario, and assembled a group of scientists who presented data on the monitoring of trends in the concentrations of persistent toxic substances in biotic and abiotic samples from the Great Lakes basin. Other scientists presented evidence of the incidence and trends in effects of persistent toxic substances on populations of organisms, including humans.

The research and monitoring on persistent toxic substances in the Great Lakes Basin is among the most extensive that has been undertaken anywhere in the world. This volume of the *Environmental Monitoring and Assessment Journal* contains several of the papers that were presented at the Workshop on Environmental Results. The purpose of publishing these papers in the peer-reviewed literature is not only to bring this monitoring information to the attention of scientists and administrators in other parts of the world, but also to draw attention to the continuing need, in these days of budgetary constraints, to focus attention on a core group of indicators by which the Parties can reliably demonstrate whether their policy concerning the virtual elimination of discharges of persistent toxic substances is being achieved.

We should like to thank those who participated in the International Joint Commission's Workshop on Environmental Results and those who have contributed papers to this publication.

Michael Gilbertson
Glen Fox
William Bowerman
March 1998

*Environmental Monitoring and Assessment* **53:** 1, 1998.

# TRENDS IN CONCENTRATIONS AND EFFECTS OF PERSISTENT TOXIC CONTAMINANTS IN THE GREAT LAKES: THEIR SIGNIFICANCE FOR INFERRING CAUSE-EFFECT RELATIONSHIPS AND VALIDATING MANAGEMENT ACTIONS

IAN C.T. NISBET[1]

[1]*I.C.T. Nisbet & Company, 150 Alder Lane, North Falmouth, MA 02556, USA.*

**Abstract.** This paper summarizes a workshop on temporal trends in levels and effects of persistent toxic contaminants in the North American Great Lakes. Information on trends in contaminant levels is reasonably good for sediments, fish, and birds, but is scanty or absent for other ecosystem components. Information on trends in effects has been reported for birds, but is scanty or absent for other groups of organisms. In principle, information on differential trends in effects of contaminants could be used to validate or improve hypotheses about cause-effect relationships and to verify the effectiveness of management actions. However, little or no useful information on differential trends appears to be available. Use of trend data for these purposes will require collection of more detailed information and greater attention to conceptual formulation of hypotheses.

## 1. Introduction

Nineteen papers were presented at this workshop, of which ten are published in this issue of ***Environmental Monitoring and Assessment***. Although some of the papers were limited to documenting levels or effects of persistent toxic contaminants in the Great Lakes, the main topic of the workshop was trends in these levels and effects. This overview addresses the following questions: (1) Can information on these trends be useful in clarifying cause-effect relationships and the importance of specific agents and sources? (2) Can information on these trends be useful in validating or improving management actions? (3) Is the information now available on trends adequate to serve these functions?

## 2. Conceptual Framework

Figure I presents a schematic outline of the process by which information on the levels and effects of toxic contaminants is acquired, assessed, and used as the basis for remedial actions. Toxic contaminant problems are usually identified first by detection of contaminants in the environment (1), effects on humans or wildlife (2), or both. Measurement of levels and distribution of contaminants in the environment (1) leads to exposure assessment for the affected populations (3), while study of effects promotes laboratory studies to elucidate mechanisms (4). Epidemiological studies (5) relating exposure (3) to effects (2) lead to inferences of cause-effect

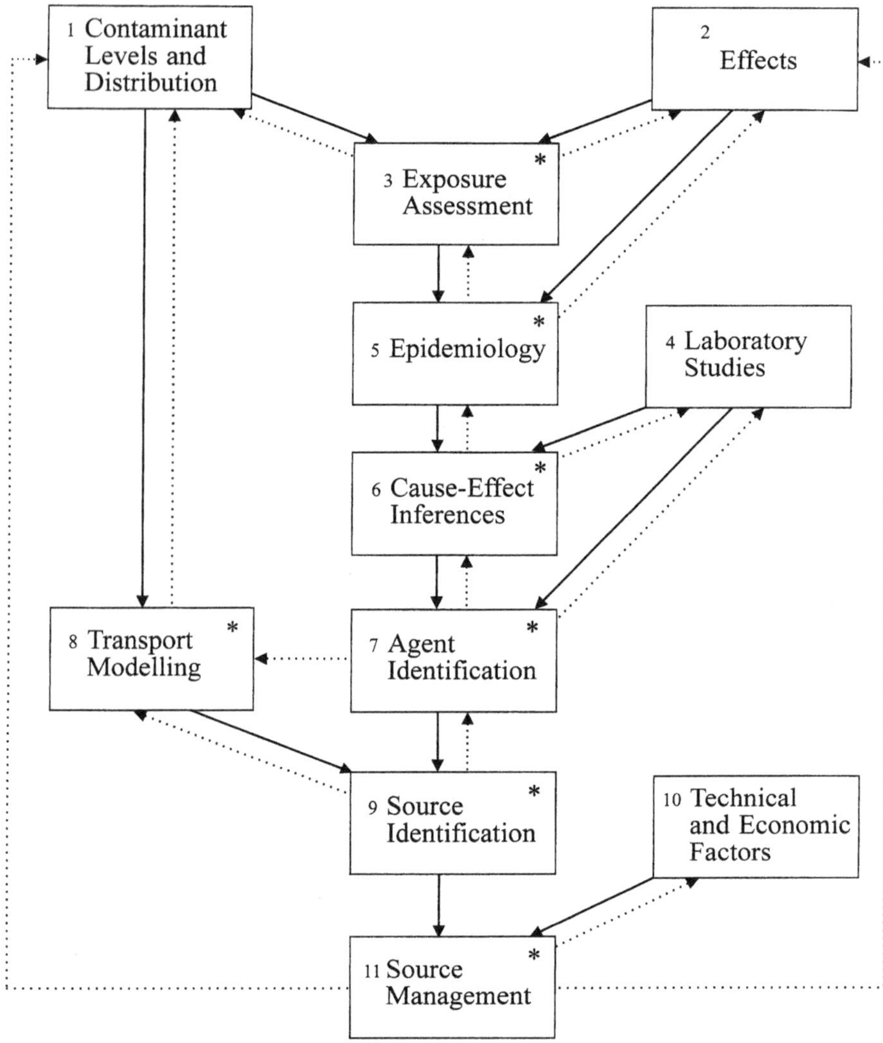


**Figure I.** Schematic Outline of the Process of Assessing and Managing an Environmental Contamination Problem

Boxes linked by solid arrows show sequential steps in the process from identification of a problem (boxes 1-2), through assessment (boxes 3-10) to management (box 11). Dotted arrows show feedback steps leading to iteration and progressive improvement of the assessments. The outer two arrows show the effect of source management in reducing levels and effects of contamination. Asterisks indicate the steps at which information on trends can be used to improve the assessments and hence improve the effectiveness of management.

relationships (6), which are verified or refuted by laboratory studies (4). This sequence of studies and inferences has been discussed in detail in previous workshops on effects of contaminants in the Great Lakes (Gilbertson and Scheider 1991, 1993, Gilbertson 1996).

Steps (7) and (9) in Figure I are less familiar and are often taken for granted, although they are important steps on the path to effective remedial action. Agent identification (7) is the process by which an initial inference of a cause-effect relationship between an adverse effect and environmental contamination (6) is made specific to an individual agent and an identified mechanism of action. This requires differential epidemiological studies (5) or differential laboratory studies (4) to distinguish the effects of different contaminants which may occur together in the environment. One of the best known examples is the identification of DDE as the exclusive cause of eggshell-thinning in wild birds through a combination of differential laboratory and field studies (Risebrough 1986). In the Great Lakes, agent identification still poses important unresolved questions, because many of the contaminants are highly correlated with each other (Environment Canada 1991), and rigorous differential analyses have rarely been undertaken. For example, the relative importance of DDE, PCBs, and other contaminants in causing reproductive impairment in fish-eating birds continues to be investigated and debated (Nisbet 1989, Ludwig *et al.* 1995, Bowerman *et al.* 1998, Custer *et al.* 1998). Source identification (9) is the process by which contamination responsible for specific adverse effects is traced back to specific sources. This usually requires transport modelling (8) to relate levels of contamination to rates of input, output, degradation, and transfer between media. Transport models must be time-dependent to separate the effects of past and present sources. For example, the current distribution and levels of persistent toxic contaminants in the Great Lakes reflect both past and present uses and discharges, point and non-point sources, releases from localized "hot-spots" and from area-wide reservoirs of contamination, and continuing atmospheric inputs from distant sources (International Joint Commission 1993, Mackay *et al.* 1994, Mackay and Di Guardo 1995, Hoff *et al.* 1996).

Finally, source management (11) includes all actions that change rates of discharge or release from the identified sources. These include not only regulatory actions to restrict uses and discharges, but also remedial actions to clean up "hot spots", and in some cases non-regulatory changes in uses or discharges. For example, changes in industrial practices and combustion of wastes appear to have reduced rates of release of polychlorinated dibenzo-*p*-dioxins (PCDDs) into the environment before they were even recognized as widespread contaminants (Alcock and Jones 1996). Effective source management requires accurate identification of the importance of different sources (9), as well as consideration of technical feasibility and socio-economic costs and benefits (10).

Figure I shows many feedback loops through which successive steps in the process are modified and iterated. In particular, the steps of cause-effect inference (6), agent identification (7), transport modelling (8) and source identification (9) are usually repeated and refined many times. Each iteration incorporates new information from steps (1) to (5), and then suggests new studies that are required to support the next iteration. The ultimate feedback loops stem from source management, which, if effective, reduces both the levels and effects of the contaminants. The main topic of this workshop was the documentation and interpretation of these reductions.

## 3. Trends in Levels and Effects of Contaminants in the Great Lakes

Table I presents the author's assessments of the adequacy of data on trends in levels and effects of contaminants in 8 components of the Great Lakes ecosystem. "Adequacy" in this sense includes both completeness of description of spatial and temporal trends, and utility for validating cause-effect relationships, agent identification, source identification, and management actions. The assessments in Table I are based on the papers in these proceedings, other papers presented at the workshop, and other recently published papers.

### 3.1 TRENDS IN LEVELS

**Sediments.** Because sediments in some locations preserve a record of contamination at the time of deposition and can be dated using $^{210}Pb$, the historical record of contamination in sediments of the Great Lakes is unusually good, dating back in some cases to 1900 and with a resolution of a few years (Czuczwa *et al.* 1986, Golden *et al.* 1993). Although not yet fully published, data are now available for all the 14 chemicals of concern listed by the International Joint Commission (D. Swackhammer, University of Minnesota, Minneapolis, Minnesota, oral presentation). The most noteworthy finding is that deposition rates for all these chemicals peaked between 1960 and 1980, in most cases before contaminant data became available for any other medium and in many cases before any specific regulatory actions were taken to limit discharges. The main limitation of these data is that they are restricted to a few sites in Lakes Superior, Michigan and Ontario.

**Water.** Although not yet published, analyses of water concentrations are available from the St. Clair and Niagara Rivers on a weekly basis since 1986. These show declines in concentrations of at least 8 chemicals (M. Neilson, Canada Centre for Inland Waters, Burlington, Ontario, oral presentation). No data are available from earlier than 1986, except for some measurements of total PCBs in open lake waters, but these were not well designed for trend analysis.

**Air.** Data are available for 20 chemicals at 5 stations (one on each lake) since 1990 (Hoff *et al.* 1996, R. Hoff, Atmospheric Environment Service, Egbert, Ontario, oral presentation). After controlling for large effects of temperature, downward trends in concentrations of several chemicals are indicated, but are not yet statistically significant. The data are further limited by problems of intercalibration, which make estimates of gas-to-water transfer rates uncertain (Hoff *et al.* 1996).

**Table I.** Adequacy of Data on Temporal Trends in Levels and Effects of Persistent Toxic Contaminants in the Great Lakes

| | Adequacy of Data on Trends | |
|---|---|---|
| Ecosystem Compartment | Contaminant Levels | Contaminant Effects |
| Sediments | 8 | NA |
| Water | 4 | NA |
| Air | 2 | NA |
| Fish | 6 | 1 |
| Birds | 8 | 6 |
| Reptiles/Amphibians | 1 | 0 |
| Mammals | 2 | 0 |
| Biomarkers | NA | 2 |

The adequacy of data is assessed subjectively on a scale of 0-10, based on criteria and data discussed in the text.
NA - not applicable.

**Fish.** Extensive data are available on contaminant levels in several species of fish from all the major lakes since 1977 (De Vault *et al.* 1996, Huestis *et al.* 1996, 1997, Scheider *et al.* 1998). After controlling for age and fat content, these data show declines in concentrations of DDE, PCBs, PCDDs and several other organochlorines, but less consistent declines in metals. Patterns of decline vary considerably among locations (Scheider *et al.* 1998). Stow *et al.* (1995) analyzed data on PCB levels in 7 species of fish from Lake Michigan from 1974 to 1992. They reported that the data for 5 species fitted best to a non-zero asymptote model (i.e., an approach to a steady state), while data for the other two species suggested a recent slight increase. However, this analysis was based on estimates of total PCBs, and it is not clear that analytical methods were standardized across laboratories or throughout the 19-year period.

## 3.2 TRENDS IN EFFECTS

**Fish.** Effects of organochlorine contaminants on fish are disputed (Mac and Edsall 1991, Gilbertson 1996, 1997, Carpenter *et al.* 1996) and there is little information on trends in these effects, although improved reproduction in lake trout *(Salvelinus namaycush)* has been tentatively attributed to declining contaminant levels (Gilbertson 1996). In one river in Ohio, liver neoplasms in brown bullheads *(Ameiurus nebulosus)* fluctuated in association with levels of contamination with polynuclear aromatic hydrocarbons (Baumann and Harshbarger 1998).

**Birds.** The most systematic monitoring programme on the Great Lakes is that based on herring gull *(Larus argentatus)* eggs, which have been collected from 13 sites on all five lakes since 1974 (Pekarik and Weseloh 1998). Archiving of samples has permitted standardization of analytical methods throughout the period for most chemicals (Turle *et al.* 1991) as well as retrospective studies of newly-discovered contaminants (Hebert *et al.* 1994). Analysis of data through 1995 showed changes in the pattern of decline for most chemicals. Many of these were upward or downward steps in mean levels; 20% of comparisons showed reductions in the rate of decline, while 13% of comparisons showed increases in the rate of decline. Some of the changes may be attributable to local or secular changes in the diet of the gulls, i.e., changes in the relationship between levels in the environment and those in the gull eggs (Pekarik and Weseloh 1998). In an earlier study, Stow (1995) concluded that data on total PCB levels in gull eggs between 1978 and 1992 indicated an approach to a steady state in four of the five lakes, but this finding is not inconsistent with the results of the more detailed analyses by Pekarik and Weseloh (1998) for total PCBs. Data are also available on trends in contamination levels in several other bird species (Environment Canada 1991, Ewins *et al.* 1994, Pettit *et al.* 1994, Weseloh *et al.* 1995, Ryckman *et al.* 1998, Bowerman *et al.* 1998).

Several published papers have reported improvements in reproductive success of fish-eating birds in the Great Lakes in association with declining contaminant levels (Fox 1991, Ewins 1992, Ewins *et al.* 1994, Harris *et al.* 1993, Weseloh *et al.* 1995), but only a brief summary of these trends was presented at the workshop (Grasman *et al.* 1998). Two other papers in the workshop reported no significant trends in effects (Ryckman *et al.* 1998, Bowerman *et al.* 1998).

**Reptiles and Amphibians.** Reptiles and amphibians in the Great Lakes are thought to be suffering adverse effects from persistent toxic contaminants, but few data are available on trends in contaminant levels and none on trends in effects (Bishop and Gendron 1998).

**Mammals.** The only longitudinal study of contaminant levels in humans consuming fish from the Great Lakes is not yet published, but showed a small decline in total PCB and total DDT levels between 1979 and 1995 (H. Humphrey, oral presentation). Although effects of PCB exposure are now recognized (Johnson *et al.* in press), there is no information on trends in effects. Little or no information is available on trends in exposure or effects in other fish-eating mammals.

**Biomarkers.** Extensive data have been published on trends in eggshell thickness in fish-eating birds of the Great Lakes (Ewins 1992, Ewins *et al.* 1994, Harris *et al.* 1993, Wiemeyer *et al.* 1993, Weseloh *et al.* 1995), but no recent compilation or synthesis is available. The only other biomarker for which trend data were reported was enlargement of the thyroid in herring gulls (Grasman *et al.* 1998).

## 4. Use of Trend Data to Validate and Refine Cause-Effect Relationships, Agent Identification and Source Identification

Information on trends in contamination and effects has been used in support of the general hypothesis of a cause-effect relationship between adverse effects on reproduction in fish-eating birds in the Great Lakes and exposure to persistent toxic contaminants. Specifically, most of the adverse effects appeared after the introduction of these chemicals into the environment, and ameliorated or disappeared after declines in exposure levels (Fox 1991, Gilbertson *et al.* 1991, Colborn 1991, Weseloh *et al.* 1995, Ludwig *et al.* 1996, Bowerman *et al.* 1998). At a minimum, these data are consistent with the temporal sequencing required by Koch's postulates (Fox 1991). In some cases, they extend to detailed spatio-temporal correlations between trends in exposure and effects (Fox 1991, Harris *et al.* 1993, Weseloh *et al.* 1995). This general pattern of association between increasing and decreasing trends in exposure to persistent toxic contaminants and effects has been pointed out also for fish reproduction (Gilbertson 1997), but the cause-effect relationships in this case are disputed (Carpenter *et al.* 1996).

In principle, information on trends in contaminant levels and effects could also be used at several other steps in the process outlined in Figure I. Specifically, information on *differential* trends in levels of different contaminants could be used in agent identification and in source identification. One example of the former from other areas is the evidence that the decline and subsequent recovery of the European sparrowhawk *(Accipiter nisus)* in Great Britain coincided with the increase and subsequent decrease of residues of dieldrin, but that DDE concentrations remained more or less constant through both periods (Newton and Haas 1984). This suggested that DDE was not the critical factor in either the population decline or recovery of that species. Another example is the evidence that recovery in reproductive success and numbers of the osprey *(Pandion haliaetus)* in the eastern USA coincided with declines in residue levels of DDE, but that concentrations of PCBs and mercury remained more or less constant during this period(Spitzer *et al.* 1978). This suggested that PCBs and mercury were not critical factors in the prior reproductive failures and population crash of that species. In these and similar cases, information on differential trends has provided strong evidence *excluding* specific contaminants as causative agents. If only one agent remains correlated with the adverse effects in space and time (e.g. dieldrin for the sparrowhawk and DDE for the osprey), this constitutes strong evidence that that one was the primary causative agent. However, in a rigorous sense, such evidence remains correlative and additional epidemiological criteria have to be applied to support a strong inference of a cause-effect relationship (Fox 1991). Thus, use of information on differential trends can be useful in shortening the list of suspected causative agents, but additional information is usually needed to reduce the list to one.

An example of the use of information on differential trends in source identification is the fact that DDE levels in ospreys and bald eagles *(Haliaeetus leucocephalus)* in parts of North America were already declining by 1972 and reproduction was improving by 1975 (Spitzer *et al.* 1978, Grier 1982, Bowerman *et al.* 1998). Hence, the sources of DDE that were primarily responsible for the earlier reproductive failures in these species must have been the dispersive uses that were reduced or eliminated prior to 1972, not the remaining agricultural uses that were banned in that year. Again, information on differential trends was useful in reducing the list of suspected sources, but not, by itself, in pinpointing one specific source.

In the Great Lakes ecosystem, however, these approaches will be very difficult because levels of many contaminants increased and decreased at approximately the same time (Environment Canada 1991, Golden *et al.* 1993, Huestis *et al.* 1996, 1997, Pekarik and Weseloh 1998). No study reported in this workshop used *differential* trends in levels of different contaminants either in agent identification or source identification. To do so will require much more detailed analysis of trends in levels and effects of contaminants, including a search for cases where trends differ markedly among different contaminants. It will also require rigorous formulation of hypotheses about differential exposure and effects.

## 5. Use of Trend Data to Validate and Refine Management Actions

In principle, information on trends in contaminant levels and effects could also be used to validate and refine (or invalidate) management actions. For example, Stow (1995) and Stow *et al.* (1996) have argued that data on trends of contaminant levels in fish and birds indicate that PCBs in the Great Lakes are approaching steady-state levels, perhaps maintained by diffuse releases from residue pools retained in sediments. If so, regulatory action to control local sources and discharges are likely to be ineffective in reducing ecosystem-wide levels and effects of contamination (although efforts to remediate local "hot-spots" may be justified to reduce local risks). Similarly, Cooper (1995) argued that effects of contaminants on birds in the Great Lakes have already been ameliorated and that further attempts to reduce lake-wide contaminant levels would not be cost-effective.

Except for some of the data presented by Pekarik and Weseloh (1998) and Bowerman *et al.* (1998), none of the data presented in this workshop had sufficient spatial or temporal resolution to address this issue. Even the data presented by Stow (1995), Stow *et al.* (1996) and Cooper (1995) were lake-wide averages and were inadequate to support conclusions about the relative importance of local and area-wide sources in maintaining lake-wide average levels of contaminants. If trend data are to be useful in guiding future management actions, they will need to differentiate among areas within lakes that are influenced by different sources.

## 6. Summary and Conclusions

1. Information on temporal trends in levels of persistent toxic contaminants in the Great Lakes ecosystem is reasonably good for sediments, fish and birds, but scanty or absent for other compartments (Table I).
2. Information on temporal trends in effects has been reported for birds, but has not been compiled or assessed recently and was not well reported in this workshop. Little or no information is available for other groups of organisms (Table I).
3. In principle, information on differential trends in levels and effects of contaminants could be used to validate or improve hypotheses about sources, agents, and cause-effect relationships. In practice, this will be difficult for the Great Lakes because temporal trends in levels and effects have been closely parallel among different contaminants. Little or no useful information on differential trends was presented in this workshop.
4. Use of trend data for these purposes will require careful attention to comparability of measurements, identification of cases where trends differ significantly among different contaminants, and greater attention to conceptual formulation of hypotheses (Figure I).

## References

Alcock, R.E., and Jones, K.C.: 1996. 'Dioxins in the Environment: A Review of Trend Data', *Environ. Sci. Technol.* **30**,3133-3143.

Baumann, P.C., and Harshbarger, J.C.: 1998. 'Long Term Trends in Liver Neoplasm Epizootics of Brown Bullhead in the Black River, Ohio', *Environ. Monitor. Assess.* this issue.

Bishop, C.A., and Gendron, A.D.: 1998. 'Reptiles and Amphibians: The Shy but Sensitive Vertebrates of the Great Lakes Basin and St. Lawrence River', *Environ. Monitor. Assess.* this issue.

Bowerman, W.W., Best, D.A., Grubb, T.G., Zimmerman, G.M., and Giesy, J.P.: 1998. 'Trends of Contaminants and Effects for Bald Eagles of the Great Lakes', *Environ. Monitor. Assess.* this issue.

Carpenter, S.R., Jackson, I.J., Kitchell, J.F., and Stow, C.A.: 1996. 'Organochlorine Contaminants in the Great Lakes', *Ecol. Appl.* **6**,971-974.

Colborn, T.: 1991. 'Epidemiology of Great Lakes Bald Eagles', *J. Toxicol. Environ. Health* **33**,395-454.

Cooper, W.E.: 1995. 'Risks of Organochlorine Contaminants to Great Lakes Ecosystems are Overstated', *Ecol. Appl.* **5**,293-298.

Custer, T.W., Custer, C.M., Melancon, M.J., and Stromborg, K.L.: 1998. 'Effects of Contaminants on Double-Crested Cormorant Reproduction in Green Bay, Wisconsin',

Czuczwa, J.M., McVeety, B.D., and Hites, R.A.: 1984. 'Polychlorinated Dibenzo-*p*-Dioxins and Dibenzofurans in Sediments from Siskiwit Lake, Isle Royale', *Science* **226**,568-569.

De Vault, D.S., Hesselberg, R., Rodgers, P.W., and Feist, T.J.: 1996. 'Contaminant Trends in Lake Trout and Walleye from the Laurentian Great Lakes', *J. Great Lakes Res.* **22**,884-895.

Environment Canada: 1991. 'Toxic Chemicals in the Great Lakes and Associated Effects, Vols. I-III. Environment Canada, Toronto.

Ewins, P.J.: 1992. 'Ospreys and Contaminants on the Canadian Great Lakes, 1991', Interim Report to the Canadian Wildlife Service (Ontario Region) and the Ontario Ministry of Natural Resources.

Ewins, P., Weseloh, D.V., Norstrom, R.J., Legierse, K., Auman, H.J., and Ludwig, J.P.: 1994. 'Caspian Terns on the Great Lakes: Organochlorine Contamination, Reproduction, Diet, and Population Changes 1972-91', Can. Wildl. Serv. Occas. Paper 85,1-34.

Fox, G.A.: 1991. 'Practical Causal Inference for Ecoepidemiologists', *J. Toxicol. Environ. Health* **33**,359-373.

Fox, G.A., Trudeau, S., Won, H., and Grasman, K.A.: 1998. 'Monitoring the Elimination of Persistent Toxic Substances from the Great Lakes: Chemical and Physiological Evidence from Adult Herring Gulls', *Environ. Monitor. Assess.* this issue.

Gilbertson, M.: (Ed.). 1996. 'Special Section on Cause-Effect Linkages', *J. Great Lakes Res.* **22**,128-212.

Gilbertson, M.: 1997. 'Are Causes Knowable? Some Consequences of Successional Versus Toxicological Interpretations of the Great Lakes Water Quality Agreement', *Can. J. Fish. Aquat. Sci.* **54**,483-495.

Gilbertson, M., and Schneider, R.S. (Eds.): 1991. 'Special Issue: International Joint Commission Workshop on Cause-Effect Linkages', *J. Toxicol. Environ. Health* **33**,359-639.

Gilbertson, M., and Schneider, R.S. (Eds.): 1993. 'Special Section on Cause-Effect Linkages', *J. Great Lakes Res.* **19**,720-807.

Gilbertson, M., Kubiak, T.J., Ludwig, J.P., and Fox, G.A.: 1991. 'Great Lakes Embryo Mortality, Edema, and Deformities Syndrome (GLEMEDS) in Colonial Fish-Eating Birds: Similarity to Chick Edema Disease', *J. Toxicol. Environ. Health* **33**,455-520.

Golden, K.A., Wong, C.S., Jeremiason, J.D., Eisenreich, S.J., Sanders, G., Swackhammer, D.L., Engstrom, D.R., and Long, D.T.: 1993. 'Accumulation and Preliminary Inventory of Organochlorines in Great Lakes Sediments', Preprint from First International Specialized Conference on Contaminated Aquatic Sediments: Historical Records, Environmental Impact, and Remediation:13-22. International Association on Water Quality, Milwaukee.

Grasman, K.A., Scanlon, P.F., and Fox, G.A.: 1998. 'Reproductive and Physiological Effects of Environmental Contaminants in Fish-Eating Birds of the Great Lakes: A Review of Historical Trends', *Environ. Monitor. Assess.* this issue.

Grier, J.W.: 1982. 'Ban of DDT and Subsequent Recovery of Reproduction in Bald Eagles', *Science* **218**,1232-1235.

Harris, H.J., Erdman, T.C., Ankley, G.T., and Lodge, K.B.: 1993. 'Measures of Reproductive Success and Polychlorinated Biphenyl Residues in Eggs and Chicks of Forster's Terns on Green Bay, Lake Michigan, Wisconsin -- 1988', *Arch. Environ. Contam. Toxicol.* **25**,304-314.

Hebert, C.E., Weseloh, D.V., Kot, L., and Glooschenko, V.: 1994. 'Temporal Trends and Sources of PCDDs and PCDFs in the Great Lakes: Herring Gull Egg Monitoring, 1981-1991', *Environ. Sci. Technol.* **28**,1268-1277.

Hoff, R.M., Strachan, W.M.J., Sweet, C.W., Chan, C.H., Shackleton, J.M., Bidleman, T.F., Brice, K.A., Burnston, D.A., Cussion, S., Gatz, D.F., Harlin, K., and Schroeder, W.H.: 1996. 'Atmospheric Deposition of Toxic Chemicals to the Great Lakes: a Review of Data through 1994', *Atmos. Environ.* **30**,3505-3527.

Huestis, S.Y., Servos, M.R., Whittle, D.M., and Dixon, D.G.: 1996. 'Temporal and Age-Related Trends in Levels of Polychlorinated Biphenyl Congeners and Organochlorine Contaminants in Lake Ontario Lake Trout *(Salvelinus namaycush)*', *J. Great Lakes Res.* **22**,310-330.

Huestis, S.Y., Servos, M.R., Whittle, D.M., Van Den Heuvel, M., and Dixon, D.G.: 1997. 'Evaluation of Temporal and Age-Related Trends of Chemically and Biologically Generated 2,3,7,8-Tetrachloro-*p*-Dioxin Equivalents in Lake Ontario Lake Trout, 1977 to 1993', *Environ. Toxicol. Chem.* **16**,154-164.

International Joint Commission: 1993. 'Source Investigation for Lake Superior', International Joint Commission, Windsor, Ontario.

Johnson, B.L., Hicks, H.E., Jones, D.E., Cibulas, W., and De Rosa, C.T.: in press. 'Public Health Implications of Persistent Toxic Substances in the Great Lakes and St. Lawrence Basins'. *J. Great Lakes Res.*

Ludwig, J.P., Kurita-Matsuba, H., Auman, H.J., Ludwig, M.E., Summer, C.L., Giesy, J.P., Tillitt, D.E., and Jones, P.D.: 1996. 'Deformities, PCBs and TCDD Equivalents in Double-Crested Cormorants *(Phalacrocorax auritus)* and Caspian Terns *(Hydroprogne caspia)* of the Upper Great Lakes 1985-1991:Testing a Cause-Effect Hypothesis', *J. Great Lakes Res.* **22**,172-197.

Mac, M.J., and Edsall, C.C.: 1991. 'Environmental Contaminants and the Reproductive Success of Lake Trout in the Great Lakes: An Epidemiological Approach', *J. Toxicol. Environ. Health* **33**,375-394.

Mackay, D., and Di Guardo, A.: 1995. 'Organochlorines in the Great Lakes Ecosystem: Sources, Partitioning, and Control', *Ecol.Appl.* **5**,301-304.

Mackay, D., Sand, S., Vlahos, P., Diamond, M., Gobas, F., and Dolan, D.: 1994. 'A Rate Constant Model of Chemical Dynamics in a Lake Ecosystem: PCBs in Lake Ontario', *J. Great Lakes Res.* **20**,625-642.

Newton, I., and Haas, M.B.: 1984. 'The Return of the Sparrowhawk', *Brit. Birds* **77**,47-70.

Nisbet, I.C.T.: 1989. 'Organochlorines, Reproductive Impairment, and Declines in Bald Eagle *Haliaetus leucocephalus* Populations: Mechanisms and Dose-Response Relationships', In Meyburg, B.-U., and Chancellor, R.D. (Eds.), Raptors in the Modern World, 483-489. World Working Group on Birds of Prey, Berlin.

Pekarik, C., and Weseloh, D.V.: 1998. 'Organochlorine Contaminants in Herring Gull Eggs from the Great Lakes, 1974-1995: Change Point Analysis and Short Term Regression', *Environ. Monitor. Assess.* this issue.

Pettit, K.E., Bishop, C.A., Weseloh, D.V., and Norstrom, R.J.: 1994. 'An Atlas of Contaminants in Eggs of Fish-Eating Colonial Birds of the Great Lakes (1989-1992), Vol.I',Can. Wildl. Serv. Tech. Rep. Ser. 193. Canadian Wildlife Service, Toronto.

Risebrough, R.W.: 1986. 'Pesticides and Bird Populations', *Curr. Ornithol.* **3**,397-427.

Ryckman, D.P., Weseloh, D.V., Hamr, P., Fox, G.A., Collins, B., Ewins, P.J., and Norstrom, R.J.: 1998. 'Spatial and Temporal Trends in Organochlorine Contamination and Bill Deformities in Double-Crested Cormorants *(Phalacrocorax auritus)* from the Great Lakes', *Environ. Monitor. Assess.* this issue.

Scheider, W.A., Cox, C., Hayton, A., Hitchin, G., and Vaillancourt, A.: 1998. 'Current Status and Temporal Trends in Concentrations of Persistent Toxic Substances in Sport Fish and Forage Fish in the Canadian Great Lakes', *Environ. Monitor. Assess.* this issue.

Spitzer, P.R., Risebrough, R.W., Walker, W., II, Hernandez, R., Poole, A., Puleston, D., and Nisbet, I.C.T.: 1978. 'Productivity of Ospreys in Connecticut-Long Island Increases as DDE Residues Decline', *Science* **202**,333-335.

Stow, C.A.: 1995. 'Great Lakes Herring Gull Egg PCB Concentrations Indicate Approximately Steady-State Conditions', *Environ. Sci. Technol.*, **29**,2893-2897.

Stow, C.A., Carpenter, S.R., Eby, L.A., Amrhein, J.F., and Hesselberg, R.J.: 1995. 'Evidence that PCBs are Approaching Stable Concentrations in Great Lakes Fishes', *Ecol. Appl.* **5**,248-260.

Turle, R., Norstrom, R.J., and Collins, B.: 1991. 'Comparison of PCB Quantitation Methods: Re-analysis of Archived Specimens of Herring Gulls from the Great Lakes', *Chemosphere* **22**,201-213.

Weseloh, D.V., Ewins, P.J., Ewins, J., Mineau, P., Bishop, C.A., Postupalsky, S., and Ludwig, J.P.: 1995. 'Double-Crested Cormorants of the Great Lakes: Changes in Population Size, Breeding Distribution and Reproductive Output, 1913-1991', In Nettleship, D.N., and Duffy, D.C. (Eds.), The Double-Crested Cormorant: Biology, Conservation, and Management. *Colon. Waterbirds* **18**(Spec. Publ. 1),48-59.

Wiemeyer, S.N., Bunck, C.M., and Stafford, C.J.: 1993. 'Environmental Contaminants in Bald Eagles -- 1980-84 -- and Further Interpretations of Relationships to Productivity and Shell Thickness', *Arch. Environ. Contam. Toxicol.* **24**,213-277.

# DESIGNING THE ENVIRONMENTAL RESULTS WORKSHOP: HISTORICAL CONTEXT, CAUSALITY AND CANDIDATE SPECIES

MICHAEL GILBERTSON[1], GLEN FOX[2], WILLIAM BOWERMAN[3]
[1]*International Joint Commission, 100 Ouellette Avenue, Windsor, Ontario, N9A 6T3, Canada,* [2]*Canadian Wildlife Service, National Wildlife Research Centre, Environment Canada, Ottawa, Ontario, K1A 0H3, Canada,* [3]*Gale Gleason Environmental Institute, School of Science and Natural Resources, Lake Superior State University, Sault Ste Marie, Michigan, 49783, USA.*

**Abstract.** During the past thirty years, researchers have been documenting the concentrations and effects of persistent toxic substances, such as DDT, PCBs, dioxins and furans, in populations of Great Lakes organisms. In designing the International Joint Commission's Workshop on Environmental Results, the organizers on the Great Lakes Science Advisory Board started from the premise that the selection of indicator organisms for long-term trend analysis requires that the causal relationships between the observed effects and the putative cause should be demonstrated and believed. However, the causal relationships that have been documented have generally been met with skepticism by fellow scientists and by regulatory officials resulting in no agreement on valid indicator organisms. To overcome this skepticism, wildlife, fisheries and human health researchers have adapted the epidemiological criteria used by medical and veterinary researchers for synthesizing the causal evidence. Brief reviews of several candidate species and of their suitability as long term monitors are presented, and the adequacy of monitoring programs for determining trends in the incidence of chemically-induced effects is assessed.

## 1. Introduction

The September 1996 Workshop on Environmental Results was hosted by the Great Lakes Science Advisory Board, and represented part of the long tradition of involvement of the International Joint Commission in reporting on monitoring of water quality in the Great Lakes Basin. Since the signing of the 1972 Great Lakes Water Quality Agreement between the United States and Canada (the Parties), the International Joint Commission has been involved in collating a variety of different kinds of evidence on various pollutants including persistent toxic substances and in advising the Parties of the related trends and of the adequacy of their regulatory programs. Much of this information was concerned with the concentrations of these pollutants in effluents, environmental media and in tissues of fish and wildlife, and occasionally of humans. For example, since 1975, the Great Lakes Water Quality Board published (International Joint Commission 1975) the first of several annual compilations of surveillance data collected by the Parties and issued as Appendix B to the Board's reports. The most recent of these surveillance reports was issued in

1989 (International Joint Commission 1989a) at which time the Parties took over the responsibility for collating and reporting surveillance information under the 1987 Protocol to amend the 1978 Great Lakes Water Quality Agreement. These reports show the declines that occurred in the late 1970s in the quantities of persistent toxic substances released into the Great Lakes and the concentrations in environmental samples and in organisms collected from the Great Lakes Basin. More recent reports have shown that since about 1980 there has been almost no further progress. It seemed that further progress would only be achieved from large expenditures for remedial actions and that these expenditures would only be forthcoming if there was evidence of actual injury to health and property from existing levels of persistent toxic substances in the Great Lakes.

## 2. Evidence of Injury to Health and Property

At the same time that this chemical analytical data was being compiled, the International Joint Commission was involved in efforts to integrate the science relating to the effects of persistent toxic substances on Great Lakes organisms exposed to these compounds. The acceptance of the evidence concerning effects has been much slower than that for the chemical analytical data and there has been a prevailing skepticism about the causal relationships and even recently a belief that the causes were unknowable (Gilbertson 1997). Despite this skepticism, the specific chemical(s) that have caused the reproductive failure of colonies of Lake Ontario herring gulls *(Larus argentatus)* and Green Bay Forster's terns *(Sterna forsteri)* have been identified, respectively, as 2,3,7,8-tetrachlorodibenzo-*p*-dioxin and hexachlorobenzene, and two pentachloro- biphenyls (Gilbertson 1983, Boersma *et al.* 1986, Kubiak *et al.* 1989, Gilbertson *et al.* 1991).

The International Joint Commission has had an interest in the human health aspects of pollutants for a long time. In relation to the Great Lakes Water Quality Agreement, the Commission established a Committee on the Assessment of Human Health Effects of Great Lakes Water Quality in early 1978, that reported to both the Great Lakes Water Quality Board and to the Great Lakes Science Advisory Board (International Joint Commission 1979). The Commission was responding to the growing concern about the number of chemicals that were being reported, and urged the Parties to collaborate and develop a program which would establish a running inventory of toxic chemicals used, manufactured and imported into the Great Lakes Basin and evaluate their risk to human health and the environment. The Committee therefore initially took a toxicological rather than epidemiological approach and, through the 1980s, developed extensive documentation of the toxicological profiles including exposure assessments for the compounds that had been detected, and commented on the adequacy of water quality criteria to protect human health.

In 1983, the Great Lakes Science Advisory Board established the Health of Aquatic Communities Task Force to integrate information on the effects of toxic substances on the Great Lakes aquatic community, to explore improved methods for assessing effects of toxic substances on the biota, and to develop recommendations for further research. The Task Force undertook a literature review of the effects of persistent toxic substances on Great Lakes biota (Fitchko 1986).

In June 1985, the Task Force held a symposium and workshop as part of the annual meeting of the American Society of Limnology and Oceanography and published the proceedings (Evans 1988). The editor noted that despite the multitude of experimental studies to investigate the effects of toxic substances on various components of the aquatic ecosystem, including phytoplankton, benthos, fish, birds and mammals including humans, researchers had a poor understanding of how toxic substances had affected these components. She concluded that even where toxic substances have been implicated as having had significant effects on the biota, the specific causal factor has been difficult to ascertain, and that this was well illustrated by the confusion in the early bird and mammal studies. She further concluded that despite these uncertainties, it was generally thought that toxic substances had the potential to adversely affect organisms in areas where toxic chemical concentrations were relatively high and in those long-lived organisms (including humans) that bioaccumulate certain toxic substances. It is interesting to note the tentative tone in these conclusions even as late as 1988.

## 3. Skepticism and Developments in Causality

The Council of Great Lakes Research Managers initiated a project in 1988 on how scientists "prove" causal relationships to overcome the apparent uncertainties in the science. The Council held the first Cause-Effect Linkages Workshop in 1989 and used an approach based on epidemiological criteria (Fox 1991) to review several case studies (International Joint Commission 1989b). In a poll of the participants, it was interesting to note that there was fairly widespread skepticism about whether any causal relationships had been demonstrated and even some participants who voted that there was no relationship between the effects reported in some case studies and exposures to persistent toxic substances. The papers from this First Cause-Effect Linkages Workshop were published in the August 1991 issue of the *Journal of Toxicology and Environmental Health* volume 33, number 4. Subsequent Cause-Effect Linkages Workshops have been held in association with the 1991 Biennial Meeting of the International Joint Commission, at Traverse City, and with the 1995

Conference of the International Association of Great Lakes Research at Ann Arbor, Michigan. The papers from these workshops were published, respectively, in the *Journal of Great Lakes Research* in volume 19, number 4, 1993, and in volume 22, number 2, 1996.

This skepticism is not confined to the scientific community but was expressed by the Great Lakes Water Quality Board in a report to the International Joint Commission (1993). In its report it stated that "the crux of the matter for regulatory officials was how to make policy decisions in the face of uncertainty, given the implications for their long-term credibility of making a decision without a proven cause and effect". Ironically, there is a similar skepticism about the necessity of including information concerning the effects of persistent toxic substances on populations of Great Lakes organisms in the preparation of the Remedial Action Plans and the Lakewide Management Plans as required in Annex 2 of the 1987 Protocol to the 1978 Great Lakes Water Quality Agreement.

In 1988, the Committee for the Assessment of Human Health Effects of Great Lakes Water Quality expressed its concern about the lack of epidemiological studies to directly assess the effect of environmental contaminants in water and fish on human health since most assessments were based on extrapolations from animal toxicology studies and from estimates of human exposure (International Joint Commission 1990a). The Committee held a workshop on "The Role of Epidemiology in Assessing the Effects of Great Lakes Water Quality on Human Health," in March, 1988, in Scarborough, Ontario. In a preface, the co-chairs noted that there had been a substantial evolution in the thinking about the risks posed by chemicals and noted the shift in orientation from cancer and birth defects to attention on reproductive, developmental and metabolic processes. This shift in orientation, they stated, was a significant outcome of the workshops sponsored by the International Joint Commission. These included the Cause-Effect Linkages Workshop, held by the Council of Great Lakes Research Managers, and a subsequent workshop of the Human Health Effects Committee to develop a research prospectus for human health in the Great Lakes basin. The Fourth Biennial Meeting of the International Joint Commission, held in Hamilton, Ontario, in October, 1989, was the occasion when the ideas from these workshops and from the publication of *Great Lakes: Great Legacy?* (Colborn *et al.* 1990), coincided with the activism among the public interest groups in influencing the message of the International Joint Commission in preparing the strong message to the two governments contained in the Fifth Biennial Report (International Joint Commission 1990b).

In 1991, the International Joint Commission set up a Virtual Elimination Task Force to "provide advice and recommendations to the Commission about what a virtual elimination strategy should contain and how the strategy could be implemented". The Task Force noted that the attainment of the policy on virtual elimination of discharges of persistent toxic substances would be characterized by an absence of injury to living organisms and to society (Virtual Elimination Task Force 1993). The Task Force held a workshop on indicators, in Ann Arbor, Michigan on April 28-29, 1992 to provide advice to the International Joint Commission on specific indicators relevant to this policy on virtual elimination (Fox 1994).

## 4. Design of the Workshop on Environmental Results

In 1995, the International Joint Commission asked its Great Lakes Science Advisory Board to undertake an evaluation of any progress made in reducing or eliminating the effects of persistent toxic substances in wildlife and in humans. The Board hosted the Workshop on Environmental Results to assemble scientists with information on trends in the concentrations of persistent toxic substances in the Great Lakes, and on trends in the incidence of effects in populations of Great Lakes organisms. The workshop was designed with an awareness of this historical context of the Commission's work on effects and of the continuing skepticism within the scientific and regulatory communities about the causal relationships between injury to health and property and exposures to persistent toxic substances. There was an assumption that the most useful indicators were those for which there was a known cause and effect relationship between exposures to specific persistent toxic substances and injury observed in populations of wild organisms or in subpopulations of humans.

The designers of the Great Lakes Science Advisory Board's workshop made several other assumptions . First, it was assumed that the primary interest of the Commission is in trends in the incidence of effects of Great Lakes pollutants on human health. While there is a little long-term trend data on the concentrations of contaminants in human tissues or fluids, there are no data on the long-term trends in effects. Second, in the absence of this long-term trend data on effects on human health, the designers assumed that data on the incidence of effects in wildlife that was critically exposed to high concentrations of pollutants could be used as a surrogate. Third, because persistent toxic substances disperse over a large area far from the site of release and are biologically available and active over a long period of time, the workshop was designed primarily concerning species that integrate pollutants over large temporal and spatial scales. While the main interest was assumed to be in trend data for large scale integrators, selected data on species that show trends at the local scale were also included.

## 5. Causality and Candidate Species

There have been significant advances, in the past 150 years, in the process of establishing (causal) relationships between observations of disease in organisms and specific putative causal agents, since the time that Robert Koch and Jacob Henle formulated their postulates (Evans 1976). Renewed interest in the criteria for inferring causal relationships, particularly in relation to exposures to substances, arose from the United States Surgeon General's investigations into lung cancer and tobacco (US Department of Health, Education and Welfare, 1964). Since that time, there has been considerable interest in the epidemiological criteria and their application to a variety of diseases (Susser 1986). The particular formulation of the epidemiological criteria used by the Council of Great Lakes Research Managers in the First Cause-Effect Linkages Workshop included the following (Fox 1991):-

- **Time-Order**. Does the exposure to supposed cause precede the effect in time? This criterion is useful for excluding spurious putative factors but may be difficult to establish in systems with little historic data.
- **Strength of the Association**. Do cause and effect coincide in their distribution? Is the effect large relative to the cause? Is the prevalence of the effect in exposed populations large relative to unexposed populations?
- **Consistency of the Association**. Has the association been repeatedly observed in different places, circumstances, times, and species, or by other investigators with different research designs?
- **Coherence of the Association**. Is the cause-effect interpretation consistent with our current understanding of the biological mechanism(s) underlying the effect? Are there plausible routes of exposure? Is an exposure-response relationship present? Do laboratory studies support the proposed relationship? Do remedial actions lead to altered frequency or severity of the effects?
- **Specificity of the Association**. Could the effect be due to a different cause? Could the proposed cause produce other effects? Can alternative hypotheses be eliminated? In the context of the Great Lakes, where a multiplicity of persistent toxic substances and ecological perturbations are present, specificity may be complicated by chemical interactions, commonality of mode of action, and interspecific differences.
- **Predictive Power of the Association**. A hypothesis drawn from an observed association predicts a previously unknown fact or consequence, and must in turn be shown to lead to that consequence.

These criteria were used as a template by the authors of the papers from the First Cause-Effect Linkages Workshop to review a series of case studies on chemically-induced syndromes in a variety of St Lawrence - Great Lakes organisms exposed to high concentrations of persistent toxic substances. The criteria were less strictly applied in the preparation of papers from the subsequent Cause-Effect Linkages Workshops. The various case studies include syntheses of the evidence on humans (Swain 1991, Jacobson and Jacobson 1993, Lonky *et al.* 1996), mink *(Mustela vison)* and otters *(Lutra canadensis)* (Wren 1991), beluga whales *(Delphinapterus leucas)* in the Gulf of St Lawrence (Beland *et al.* 1993), bald eagles *(Haliaeetus leucocephalus)* (Colborn 1991), colonial fish-eating birds (Gilbertson *et al.* 1991, Fox 1993, Ludwig *et al.* 1996), snapping turtles *(Chelydra serpentina)* (Bishop *et al.* 1991), lake trout *(Salvelinus namaycush)* (Mac and Edsall 1991, Mac *et al.* 1993), Pacific salmon *(Oncorhynchus sp.)* (Leatherland 1993) and brown bullhead *(Ameiurus nebulosus)* (Baumann *et al.* 1996). There was, thus, evidence of chemically-induced syndromes on a large number of Great Lakes organisms. The available case studies yielded candidate species that were evaluated at the 1992 workshop on indicators in Ann Arbor, Michigan (Fox 1994). All case studies available by 1996 were evaluated in detail in designing this Workshop on Environmental Results. These detailed evaluations of the candidate species are contained in the text and tables in Appendix 1 to this paper and extensively draw on the original descriptions in the proceedings of the 1992 workshop without major revisions or updating with more recent references (Fox 1994).

Though these epidemiological criteria have not been institutionalized within agencies addressing chemically-induced injury to fish, wildlife and human populations, industry does seem to have supported the methodology. Shortly after the publication of these epidemiological criteria and the associated case studies, the Chlorine Coordinating Council, and the Chlorine Institute sponsored a review of the evidence (Willes *et al.* 1993). The review concluded that "If the application of sound scientific principles is to be followed, these (epidemiological) criteria must be applied before a causal relationship is inferred between effects and exposures to specific chemicals in the environment."

There has been extensive work undertaken on the development of criteria for selecting monitoring species. In the mid-1960s, Moore (1966) outlined five criteria for choosing indicator species for monitoring the concentrations and effects of pesticides. These criteria included consideration that :-

a) The species should be easily available with a wide distribution, relative abundance and ease of collection;
b) The species should be large enough to provide sufficient tissues or samples for individual analyses;

c) The age of the specimens at sampling should be able to be determined to infer whether contamination was recent or over a long period of time;
d) The species should accumulate sufficient of the residues to be analytically detectable, but not so high that toxicological effects are occurring in the population;
e) There should be knowledge of the range of the species so that inferences can be made of the geographic scale of contamination.

These criteria were used in the early 1970s to select the herring gull as a monitor species to establish the geographic and temporal trends of the concentrations and effects of organochlorine pollutants in the Great Lakes basin (Gilbertson 1974).

More recently, there have been attempts by fisheries biologists to develop criteria for the development of a General Objective for Ecosystem Quality "for the restoration and maintenance of Great Lakes biological communities similar to those present before the modification by human intervention". The underlying assumption is that "deviations from the objective would indicate the existence of an ecosystem stress, the precise nature of which might need to be determined for remedial action to be taken" (International Joint Commission 1985). Some of the criteria were similar to those outlined above, but in addition the selected indicator was to:-

a) Be indigenous and maintain itself through natural reproduction;
b) Interact directly with many components of its ecosystem;
c) Have well documented and quantified niche dimensions expressed in terms of metabolic and behavioural responses;
d) Exhibit a gradual response to a variety of human induced stresses;
e) Serve as a diagnostic tool for specific stresses of many sorts;
f) Respond to stresses in a manner that is both identifiable and quantifiable;
g) Be a suitable species for laboratory investigations;
h) Be generally recognized as important to humans; and
i) Serve to indicate aspects of ecosystem quality other than those represented by presently accepted parameters.

These rather unwieldy criteria subsequently became the criteria for evaluation of candidate species as indicators of trends in persistent toxic substances as well as general ecosystem stress by any and all human activities. Based on an evaluation of the characteristics of lake trout and *Pontoporeia hoyi* against these criteria, these species were included as Lake Ecosystem Objectives for Lake Superior in the 1987 Protocol amending the 1978 Great Lakes Water Quality Agreement. These criteria have also been applied to walleye *(Stizostedion vitreum)* and *Hexagenia* spp. as proposed indicators in relation to the general condition of the central and western basins of Lake Erie but these proposed indicators have not yet been incorporated into the Agreement (International Joint Commission 1990c). There have been several informal workshops convened by the International Joint Commission to evaluate the suitability of several candidate species as specific indicators of pollution by persistent toxic substances. The criteria have been applied to the evidence on the bald eagle and this species has been recommended as an indicator but has not been formally adopted and incorporated into the Agreement (International Joint Commission 1990c). It would seem that the 1985 proposals for development of a General Objective for Ecosystem Quality have brought some degree of ambiguity to the process of developing indicators of water quality for the Great Lakes Basin.

## 6. Designing the Workshop on Environmental Results

In designing the Workshop on Environmental Results the organizers in the Great Lakes Science Advisory Board recognized the need to start with presentations on the trends in the concentrations of persistent toxic substances in the abiotic (sediment, water and atmospheric), and in biotic (fish, gulls and humans) compartments. Presentations were then given on trends in the incidence of effects in birds (population status of bald eagle and osprey *(Pandion haliaetus)*, Great Lakes Embryo Mortality, Edema and Deformities Syndrome in herring gull and double-crested cormorant *(Phalacrocorax auritus)*, recruitment in Caspian terns *(Sterna caspia)*, as well as biochemical markers and a vertebral somite assay), deformities in snapping turtles, tumors in brown bullheads, and reproductive anomalies in lake trout. This issue of the *Environmental Monitoring and Assessment Journal* contains the peer-reviewed papers for about a half of the presentations at the workshop. A detailed commentary of the adequacy of the monitoring projects for inferring trends in the concentrations and effects of persistent toxic substances have been published (Nisbet, this volume).

## 7. Conclusions and Recommendation

The main finding of the Great Lakes Science Advisory Board from conducting the Workshop on Environmental Results was that, while there are adequate long-term monitoring projects to document gross trends in the concentrations of organochlorine pollutants in the Great Lakes environment, ***there is no formal existing program in either country for gathering long-term evidence for determining trends in the incidence of the effects of persistent toxic substances on organisms***. There have been occasional surveys at the local level to investigate the incidence of tumours in fish exposed to polynuclear aromatic hydrocarbons (Baumann and Harshbarger, this volume) and from which trends in the incidence of effects can be inferred. Similarly, at the local level, surveys of the variation in hatchability and in the incidence of deformities in snapping turtles have been undertaken ( Bishop and Gendron, this volume). The most reliable long term trend data at the Great Lakes regional scale concerns the evidence of the population collapse, reproductive failure and subsequent partial recovery of bald eagles (Bowerman *et al.* this volume). This evidence has been collected through the dedication of a few individuals, notably Sergej Postupalsky, since the early 1960s, with financial resources supplied on an ad hoc basis from year to year. Surveys, during the past decade, of the incidence of bill deformities in double-crested cormorants have provided a baseline of the geographic and temporal variation in teratogenic activity in the Great Lakes basin. Similarly, the occasional surveys to monitor biochemical markers, such as porphyrins, vitamin A storage, and thyroid status in adult herring gulls (Fox *et al.* this volume) and in embryos and chicks of fish eating birds (Grasman *et al.* this volume), has shown the spatial variation in the effects of persistent toxic substances and an indication of temporal trends.

The Great Lakes Science Advisory Board concluded that there was a need for the Parties to the Great Lakes Water Quality Agreement to formally name species that would be used as indicators to evaluate whether they have been successful in implementing their policy concerning the virtual elimination of discharges of persistent toxic substances, and in assessing whether resources and human health continue to be injured by transboundary pollution. The Board recommended that the Parties establish a formal bilateral mechanism to select indicator organisms and commit funding on a long-term basis.

## 8. Appendix 1: Evaluation of Candidate Species

There are large numbers of Great Lakes species that have been sampled for residue analysis. There are, however, very few for which there are any data available concerning effects of persistent toxic substances. A variety of effects caused by exposures to persistent toxic substances have been observed in populations of several species in the Great Lakes basin. The process of reviewing this extensive evidence, to select the anomalies in different species that can be reliably used to report whether there has been a reduction in the incidence of effects, has been aided by previous reviews compiled by the Government of Canada (1991) and the International Joint Commission (Fox 1994). Those members of the Great Lakes Science Advisory Board who designed the Workshop on Environmental Results drew extensively on these sources to compile the following brief accounts of the various species that have been investigated and of the effects that have been documented. While there have been several papers concerning the concentrations in, and effects of persistent toxic substances on organisms in the Great Lakes Basin, that have been published since these original works, these have not been included in this Appendix because they were unavailable to the Great Lakes Science Advisory Board at the time that the Workshop on Environmental Results was being designed. Many of these more recent references can be found in the detailed papers contained in this issue of *Environmental Monitoring and Assessment.*

## 8.1 BIRDS

In the past thirty-five years, the research on the effects of persistent toxic substances on Great Lakes birds has yielded the most comprehensive and indisputable causal relationships. There are several reasons for the success of this research. First, there have been many knowledgeable observers including naturalists, bird banders, as well as professional ornithologists, who have noticed anomalies in avian populations soon after they occurred. Second, birds and eggs that are affected by exposures to chemicals are visible and can be recovered for diagnostic and experimental purposes. Populations of birds and their eggs can be sampled, and identifiable individuals followed over time. Third, there is a large body of knowledge about avian biology, physiology, pathology and toxicology, partly deriving from the economic importance of poultry and wild water fowl, and partly from extensive laboratory testing with representatives of wild species. Finally, many species that have been affected are of no value as game and evidence concerning the status of their populations is not confounded by annual mortalities from hunting.

The use of Great Lakes fish-eating birds as monitors of the effects of exposures to persistent toxic substances has been extensively reviewed (Mineau *et al.* 1984, Fitchko 1986, Fox and Weseloh 1987, Peakall and Fox 1987, Peakall 1988, Gilbertson 1988, Government of Canada 1991, and Fox 1993). Tables I and II are matrices of the references to gross effects, and histological and biochemical effects, respectively, that have been observed in various species of fish-eating birds, and that have been reliably related to exposures to persistent toxic substances. The designers of the program for the Workshop on Environmental Results considered data on the incidence of these effects in the following candidate species to establish the suitability of the species for establishing trends:

**Bald Eagle *(Haliaeetus leucocephalus)***

The bald eagle was probably the first species that was widely affected by exposure to a persistent toxic substance with the introduction of DDT (Broley 1958) with declines and extirpation of most populations throughout the range of this species. According to the U.S. Fish and Wildlife Service, bald eagle numbers in the United

**Table I.** Candidate Species: Chemically-Induced Anomalies Observed in Great Lakes Fish-Eating Birds

| Species | Abundance and Distribution | Production | Eggshell Thinning | Embryo Mortality | Deformity | Wasting | Feminization Female Pairing | Behaviour |
|---|---|---|---|---|---|---|---|---|
| Bald Eagle | Postupalsky 1971 Sprunt *et al*. 1973. Best *et al*. 1994 McKeating 1985 Bowerman 1993 | Postupalsky 1971 Sprunt *et al*. 1973. Nisbet 1989 Best *et al*. 1994 Giesy *et al*.1996 | Postupalsky 1971 Wiemeyer *et al*. 1984, 1993. | Postupalsky 1971 | Bowerman *et al*. 1994 | | | |
| Osprey | Postupalsky 1971 Postupalsky 1977a,b, Ewins *et al*. 1995 | Postupalsky 1971 Postupalsky 1977a,b, Ewins 1992 Ewins *et al*. 1995 | Anderson and Hickey 1972 Ewins *et al*. 1995 | | | | | |
| Herring Gull | Ludwig and Ludwig 1969 Gilman *et al*. 1977 Weseloh 1984 Blokpoel and Scharf 1991 | Keith 1966 Ludwig and Tomoff 1966 Gilbertson 1974 Gilbertson and Hale 1974a Gilman *et al*. 1977 Teeple 1977 Mineau *et al*. 1984 Government of Canada 1991 | Hickey and Anderson 1968 Faber and Hickey 1973 Gilbertson 1974 Weseloh *et al*. 1994 | Keith 1966 Ludwig and Tomoff 1966 Gilbertson and Hale 1974b Gilbertson and Fox 1977 Peakall and Fox 1987. | Gilbertson *et al*. 1976 Scharf and Buckingham 1974. | Grasman *et al*. 1996 | Shugart 1980 Fox and Weseloh 1987 | Fox *et al*. 1978 Peakall and Fox 1987. |
| Ring-billed Gull | Ludwig 1974 Blokpoel and Scharf 1991 | Ludwig 1974Ludwig 1974 | | | Ryder and Chamberlain 1972. Gilbertson *et al*. 1976 | | | |

| | | | | | | | |
|---|---|---|---|---|---|---|---|
| Double-crested Cormorant | Postupalsky 1971<br>Weseloh and Ewins 1994<br>Weseloh *et al.* 1983<br>Ludwig 1984.<br>Government of Canada 1991<br>Blokpoel and Scharf 1991<br>Weseloh *et al.* 1995 | Postupalsky 1971<br>Weseloh and Ewins 1994<br>Weseloh *et al.* 1983<br>Ludwig 1984<br>Weseloh *et al.* 1995 | Postupalsky 1971<br>Anderson and Hickey 1972<br>Weseloh *et al.* 1983 | Tillitt *et al.* 1992,<br>Ludwig *et al.* 1996<br>Weseloh *et al.* 1983 | Fox *et al.* 1991.<br>Yamashita *et al.* 1993.<br>Henshel *et al.*, 1997 | | |
| Black-crowned Night Heron | Government of Canada 1991 | Government of Canada 1991 | Faber and Hickey 1973<br>Heinz *et al.* 1985<br>Government of Canada 1991 | Government of Canada 1991 | Gilbertson *et al.* 1976<br>Government of Canada 1991 | | Government of Canada 1991 |
| Caspian Tern | Ludwig 1979.<br>L'Arrivee and Blokpoel 1988.<br>Blokpoel and Scharf 1991 | Ludwig 1979.<br>Mora *et al.* 1993.<br>Ludwig *et al.* 1993b. | | Ludwig *et al.* 1993b<br>Ludwig *et al.* 1996 | Yamashita *et al.* 1993<br>Ludwig *et al.* 1996 | Ludwig *et al.* 1993b<br>Grasman *et al.* 1996 | Mora *et al.* 1993. |
| Common Tern | Courtney and Blokpoel 1983<br>Blokpoel and Scharf 1991 | Gilbertson and Reynolds 1972<br>Government of Canada 1991 | | Government of Canada 1991 | Gilbertson *et al.* 1976<br>Government of Canada 1991 | | |
| Forster's Tern | Scharf and Shugart 1984<br>Blokpoel and Scharf 1991 | Kubiak *et al.* 1989<br>Harris *et al.* 1993 | | Hoffman *et al.* 1987<br>Kubiak *et al.* 1989 | Hoffman *et al.* 1987<br>Kubiak *et al.* 1989 | Kubiak *et al.* 1989<br>Harris *et al.* 1993 | Kubiak *et al.* 1989. |

**Table II.** Candidate Species: Histological and Biochemical Anomalies in Great Lakes Fish-Eating Birds

| Species | Liver Histology | Porphyria | Vitamin A and Retinoids | Cytochrome P4501A1 | Thyroid T4/T3 | Immunity |
|---|---|---|---|---|---|---|
| Bald Eagle | | | | Feyk *et al.* 1995 | | |
| Osprey | | | | | | |
| Double-crested Cormorant | | | (Fox *et al.* in prep.) (Bishop *et al.* in prep.) | Hoffman,cited in Fox 1993. Feyk *et al.* 1995 | | |
| Herring Gull | Gilbertson 1983. | Gilbertson and Fox, 1977. Fox *et al.* 1988. Kennedy and Fox, 1990. Fox *et al.* This volume | Spear *et al.* 1986 Spear *et al.* 1990 Government of Canada 1991 Grasman *et al.* 1996. Fox *et al.* This volume. | Ellenton *et al.* 1985. Boersma *et al.* 1986. Feyk *et al.* 1995 | Moccia *et al.* 1986. Grasman *et al.* 1996. Fox *et al.* This volume. | Grasman *et al.* 1996 |
| Ring-billed Gull | | | | Feyk *et al.* 1995 | | |
| Black-crowned Night Heron | | | | Hoffman *et al.* 1993 Rattner *et al.* 1993 | | |
| Caspian Tern | | | Grasman *et al.* 1996 | | Grasman *et al.* 1996 | Grasman *et al.* 1996 |
| Common Tern | | | | Hoffman *et al.* 1993 | | |
| Forster's Tern | | | | Hoffman *et al.* 1987 | | |

States have recovered from a low of approximately 400 pairs nationwide in 1964 to 2,700 pairs in 1989. The Great Lakes population of bald eagles has been extensively studied since the mid-1960s (Postupalsky 1971, McKeating 1985). In the 1970s, when contaminant levels were the highest in the Great Lakes, the population along the Great Lakes shoreline and environs dwindled to 16 nesting pairs and reproductive success was the lowest of six populations surveyed throughout the continent (Sprunt *et al.* 1973). Presently, the population is up to about 120 pairs, and the bald eagle is slowly reoccupying the Great Lakes coastal territories (Bowerman 1993). However, reproduction is still impaired along Great Lakes shorelines of Michigan and Ohio (Best *et al.* 1994) and at inland sites accessible to runs of anadromous Great Lakes fish (Giesy *et al.* 1995). With the reestablishment of Great Lakes territories in contaminated areas, congenital malformations are being found more frequently (Bowerman *et al.* 1994, this volume). Although suitable habitat has existed since the bald eagle was extirpated from the Lake Ontario shoreline, there are as yet no territories that have been reestablished next to that lake.

The demise and subsequent partial recovery of the bald eagle in North America has been causally related to the concentrations of DDT and metabolites in the environment (Wiemeyer *et al.* 1984, 1993, Nisbet 1989). Concentrations of DDT and metabolites have declined in the Great Lakes Basin, but the recovery of the bald eagle populations in Great Lakes shoreline territories has been constrained by the continuing presence of PCBs and related compounds (Bowerman 1993, Bowerman *et al*, 1995, Bowerman *et al.* this volume). The causal relationship between the status and reproductive success of the bald eagle in the Great Lakes Basin has been reviewed by Colborn (1991).

There has been extensive consultation and review of proposals to use the bald eagle as an indicator of the virtual elimination of discharges of persistent toxic substances to the Great Lakes (International Joint Commission 1991, 1992) and the species was recommended as an indicator by the Great Lakes Science Advisory Board in its 1989 and 1991 reports to the International Joint Commission. Clearly, data on the population status and reproductive success of this candidate species might provide evidence of trends in the incidence of effects caused by persistent toxic substances.

**Osprey *(Pandion haliaetus)***

The status of osprey populations in the United States in relation to the effects of persistent toxic substances has been comprehensively reviewed (Poole 1989) and, as with the bald eagle, many populations have been extirpated or declined with characteristic signs of the effects of organochlorine compounds on reproduction, including eggshell thinning and embryo mortality. Ospreys have not been one of the principal bird species that have been studied in relation to contaminants in the Great

Lakes basin, but there are reports on the distribution and reproductive success in parts of the Great Lakes Basin (Postupalsky 1971, 1977a&b), though not on the main lake shorelines. The historic distribution of the osprey around the Great Lakes shorelines is somewhat anecdotal and there is uncertainty about the former range in the lower Great Lakes and the possible influence of bald eagles on osprey distribution. Similarly, there is only limited material, such as eggshells for measuring thickness, and few analytical results from which to attempt to reconstruct the effects of organochlorine chemicals on Great Lakes osprey populations.

More recently, the reproductive success of populations of ospreys in the Kawartha Lakes region have been compared with those in the regions of Georgian Bay and the St Mary's River (Ewins 1992). The story of the recovery of the osprey population in the Georgian Bay and North Channel (Ewins *et al.* 1995) is a remarkable example of the way in which an avian top predator can be used to document the restoration of a Great Lake from the effects of specific organochlorine compounds (Environment Canada 1994).

**Herring Gull *(Larus argentatus)***

The herring gull has been extensively studied in the Great Lakes basin since the mid-1960s (Keith 1966, Ludwig and Tomoff 1966). This widespread species is a year-round resident in the basin (Gilman *et al.* 1977, Weseloh 1984). Both field studies (Fox *et al.* 1990, Ewins *et al.* 1994) and a bioenergetics-based model (Clark *et al.* 1987) indicate that Great Lakes herring gulls are opportunistic piscivores, and when forage fish are available, 80% of their diet is composed of fish. In 1974, the Canadian Wildlife Service initiated a surveillance project to monitor chemical contaminants and their effects in this species (Gilbertson 1974). Since its inception, the Great Lakes Herring Gull Monitoring Program has acquired one of the most complete and continuous databases in the world for contaminant residues and their associated effects.

The biological measures utilized include egg viability (embryo mortality), chick growth and survival, incidence of deformities, and other components of GLEMEDS (Great Lakes Embryo Mortality Edema and Deformity Syndrome) (Gilbertson *et al.* 1991) and population size. Histological examination of gonads of developing embryos have shown the occurrence of chemically-induced feminization (Fox and Weseloh 1987). The species is sufficiently well characterized that nutritional stress can be differentiated from toxic stress. Basinwide biomarker studies involving vitamin A storage, retinoid and thyroid status and the induction of cytochrome P4501A1 have been conducted for over 10 years (reviewed in Fox 1993) and provide an existing historical database against which to measure progress toward virtual elimination. Studies of immune function and weight gain of chicks between

21 and 28 days of age have been successfully completed in this species (Grasman *et al.* 1996) so that measurements of immune competence and the incidence of wasting syndrome can be assessed.

Environment Canada has prepared a State of the Environment Fact Sheet on the long-term trend data and the relationship to reproductive anomolies in this species (Environment Canada 1990). Clearly, this is a third candidate species to document the trends in the incidence of effects caused by persistent toxic substances.

**Ring-billed Gull *(Larus delawarensis)***

There is extensive data on the status of populations of ring-billed gulls in the Great Lakes basin (Ludwig 1974, Blokpoel and Scharf 1991). There does not seem to be any evidence that the population has been affected by exposures to persistent toxic substances and thus it is unlikely to be a suitable indicator of the virtual elimination of discharges of persistent toxic substances, even though there have been occasional reports of deformities in chicks (Ryder and Chamberlain *et al.* 1972, Gilbertson *et al.* 1976) and mortality of juveniles and adults in fall migrations (Sileo *et al.* 1977.). Thus the ring-billed gull was not considered further as a candidate species in designing the Workshop on Environmental Results.

**Double-crested Cormorant *(Phalacrocorax auritus)***

The double-crested cormorant is a widespread obligate piscivore and the Great Lakes population of cormorants has been studied extensively, particularly in relation to the effects of persistent toxic substances (Postupalsky 1971, Blokpoel and Scharf 1991, Weseloh *et al.* 1995). Cormorants that nest in the Great Lakes basin, winter on the Atlantic or Gulf of Mexico coasts of Florida and thus there is a potential for ambiguity about the locations in which particular individuals became contaminated. This species is more sensitive to DDE-induced eggshell thinning (Postupalsky 1971, Weseloh *et al.* 1983) and PCB/TCDD-induced developmental abnormalities (Yamashita *et al.* 1993), particularly crossed bills (Fox *et al.* 1991), than is the herring gull (Ludwig *et al.* 1993a). There is extensive research that has been undertaken to demonstrate the causal linkage between the incidence of abnormalities and embryo mortality in colonies of double-crested cormorants in the Great Lakes and exposures to persistent toxic substances and particularly PCBs (Ludwig 1995, Ludwig *et al.* 1996). There is unpublished evidence of an association between the cross-billed phenomenon and a skewed ratio of females in cormorant chicks that might have potential as a biomarker. Henshel *et al.* 1997 have shown correlations between the degree and frequency of morphometric abnormalities of the brains of double-crested cormorants and exposure to elevated mixtures of PCBs, PCDDs, and PCDFs.

Bioeffects that are monitored in this species are eggshell thickness, and the incidence of bill abnormalities (Fox *et al.* 1991) and other abnormalities in chicks.

Recent outbreaks of Newcastle Disease, a highly pathogenic viral disease of poultry, suggests that this virus is endemic in Great Lakes cormorants and its presence could confound the search for potential toxic effects. Despite this drawback, the double-crested cormorant was included as a candidate species for indicating the reduction of effects.

**Black-crowned Night Heron *(Nycticorax nycticorax)***

The Government of Canada (1991) has reviewed the evidence for the role of persistent toxic substances in affecting the status of Great Lakes populations of black-crowned night herons. There is unpublished evidence of a decline in the population on Pigeon Island in Lake Ontario in the late 1960s and early 1970s associated with poor reproductive success and eggshell thinning and embryo mortality. Eggshell thinning was documented in the population in Green Bay, Wisconsin (Heinz *et al.* 1985) and there was a possibility that the high rate of egg breakage might have been caused by aberrant behaviour of the adult birds in piercing the eggs with their bills.

The U.S. Fish and Wildlife Service has chosen this species as its estuarine and freshwater wetland sentinel species (Custer *et al.* 1991) and biochemical studies have been ongoing for several years at their Patuxent laboratory in Maryland. This species will be used as part of the nationwide BEST (Biomonitoring of Environmental Status and Trends) program. To date, the main factors being studied in the black-crowned night heron include reproductive success, eggshell thinning, mixed function oxidase induction and other biochemical markers, in pipping embryos and chicks (Hoffman *et al.* 1993, Rattner *et al.* 1993), and a small amount of work on DNA damage.

Despite the apparent suitability of this species as an indicator and the plans and decisions that have been made for the use of this species, the existing evidence concerning populations on the Great Lakes was too limited, and the species was therefore not considered further in designing the Workshop on Environmental Results.

**Caspian Tern *(Sterna caspia)***

The Caspian tern is the largest tern in North America. This migratory species is a strict piscivore and nests in Lakes Ontario, Huron, Michigan, and Superior (Blokpoel and Scharf 1991). It is considered threatened or endangered. The Caspian tern is a long-lived species and its populations within the Great Lakes basin have been the subject of large-scale, long-term banding efforts that suggest differential recruitment patterns between colonies in Canadian and United States waters (Ludwig 1979; L'Arrivee and Blokpoel, 1988), and a significant negative correlation between mean concentration of PCBs in the plasma of breeding adults and the proportion of young recruiting to that natal region has been reported by Mora *et al.* (1993).

Bioeffect measures include embryonic viability and the incidence of terata in dead eggs and abnormalities in chicks in relation to concentrations of persistent toxic substances (Ludwig *et al.* 1996). Recent studies suggest that, in terms of reproductive effects, this species is considerably more sensitive to PCBs than is the herring gull. Studies of immune function and weight gain of chicks between 21 and 28 days of age have also been successfully conducted in this species (Grasman *et al.* 1996).

The species has been sufficiently well studied in relation to the effects of persistent toxic substances that it was included in planning for the Workshop on Environmental Results.

**Common Tern *(Sterna hirundo)***

The Great Lakes population of common terns has undergone a serious decline in the past thirty years as a result of a complex of factors including interspecific competition, flooding, predation and the encroachment of vegetation, as well as the possibility of effects of contamination by persistent toxic substances (Courtney and Blokpoel 1983, Blokpoel and Scharf 1991, Government of Canada 1991). Common terns have been studied intensively at a limited number of locations such as at Toronto Harbour, Hamilton Harbour and Port Colborne (Morris *et al.* 1976, 1980). Though the Government of Canada concluded that toxic contaminants have had a significant negative effect on the reproduction of this species in the lower Great Lakes, it has proved much more difficult than in other species to demonstrate causal relationships between observed effects and exposures to persistent toxic substances. Thus this limited database on the incidence of effects in this species is not further considered in designing the workshop.

**Forster's Tern *(Sterna forsteri)***

The distribution of the Forster's tern, a migratory piscivorous species which occupies eutrophic marshy habitats, is more-or-less limited in the Great Lakes Basin to areas in Green Bay, Saginaw Bay, and Lake St. Clair (Scharf and Shugart 1984, Blokpoel and Scharf 1991). Despite this limited distribution, the species has been used by researchers to provide one of the most complete and compelling case studies of the effects of specific persistent toxic substances on the reproduction of any species of colonial fish-eating bird in the Great Lakes. In 1983, Kubiak *et al.* (1989) showed the relationship between reproductive impairment in a Green Bay, Wisconsin colony of Forster's terns and exposures to two specific pentachlorobiphenyls. In 1988, when concentrations of PCB in Green Bay had declined, Harris *et al.* (1993) showed that levels were still high enough to reduce reproductive performance through fledgling mortality characterised by a wasting syndrome. Biological measures include hatching success, chick growth, and the incidence of deformities and Hoffman *et al.* (1987) have shown the induction of aryl hydrocarbon hydroxylase in chicks.

These birds nest in small colonies, usually on floating mats of vegetation, and are therefore rather difficult to study. Further work is necessary to establish standard methodology for assessment of reproductive measures. The species is currently designated as threatened or endangered. Its limited distribution and its endangered status militate against its widespread use as an indicator species. Thus, despite the quality of the research that has been undertaken on this species at the Green Bay location, it was not further considered in preparing for the workshop.

## 8.2 MAMMALS

There is an extensive literature on the effects of certain organochlorine compounds in Great Lakes fish on mink survival and reproduction, which has been reviewed by Fitchko (1986) and Gilbertson (1988). Most of this literature relates to the outbreaks of kit mortality in ranch mink operations in the lower Great Lakes basin (Hartsough 1965) and to the associated experimental evidence (Aulerich and Ringer 1970, 1977, Heaton *et al.* 1995a and b). At the time of the planning for the Workshop on Environmental Results, there was almost no literature on the effects of persistent toxic substances on populations of wild mink or on the otter as the other piscivorous mammalian predator. Wren (1991) concluded that there was insufficient data to demonstrate a causal relationship between the status of wild mink and otter populations and exposures to persistent toxic substances in the Great Lakes Basin. In the past two years, there has been a series of collaborative studies to investigate methods for estimating abundance and reproductive performance under field conditions (K.A. Patnode, Wisconsin Department of Natural Resources, Madison, WI, pers.comm.).

### Mink *(Mustela vison)* and Otter *(Lutra canadensis)*

The most convincing reason for using the mink as an indicator of virtual elimination of persistent toxic substances is the very high sensitivity of this species manifested as reproductive effects caused by planar halogenated aromatic hydrocarbons (HCB, some PCBs, and TCDD) (Addison *et al.* 1991; Wren 1991). Studies in ranch mink suggest that this species is the free living mammal that is most sensitive to toxic substances such as PCBs and TCDD. Its diet provides an integrated exposure to contaminants in shoreline wetlands. Thriving mink populations in suitable Great Lakes shoreline wetlands would suggest that levels of these contaminants in their diet must be low. The chemical sensitivity of the otter has not been confirmed experimentally, but its largely piscivorous diet is more directly reflective of the nearshore aquatic environment than the mink.

In 1991, the International Joint Commission sponsored a workshop on the status of these species as potential indicators of virtual elimination of persistent toxic substances (Addison *et al.* 1991). The participants at the workshop concluded that before a reliable operational biomonitoring program using mink (and otter) could be developed and employed, further research was needed to develop field survey techniques useful for the assessment of distribution, abundance and reproductive health of these species, and to determine the physiological and biochemical responses to chemical stressors which could be measured in free-living individuals (Addison *et al.* 1991). In 1995, the Great Lakes Protection Fund provided resources for a technical meeting to develop methods and techniques for the field evaluation of the potential of wild mink populations as indicators (K.A. Patnode, Wisconsin Department of Natural Resources, Madison, WI, pers.comm.). At the time of the planning of the Workshop on Environmental Results, the information for using the mink and otter as indicator species is insufficient for the purpose of determining trends in the incidence of effects and these species were not considered further.

**Beluga Whale *(Delphinapterus leucas)***

The beluga whales, resident in the estuary of the St. Lawrence River, are also candidate indicators of virtual elimination of discharges of persistent toxic substances to the Great Lakes and the St. Lawrence River. There are historic data on changes in the population of beluga whales in the Gulf of St Lawrence (Sergeant and Brodie 1975, Reeves and Mitchell 1984) and recent evidence on the reproductive success, incidence of deformities and tumors, and the relationship to persistent toxic substances (Beland *et al.* 1993). The species has high public visibility and interest, is very sensitive to several of the persistent toxic substances targeted for virtual elimination, and is high on the food chain. Measures, beside the bioaccumulation of contaminants, include age structure of the population, incidence of tumors, various reproductive parameters, evidence of immune suppression and feminization of males, and presence of benzo-a-pyrene adducts in DNA of tissues. Beland *et al.* (1993) found levels of mercury, lead, PCBs, DDT, mirex, and other contaminants were all much higher in the St. Lawrence population than in the Arctic population that was used as a reference. Their work suggests the beluga to be a viable indicator for the St. Lawrence River ecosystem. Because the beluga whales in the Gulf of St. Lawrence are classified as a threatened or endangered population, it cannot be sacrificed and only dead whales are available for study.

## 8.3 REPTILES AND AMPHIBIANS

Research and monitoring of reptiles and amphibians, in the Great Lakes basin, was initiated several years after the studies on birds. There is evidence from only one reptile species; the snapping turtle *(Chelydra serpentina serpentina)*, of the accumulation of high concentrations of persistent toxic substances associated with toxicological effects. There are suspicions that populations of other species such as the water snake *(Natrix sp.)* may have been affected, but no research seems to have been undertaken to date.

Despite the documentation of severe declines in certain populations of amphibians in various parts of the world, there are few data on Great Lakes populations (Bishop and Pettit 1992). One species that has been studied is the mudpuppy *(Necturus maculosus)*. In addition, there are reports that leopard frogs *(Rana pipiens)* from several parts of Minnesota have an anomalous incidence of deformities.

### Snapping Turtle *(Chelydra s. serpentina)*

The snapping turtle seems to be the only Great Lakes species of reptile that has been studied sufficiently to make it a candidate species for consideration as an indicator of virtual elimination. This long-lived omnivorous species commonly inhabits wetlands on the shorelines of the Great Lakes, and has a very limited home range throughout the year. These characteristics make the snapping turtle a good integrator of pollutant contamination within a local area (i.e. Areas of Concern). Hatching success of artificially incubated turtle eggs and the incidence of deformities in hatchling turtles have been measured at a number of sites within the basin (Bishop *et al.* 1991) and pilot biochemical studies have been initiated. But there are no data on changes in populations, behaviour, or histology attributable to exposures to persistent toxic substances, though some research is being undertaken on biochemical markers of effects on some endocrine effects (Bishop et al. 1998, deSolla et al. 1988).

### Mudpuppy *(Necturus maculosus)*

The mudpuppy is an aquatic salamander with a widespread distribution in northeastern North America. Research has been undertaken (Gendron *et al.* 1994) on a population in the St Lawrence River and the Ottawa River near Montreal to investigate the suitability of this amphibian as an indicator of pollution. Research was undertaken on the levels of pollutants accumulated and on the circulating levels of steroids, fecundity, gonado-somatic index, and egg diameter. The main findings were the high prevalence of skeletal deformities in areas with high levels of organochlorine pollutants consistent with an hypothesis of developmental toxicity (Bonin *et al.* 1995, Gendron *et al.* 1997).

## 8.4 FISH

There has been an extensive amount of research undertaken on effects on fish, but this investment has produced remarkably few case studies relating the observed anomalies to exposures to persistent toxic substances. For example, Leatherland (1993) reviewed the evidence of observed anomalies in endocrine, reproductive, and developmental conditions in Great Lakes salmonids. These anomalous conditions included thyroid lesions, lowered egg thyroid hormone, precocious sexual maturation, reduced levels of gonadal steroid hormone secretion, reduced egg fertility and high embryo mortality and deformity. He concluded that the evidence in support of contaminants causing these dysfunctional states in Great Lakes fish is not as convincing as the evidence in fish-eating mammals and birds.

Similarly, the case for an environmental etiology for some of the extinctions of Great Lakes lake trout and for current and past reproductive failures, known as early mortality syndrome, is contentious (Willford 1984, Mac and Edsall 1991, Mac *et al.* 1993). For example, Hnath (1994) has reviewed the evidence concerning the occurrence and putative causes of early mortality syndrome in Great Lakes salmonids and Carpenter *et al.* (1996) noted that the consensus within fisheries agencies that changes in populations of Great Lakes fish species could be explained without invoking exposures to environmental pollutants. A strong prima facie case can be made for the involvement of persistent toxic substances in the extinction of lake trout from Lake Ontario and releases of 2,3,7,8-tetrachlorodibenzo-*p*-dioxins based on:- the timing and distribution of the exposures and the onset of the population decline; the exquisite sensitivity of lake trout embryos to this compound; and the consistency of the suite of pathologies in the affected organisms with those exposed experimentally. Though the causal relationship is contentious, the lake trout is likely the only species that could be considered as a candidate species as an indicator of the reduction of effects of persistent toxic substances and that was therefore included in the workshop.

There has been extensive monitoring of the incidence and severity of papillomas and liver tumors in subpopulations of brown bullheads and white suckers *(Catostomus commersoni)* in the Great Lakes basin. In addition, there has been extensive research on the causal relationship with exposures to pollutants and particularly polynuclear aromatic hydrocarbons (Baumann *et al.* 1996). Thus the availability of trend data in the incidences of tumors in these species were considered in designing the workshop. In addition, there were several other endpoints considered, including mixed function oxidase enzymes (Luxon *et al.* 1987), and fin ray asymmetry (Whittle *et al.* 1987).

## 8.5 INVERTEBRATES

Invertebrates have been used extensively as indicators of water pollution since Kolkwitz and Marson developed the saprobien system for stream classification in the 1930s. Great Lakes surveys of invertebrate communities and of populations of particular species, such as the fish fly *(Hexagenia limbata)*, have been undertaken and various species have been used for toxicity testing and developed as bioassays. There is, however, no systematic research that has been performed relating the status of communities or of specific populations to exposures to persistent toxic substances. Thus the invertebrates were not considered further in relation to the planning for the workshop.

## 8.6 HUMANS

Concern about the effects of persistent toxic substances on human health arose in the early 1970s as a result of the chemical analytical findings of persistent toxic substances in fish and human samples. Humphrey (1988) wrote "The existence of toxic substances in the Great Lakes, which are used by hundreds of thousands of persons for recreational and subsistence fishing creates an opportunity and an obligation to scientifically address questions concerning the health of this ecosystem including its human population". Since then, there has been extensive epidemiological research on the effects of persistent toxic substances, particularly on development of infants.

Within the Great Lakes Basin, there have been two cohorts of infants whose mothers ate appreciable quantities of Great Lakes fish that were contaminated with persistent toxic substances. In western Michigan, a cohort was set up in 1980 and the researchers have reported associations between the amount of fish eaten, the concentrations of PCBs in cord blood, and inverse relationships with neuromuscular development at birth, cognitive development at 7 months, 4 years and at 11 years, and irreversible growth retardation (Fein *et al.* 1984, Jacobson *et al.* 1984,1985,1990, Jacobson and Jacobson 1988, 1993, 1996).

In a subsequent study to replicate the findings in the western Michigan cohort, the late Dr Helen Daly set up a cohort in 1990 in upper New York State, in which the mothers had eaten significant quantities of Lake Ontario fish prior to pregnancy. The results of the study showed that the infants whose mothers had eaten the largest amount of Lake Ontario fish had poorer reflex functioning and responses to stress.

In addition these infants were less able to habituate to a mildly noxious stimulation (Lonky *et al.* 1996). Although the manifestations of toxicity documented in the various Great Lakes cohorts of infants appear to be highly sensitive, they do not lend themselves to routine monitoring.

These findings are consistent with those from studies on cohorts of infants from outside the Great Lakes Basin (Rogan and Gladen 1985, Rogan *et al.* 1988). The studies of Rogan *et al.* (1988) and Tretjak (1989) of communities exposed incidentally, accidentally or occupationally, provide considerable insight into what other physiological systems and types of measures might be most sensitive and/or specific to PCB toxicity in adults. Humans can be used as monitors of persistent toxic substances by measuring the pollutants in their blood, lipid, and breast milk, but it is difficult to link adverse health effects in the general population specifically to persistent toxic substances.

There is considerable research and monitoring on critical subpopulations of humans exposed to persistent toxic substances in the Great Lakes Basin and this was reported at the Great Lakes /St. Lawrence Health Conference held in Montreal 12-15 May, 1997 in Montreal (Johnson *et al.* in press) Currently the EAGLE (Effects on Aboriginals of the Great Lakes Environment) Project, jointly funded by Health Canada and the Assembly of First Nations, is examining the effects of contaminants on health of aboriginals in over 60 communities in the Canadian portion of the Great Lakes basin. One phase of this project will include the collection of blood, urine and hair and the application of the caffeine breath test to a subgroup of the aboriginal population. This will allow a similar suite of biochemical markers to be measured in humans as is proposed for shoreline nesting bald eagles. Canadian and American studies are underway on cohorts of sports fishermen, and biochemical markers will be measured in some of these studies.

The Ecosystem Health Work Group of the Commission's Science Advisory Board held a workshop in September 1992 which considered humans as bioindicators (International Joint Commission 1994). The report, "A Prescription for Healthy Great Lakes" (NWF/CIELAP, 1991) has also addressed human indicators as a measure of the state of well-being of the Great Lakes.

Until recently there were very few studies of the effects on Great Lakes populations of humans of exposures to persistent toxic substances. Most of the research in the past has been concerned with monitoring exposures and assessment of the results in relation to laboratory based toxicology. With the passage of the Great Lakes Critical Programs Act in 1990, Congress mandated the preparation of a report on the adverse effects of water pollution on the health of people in the Great Lakes States (United States Environmental Protection Agency 1995).

In preparing for the Workshop on Environmental Results, the organizers were aware that there was particular interest in the monitoring of effects of persistent toxic substances in humans as well as the concentrations in samples from humans. The organizers therefore included consideration of human health effects in designing the workshop.

## Acknowledgments

Some of the descriptions of the suitability of various species as indicators of the virtual elimination of persistent toxic substances were included in the proceedings of the 1992 Indicators Workshop hosted by the International Joint Commission's Virtual Elimination Task Force (Fox, 1994). We are grateful to Dr. Milagros Simmons and to Dr Marlene Evans for reviewing a previous version of this paper.

## References

Addison, E.M., Fox, G.A.,and Gilbertson, M. (Eds).: 1991, 'Proceedings of the Expert Consultation Meeting on Mink and Otter', March 5 and 6, 1991, International Joint Commission, Windsor, Ontario. 30p.

Anderson, D.W. and Hickey, J.J.: 1972, 'Eggshell changes in certain North American birds', In. Voous, K.H. (Ed.). Proceedings of the XVth International Ornithological Congress. Brill, Leiden, Netherlands. pp 514-540.

Aulerich, R.J. and Ringer, R.K.: 1970, 'Some effects of chlorinated hydrocarbon pesticides on mink', *Am. Fur Breeder* **43**,10-11.

Aulerich, R.J. and Ringer, R.K.: 1977, 'Current status of PCB, toxicity to mink, and effect on their reproduction', *Arch. Environ. Contam. Toxicol.* **6**,279-292.

Baumann, P.C., Smith, I.R., and Metcalfe, C.D.: 1996, "Linkages between chemical contaminants and tumors in benthic Great Lakes fish", *J. Great Lakes Res.* **22**,131-153.

Beland, P., DeGuise, S., Girard, C., Lagace, A., Martineau, D., Michaud, R., Muir, D.C.G., Norstrom, R.J., Pelletier, E., Ray, S., and Shugart, L.R.: 1993, 'Toxic compounds and reproductive effects in St. Lawrence beluga whales', *J. Great Lakes Res.* **19**,766-775.

Best, D.A., Bowerman, W.W., Kubiak, T.J., Winterstein, S.R., Postupalsky, S., Shieldcastle, M.C., and Giesy, J.P.: 1994, 'Reproductive impairment of bald eagles *Haliaeetus leucocephalus* along the Great Lakes shorelines of Michigan and Ohio', In: Meyburg,B.-U. and R.D. Chancellor. (Eds.). *Raptor Conservation Today.* WWGBP/ The Pica Press.

Bishop, C.A., Brooks, R.J., Carey, J.H., Ng, P., Norstrom, R.J., and Lean, D.R.S.: 1991, 'The case for a cause-effect linkage between environmental contamination and development in eggs of the common snapping turtle (*Chelydra s. serpentina*) from Ontario, Canada', *J. Toxicol. Environ. Health* **33**,521-548.

Bishop, C.A. and Pettit, K.E. (Eds).: 1992, 'Declines in Canadian amphibian populations: designing a national monitoring strategy', Canadian Wildlife Service. Occasional Paper No. 76.

Bishop, C.A., Ng, P., Brooks, R.J., Kennedy, S., Stegeman, J.J., and Norstrom, R.J.: 1998, 'Developmental abnormalities in common snapping turtles and their relationship to chlorinated hydrocarbon concentrations in their eggs from the Great Lakes basin. Environ. Poll. (in press)

Blokpoel, H. and Scharf, W.C.: 1991, 'Status and conservation of seabirds nesting in the Great Lakes of North America', ICBP Technical Publication, **11**,17-41.

Boersma, D.C., Ellenton, J.A., and Yagminas, A.: 1986, 'Investigation of the hepatic mixed-function oxidase system in herring gull embryos in relation to environmental contaminants', *Environ. Toxicol. Chem.* **5**,309-318.

Bonin, J., DesGranges, J.-L., Bishop, C.A., Rodrigue, J., Gendron, A., and Elliott, J.E.: 1995, Comparative study of contaminants in the mudpuppy (Amphibia) and the common snapping turtle (Reptilia), St Lawrence River, Canada, *Archiv. Environ. Comtam. Toxicol.* **28**,184-194.

Bowerman, W.W.: 1993, 'Regulation of bald eagle *(Haliaeetus leucocephalus)* productivity in the Great Lakes basin: An ecological and toxicological approach', PhD Dissertation. Michigan State University.

Bowerman, W.W., Kubiak, T.J., Holt, J.B., Evans, D.L., Eckstein, R.G., Sindelar, C.R., Best, D.A., and Kozie, K.D.: 1994, 'Observed abnormalities in mandibles of nestling bald eagles *Haliaeetus leucocephalus*', *Bull. Environ. Contam. Toxicol.* **53**,450-457.

Bowerman, W.W., Giesy, J.P., Best, D.A., and Kramer, V.J.: 1995, 'A review of factors affecting productivity of bald eagles in the Great Lakes region: Implications for recovery', *Environ. Health Perspect.* 103,51-59.

Broley, C.L.: 1958, 'Plight of the American bald eagle', *Audubon Magazine* **60**,162-171.

Carpenter, S.R., Jackson, L.J., Kitchell, J.F., and Stow, C.A.: 1996, "Organochlorine contaminants in the Great Lakes: Response", *Ecological Applications*. **6**,971-974.

Clark, P.C., Norstrom, R.J., Fox, G.A., and Won, H.T.: 1987, 'Dynamics of organochlorine compounds in herring gulls *(Larus argantatus)*: II. A two-compartment model and data for ten compounds', *Environ. Toxicol. Chemistry* **6**,547-559.

Colborn, T.: 1991, 'Epidemiology of Great Lakes bald eagles, *J. Toxicol. Environ. Health* **33**,395-454.

Colborn, T., Davidson, A., Green, S.N., Hodge, R.A., Jackson, C.I., and Liroff, R.A.: 1990. '*Great Lakes Great Legacy?*' The Conservation Foundation, Washington, D.C., The Institute for Research on Public Policy, Ottawa, Ontario.

Courtney, P. and Blokpoel, H.:1983, 'Distribution and numbers of common terns on the lower Great Lakes during 1900-1980: A review', *Colonial Waterbirds* **6**,107-120.

Custer, T.W., Rattner, B.A., Ohlendorf, H.M., and Melancon, M.J.: 1991, 'Herons and egrets as proposed indicators of estuarine contamination in the United States', Acta 20th Congressus Internationalis Ornithologici: 2474-2478.

deSolla, S., Bishop, C.A., Van Der Kraak, G.J., and Brooks, R.J.: 1998, The effects of chlorinated hydrocarbons on sex hormones and sexual morphology of the common snapping turtle (Chelydra serpentina) in the Great Lakes basin. Environ. Health Perspect.. **106**

Ellenton, J.A., Brownlee, L.J., and Hollebone, B.R.: 1985, 'Aryl hydrocarbon hydroxylase levels in herring gull embryos from different locations on the Great Lakes', *Environ. Toxicol. Chem.* **4**,615-622.

Environment Canada.: 1990, 'Contaminants in Herring Gull Eggs from the Great Lakes', State of the Environment Fact Sheet No. 90-2.

Environment Canada.: 1994, 'The Fall and Rise of Osprey Populations in the Great Lakes Basin', Ontario Region, Catalogue No. EN40-222/1-1994E

Evans, A.S.: 1976, 'Causation and disease: The Henle-Koch postulates revisited', *Yale J. Biol. Med.*, **49**,175-195.

Evans, M.S. (Ed.), 1988, '*Toxic Contaminants and Ecosystem Health*', John Wiley, New York. 602pp.

Ewins, P.J.: 1992, 'Ospreys and contaminants on the Canadian Great Lakes, 1991', An Interim Report to the Canadian Wildlife Service (Ontario Region) and the Ontario Ministry of Natural Resources. 194p.

Ewins, P,J., Weseloh, D.V., Groom, J.H., Dobos, R.Z., and Mineau, P.: 1994, 'The diet of herring gulls *(Larus argentatus)* during the winter and early spring on the lower Great Lakes', In: J.J. Kerekes, (Ed.), Aquatic Birds in the Trophic Web of Lakes, *Hydrobiologia*, **279/280**,39-55.

Ewins, P.J., Postupalsky, S., Weise, T., and Addison, E.M.: 1995, 'Changes in the status, distribution and biology of ospreys *(Pandion haliaetus)* breeding on Lake Huron', In: Munawar, M., Edsall, T., and J. Leach, J. (Eds.), *The Lake*

*Huron Ecosystem: Ecology, Fisheries and the Management*. SPB Academic, Amsterdam.

Faber, R.A. and Hickey, J.J.: 1973, 'Eggshell thinning, chlorinated hydrocarbons, and mercury in inland aquatic bird eggs, 1969 and 1970', *Pestic. Monit. J.* **7**,27-36.

Fein, G.G., Jacobson, J.L., Jacobson, S.W., Schwartz, P.M., and Dowler, J.K.: 1984, 'Prenatal exposure to polychlorinated biphenyls: Effects on birth size and gestational age', *J. Pediatrics* **105**,315-320.

Feyk, L.A., Giesy, J.P., and Lambert, G.H.: 1995, 'The caffeine breath test: A less invasive method to measure cytochrome P450-1A activity in birds', In: Butterworth, F.M., Corkum, L.D., and Guzman-Rincon, J. (Ed.). *Biomonitors and Biomarkers as Indicators of Environmental Change.* Plenum Press, New York.

Fitchko, J., (Ed.): 1986, 'Literature review of the effects of persistent toxic substances on Great Lakes Biota', Report to the Great Lakes Science Advisory Board. Windsor, Ontario: International Joint Commission. 256pp.

Fox, G.A.: 1991, 'Practical causal inference for ecoepidemiologists', *J. Toxicol. Environ. Health* **33**,359-373.

Fox, G.A.: 1993, 'What have biomarkers told us about the effects of contaminants on the health of fish-eating birds in the Great Lakes?' The theory and a literature review, *J. Great Lakes Res.* **19**,722-736.

Fox, G.A. (Ed.).: 1994, 'Bioindicators as a measure of success for virtual elimination of persistent toxic substances', International Joint Commission. Windsor, Ontario. 37p.

Fox, G.A., Gilman, A.P., Peakall, D.B., and Anderka, F.W.: 1978, 'Behavioral abnormalities of nesting Lake Ontario herring gulls', *J.Wildl. Manage.* **42**,477-483.

Fox, G.A. and Weseloh, D.V.: 1987, 'Colonial waterbirds as bio-indicators of environmental contamination in the Great Lakes', ICBP Technical Pub. No. 6. 209-216.

Fox,G.A., Kennedy, S.W., Norstrom, R.J., and Wigfield, D.C.: 1988, 'Porphyria in herring gulls: a biochemical response to chemical contamination of Great Lakes food chains', *Environ. Toxicol. Chem.* **7**,831-839.

Fox, G.A., Allan, L.J., Weseloh, D.V., and Mineau, P.: 1990, 'The diet of herring gulls during the nesting period in Canadian waters of the Great Lakes', *Can. J. Zool.* **68**,1075-1085.

Fox, G.A., Collins, B., Hayakawa, E., Weseloh, D.V., Ludwig, J.P., Kubiak, T.J., and Erdman, T.C.: 1991, 'Reproductive outcomes in colonial fish-eating birds: A biomarker for developmental toxicants in Great Lakes food chains. II. Spatial variation in the occurrence and prevalence of bill defects in young double-crested cormorants in the Great Lakes, 1979-1987', *J. Great Lakes Res.* **17**,158-167.

Gendron, A., Fortin, R., and Hontela, A.: 1994, 'Multi-level detection of toxic stress in the mudpuppy *(Necturus maculosus)* an aquatic salamander', Progress Report to the Canadian Wildlife Service and the St. Lawrence River Remedial Action Plan, Ottawa, Ontario. 41 pp.

Gendron, A., Bishop, C.A., Fortin, R., and Hontela, A.: 1997. 'In vivo testing of the functional integrity of the corticosterone-producing axis in mudpuppy (Amphibia) exposed to chlorinated hydrocarbons in the St Lawrence and Ottawa Rivers.' *Archiv. Environ. Contam. Toxicol.* **16**,1694-1706.

Giesy, J.P., Bowerman, W.W., Mora, M.A., Jones, P.D., Verbrugge, D.A., Othoudt, R.A., Newsted, J.L., Vandervoort, C., Summer, C.L., Aulerich. R.J., Bursian, S.J., Ludwig, J.P., Ludwig, M.E., Dawson, G.A., Kubiak, T.J., Best, D.A., and Tillitt, D.E.: 1995, 'Contaminants if fishes from Great Lakes-influenced sections and above dams of three Michigan rivers: III. Implications for health of bald eagles', *Arch. Environ. Contam. Toxicol.* 29,309-321.

Gilbertson, M.: 1974, 'Pollutants in breeding herring gulls in the lower Great Lakes', *Can. Field-Nat.* **88**,273-280.

Gilbertson, M.: 1983, 'Etiology of chick edema disease in herring gulls in the lower Great Lakes', *Chemosphere*, **12**,357-370.

Gilbertson, M.: 1988, 'Epidemics in birds and mammals caused by chemicals in the Great Lakes', In: Evans, M.S. (Ed.). *Toxic Contaminants and Ecosystem Health; a Great Lakes Focus*. John Wiley and Sons, Inc., pp. 133-152.

Gilbertson, M. and Reynolds, L.M.: 1972, 'Hexachlorobenzene (HCB) in the eggs of common terns in Hamilton Harbour, Ontario', *Bull. Environ. Contam. Toxicol.* **7**,371-373.

Gilbertson, M. and Hale, R.: 1974a, 'Characteristics of the breeding failure of a colony of herring gulls on Lake Ontario', *Can. Field-Nat.* **88**,356-358.

Gilbertson, M. and Hale, R.: 1974b, 'Early embryonic mortality in a herring gull colony in Lake Ontario', *Can. Field-Nat.* **88**,354-356.

Gilbertson, M., Morris, R.D., and Hunter, R.A.: 1976, 'Abnormal chicks and PCB residue levels in eggs of colonial birds on the lower Great Lakes (1971-1973), *Auk* **93**,434-442.

Gilbertson, M. and Fox, G.A.: 1977, 'Pollutant-associated embryonic mortality of Great Lakes herring gulls', *Environ. Poll.* **12**,211-216.

Gilbertson, M., Kubiak, T., Ludwig, J., and Fox,G., 1991, 'Great Lakes Embryo mortality, edema, and deformities syndrome (GLEMEDS) in colonial fish-eating birds: Similarity to chick-edema disease', *J. Toxicol. Environ. Health* **33**,455-520.

Gilbertson, M.: 1997, 'Are causes knowable? Some consequences of successional versus toxicological interpretations of the Great Lakes Water Quality Agreement', *Can. J. Fish. Aquat. Sci.* **54**,483-495.

Gilman, A.P., Fox, G.A., Peakall, D.B., Teeple, S.M., Carroll, T.R., and Haymes, G.T.: 1977, 'Reproductive parameters and egg contaminant levels of Great Lakes herring gulls', *J. Wildl. Mgmt.* **41**,458-468.

Grasman, K.A., Scanlon, P.F., Fox, G.A., and Ludwig, J.P.: 1996, 'Organochlorine-associated immunosuppression in prefledging Caspian terns and herring gulls from the Great Lakes: An epidemiological study', *Environ. Health Perspec.* **104**(suppl. 4),829-842.

Harris, H.J., Erdman, T.J., Ankley, G.T., and Lodge, K.B.: 1993, 'Measures of reproductive success and polychlorinated biphenyl residues in eggs and chicks of Forster's terns on Green Bay, Lake Michigan, Wisconsin - 1988', *Arch. Environ. Contam. Toxicol.* **25**,304-314.

Hartsough, G.R.: 1965, 'Great Lakes fish now suspect as mink food', *Am. Fur Breeder* **38**,27-35.

Heaton, S.N., Bursian, Giesy, J.P., Tillitt, D.E., Render, J.A., Jones, P.D., Verbrugge, D.A., Kubiak, T.J., and Aulerich, R.J.: 1995a, Dietary exposure of mink to carp from Saginaw Bay, Michigan.1. Effects on reproduction and survival, and the potential risks to wild mink populations. *Arch. Environ. Contam. Toxicol.* **28**,334-343.

Heaton, S.N., Bursian, Giesy, J.P., Tillitt, D.E., Render, J.A., Jones, P.D., Verbrugge, D.A., Kubiak, T.J., and Aulerich, R.J.: 1995a, 'Dietary exposure of mink to carp from Saginaw Bay, Michigan. 2. Hematology and liver pathology'. *Arch. Environ. Contam. Toxicol.* **29**,411-417.

Heinz, G.H., Erdman, T.C., Haseltine, S.D., and Stafford, C.: 1985, 'Contaminant levels in colonial waterbirds from Green Bay and Lake Michigan, 1975-1980'. *Environ. Monit. Assess. J.* **5**,223-236.

Henshel, D.S., Martin, J.W., Norstrom, R.J., Elliott,J., Cheng, K.M., and DeWitt, J.C.: 1997. Morphometric brain abnormalities in double-crested cormorant chicks exposed to polychlorinated dibenzo-*p*-dioxins, dibenzofurans, and biphenyls. *J. Great Lakes Res.* **23**,11-26.

Hickey, J.J., and Anderson, D.W.: 1968, 'Chlorinated hydrocarbons and eggshell changes in raptorial and fish-eating birds', *Science* **162**,271-273.

Hnath, J.G.: 1994, 'Proceedings of a workshop on EMS - Early Mortality Syndrome', Great Lakes Fish Health Committee: Great Lakes Fishery Commission. Ann Arbor, Michigan.

Hoffman, D.J., Rattner, B.A., Sileo, L., Docherty, D., and Kubiak, T.J.: 1987, 'Embryotoxicity, teratogenicity, and aryl hydrocarbon hydroxylase activity in Forster's terns on Green Bay, Lake Michigan', *Environ. Res.* **42**,176-184.

Hoffman, D.J., Smith, G.J., and Rattner, B.A.: 1993, 'Biomarkers of contaminant exposure in common terns and black-crowned night herons in the Great Lakes', *Environ. Toxicol. Chem.* **12**,1095-1103.

Humphrey, H.E.B.: 1988. 'Chemical contaminants in the Great Lakes: The human health aspect'. In: Evans, M.S. (Ed.). *Toxic Contaminants and Ecosystem Health; a Great Lakes Focus*. John Wiley and Sons, Inc., pp. 153-165.

International Joint Commission.: 1975, Appendix B: Annual Report of the Surveillance Subcommittee to the Implementation Committee, Great Lakes Water Quality Board. Windsor, Ontario, June 1975, 212 pp.

International Joint Commission.: 1979, Appendix G: Annual Report of the Committee on the Assessment of Human Health Effects of Great Lakes Water Quality to the Great Lakes Water Quality Board and Great Lakes Science Advisory Board. Windsor, Ontario, 34 pp.

International Joint Commission.: 1985, Report of the Aquatic Ecosystem Objectives Committee to the Great Lakes Science Advisory Board. Windsor, Ontario, 119pp.

International Joint Commission.: 1989a, Appendix B: Great Lakes Surveillance. Report to the Great Lakes Water Quality Board. 3 Vols. Windsor, Ontario, March 1989, 879 pp.

International Joint Commission.: 1989b, Proceedings of the Workshop on Cause-Effect Linkages held March 1989. Report of the Council of Great Lakes Research Managers to the Great Lakes Science Advisory Board. Windsor, Ontario. 51 pp.

International Joint Commission.: 1990a, Proceedings of a Workshop on the Role of Epidemiology in Assessing the Effects of Great Lakes Water Quality on Human Health, March 1-3, 1988, Scarborough, Ontario. Sponsored by the Great Lakes Water Quality Board and Great Lakes Science Advisory Board through their Committee for the Assessment of Human Health Effects. Windsor, Ontario, 30 pp.

International Joint Commission.: 1990b, Fifth Biennial Report Under the Great Lakes Water Quality Agreement of 1978 to the Governments of the United States and Canada and the State and Provincial Governments of the Great Lakes Basin, Part II. Washington, DC and Ottawa, Ontario. 60 pp.

International Joint Commission.: 1990c, Final Report of the Ecosystem Objectives Committee to the Great Lakes Science Advisory Board. Windsor, Ontario. 53pp.

International Joint Commission.: 1991, Proceedings of the Expert Consultation Meeting on Bald Eagles, February 12-13, 1990, Windsor, Ontario, Report of the Ecological Committee's Biological Effects Subcommittee to the Science Advisory Board. Best, D.A., Gilbertson, M. and Hudson, H. (Eds.). Windsor, Ontario, 33 pp.

International Joint Commission.: 1992, Proceedings of the Third Expert Consultation Meeting on Bald Eagles in the Great Lakes Basin, Report of the Ecological Committee's Biological Effects Subcommittee to the Science Advisory Board. Gilbertson, M. (Ed.). Windsor, Ontario, February 1992, 29 pp.

International Joint Commission.: 1993, Legislative and Regulatory Considerations for Virtual Elimination of Persistent Toxic Substances, Report of the Great Lakes Water Quality Board. Windsor, Ontario. 15 pp.

International Joint Commission.: 1994, Our Community, Our Health: Dialogue Between Science and Community, Report of the Workgroup on Ecosystem Health. Windsor, Ontario. 50 pp.

Jacobson, J.L., Jacobson, S.W., Fein, G.G., Schwartz, P.M., and Dowler, J.K.: 1984, 'Prenatal exposure to an environmental toxin: A test of the multiple effects model', *Dev. Psych.* **20**,523-532.

Jacobson, S.W., Fein, G.G., Jacobson, J.L., Schwartz, P.M., and Dowler, J.K.: 1985, 'The effect of intrauterine PCB exposure on visual recognition memory', *Child. Dev.* **56**,853-860.

Jacobson, J.L., and Jacobson, S.W.: 1988, 'New methodologies for assessing the effects of prenatal toxic exposure on cognitive functioning in humans'. In: Evans, M.S. (Ed.). *Toxic Contaminants and Ecosystem Health; a Great Lakes Focus*. John Wiley and Sons, Inc., pp. 373-388.

Jacobson, J.L., Jacobson S.W., and Humphrey, H.E.B.: 1990, 'Effects of exposure to PCB and related compounds on growth and activity in children', *Neurotox. Teratol.* **12**,319-326.

Jacobson, J. L. and Jacobson, S.W.: 1993, 'A 4-year follow up study of children born to consumers of Lake Michigan fish', *J. Great Lakes Res.* **19**,776-783

Jacobson, J. L. and Jacobson, S.W.: 1996. 'Intellectual impairment in children exposed to polychlorinated biphenyls in utero', *New Eng. J Med.* **335**,783-789.

Johnson, B.L., Hicks, H.E., Jones, D.E., Cibulas, W., and De Rosa, C.T.: in press. 'Public Health Implications of Persistent Toxic Substances in the Great Lakes and St. Lawrence Basins'. *J. Great Lakes Res.*

Keith, J.A.: 1966, 'Reproduction in a population of herring gulls (*Larus argentatus*) contaminated by DDT', *J. Appl. Ecol.* **3**(Suppl),57-70.

Kennedy, S.W. and Fox, G.A.: 1990, 'Highly carboxylated porphyrins as a biomarker of polyhalogenated aromatic hydrocarbon exposure in wildlife: confirmation of their presence in Great Lakes herring gull chicks in the early 1970s and important methodological details', *Chemosphere* **21**,407-415.

Kubiak, T.J., Harris, H.J., Smith, L.M., Schwartz, T.R., Stalling, D.L., Trick, J.A., Sileo, L., Docherty, D.E., and Erdman, T.C.: 1989, 'Microcontaminants and reproductive impairment of the Forster's tern on Green Bay, Lake Michigan -- 1983', *Arch. Environ. Contam. Toxicol.* **18**,706-727.

L'Arrivee, L., and Blokpoel, H.: 1988, 'Seasonal distribution and site fidelity in Great Lakes Caspian terns', *Colonial Waterbirds* **11**,202-214.

Leatherland, J.F.: 1993, 'Field observations and reproductive and developmental dysfunction in introduced and native salmonids from the Great Lakes', *J. Great Lakes Res.* **19**,737-751.

Lonky, E., Reihman, J., Darvill, T., Mather, Sr., J., and Daly, H.: 1996, 'Neonatal behavioral assessment scale performance in humans influenced by maternal consumption of environmentally contaminated Lake Ontario fish', *J. Great Lakes Res.* **22**,198-212.

Ludwig, J.P.: 1974, 'Recent changes in the ring-billed gull population and biology in the Laurentian Great Lakes', *Auk* **91**,575-594.

Ludwig, J.P.: 1979, 'Present status of the Caspian tern population of the Great Lakes', *Michigan Academician* **11**,69-77.

Ludwig, J.P.: 1984, 'Decline, resurgence and population dynamics of Michigan and Great Lakes double-crested cormorants', *Jack-Pine Warbler* **62**,91-102.

Ludwig, J.P. and Tomoff, C.S.: 1966, 'Reproductive success and insecticide residues in Lake Michigan herring gulls', *Jack-Pine Warbler* **44**,77-84.

Ludwig, J.P. and Ludwig, C.E.: 1969. 'The effect of starvation on insecticide contaminated herring gulls removed from a Lake Michigan colony', *Proc. 12th Conf. Great Lakes Res.* 53-60.

Ludwig, J.P., Giesy, J.P., Summer, C.L., Bowerman, W., Heaton, S., Aulerich, R., Bursian, S., Auman, H.J., Jones, P.D., Williams, L.L., Tillitt, D.E., and Gilbertson, M.: 1993a, 'A comparison of water quality criteria for the Great Lakes based on human and wildlife health', *J. Great Lakes Res.* **19**,789-807.

Ludwig, J.P., Auman, H.J., Kurita, H., Ludwig, M.E., Campbell, L.M., Giesy, J.P., Tillitt, D.J., Jones, P., Yamashita, N., Tanabe, S., and Tatsukawa, R.: 1993b, 'Caspian tern reproduction in the Saginaw Bay ecosystem following a 100-year flood event', *J.Great Lakes Res.* **19**,96-108.

Ludwig, J.P., Auman, H.J., Weseloh, D.V., Fox, G.A., Giesy, J.P., and Ludwig, M.E.: 1995. 'Evaluation of the effects of toxic chemicals in Great Lakes cormorants: Has causality been established?' *Colonial Waterbirds* **18**(Spec.Publ. 1),60-69.

Ludwig, J.P., Kurita-Matsuba, H., Auman,H.J., Ludwig ,M.E., Summer, C.L., Giesy, J.P., Tillitt, D.E., and Jones, P.D., 1996, 'Deformities, PCBs and TCDD-equivalents in double-crested cormorants (*Phalacrocorax auritus*) and Caspian terns (*Hydroprogne caspia*) of the upper Great Lakes 1986-1991: Testing a cause-effect hypothesis' *J. Great Lakes Res.* **22**:172-197.

Luxon, P.L., Hodson, P.V., and Borgmann, U.: 1987, 'Hepatic aryl hydrocarbon hydroxylase activity of lake trout *(Salvelinus namaycush)* as an indicator of organic pollution' *Environ. Toxicol Chem.* **6**,649-657.

Mac, M.J. and Edsall, C.C.: 1991, 'Environmental contaminants and the reproductive success of lake trout in the Great Lakes: An epidemiological approach', *J. Toxicol. Environ. Health* **33**,375-394.

Mac, M.J., Schwartz, T.R., Edsall, C.C., and Frank, A.M.: 1993, 'Polychlorinated biphenyls in Great Lakes lake trout and their eggs: relations to survival and congener composition 1979-1988', *J. Great Lakes Res.* **19**,752-765.

McKeating,G.: 1985, 'Charles Broley: Eagles then and now in Southern Ontario',In: Gerrard,J.M. and T.N.Ingram.(Eds.). *The Bald Eagle in Canada.* Hignell, Winnipeg. 25-34.

Mineau,P., Fox, G.A., Norstrom, R.J., Weseloh, D.V., Hallett, D.J., and Ellenton, J.A.: 1984, 'Using the herring gull to monitor levels and effects of organochlorine contamination in the Canadian Great Lakes', in Nriagu,J.O. and Simmons, M.S. *Toxic Contaminants in the Great Lakes.* John Wiley and Sons, Inc. New York.

Moccia, R.D., Fox, G.A., and Britton, A.: 1986, 'A quantitative assessment of thyroid histopathology of herring gulls (*Larus argentatus*) from the Great Lakes and a hypothesis on the causal role of environmental contaminants', *J. Wildl. Dis.* **22**,60-70.

Moore, N.W.: 1966, 'A pesticide monitoring system with special reference to the selection of indicator species', *J. Appl. Ecol.* **3(Suppl.)**:261-269.

Mora, M.A., Auman, H.J., Ludwig, J.P., Giesy, J.P., Verbrugge, D.A., and Ludwig, M.E.: 1993, 'Polychlorinated biphenyls and chlorinated insecticides in plasma of Caspian terns: Relationships with age, productivity, and colony site tenacity in the Great Lakes', *Arch. Environ. Contam. Toxicol.* **24**,320-331.

Morris, R.D., Hunter, R.A., and McElman, J.F.: 1976, 'Factors affecting the reproductive success of common terns *(Sterna hirundo)* colonies on the lower Great Lakes during the summer of 1972', *Can. J. Zool.* **54**,1850-1862.

Morris, R.D., Kirkham, I.R., and Chardine, J.W.: 1980, 'Management of a declining common tern colony', *J. Wildl. Mgmt.* **44**,241-245.

Nisbet, I.C.T.: 1989, 'Organochlorines, reproductive impairment, and declines in bald eagle *Haliaeetus leucocephalus* populations: mechanisms and dose-response relationships', In: Meyburg,B.-U. and R.D.Chancellor (Eds.). *Raptors in the*

*Modern World.* World Working Group on Birds of Prey: Berlin, London and Paris. Pp.483-489.

NWF/CIELAP.: 1991. *'A Prescription for Healthy Great Lakes'*, National Wildlife Federation, Ann Arbor, Michigan and Canadian Institute for Environmental Law and Policy, Toronto. 65 pp.

Peakall, D.B.: 1988, 'Known effects of pollutants on fish-eating birds in the Great Lakes of North America', In: Schmidke,N.W. (Ed.) Toxic Contaminants in Large Lakes. Lewis Publishers, Inc. Chelsea, Michigan.

Peakall, D.B. and Fox, G.A.: 1987, 'Toxicological investigations of pollutant-related effects in Great Lakes gulls', *Environ. Health Perspec.* **71**,187-193.

Pekarik, C., and Weseloh, D.V.: 1998. 'Organochlorine Contaminants in Herring Gull Eggs from the Great Lakes, 1974-1995: Change Point Analysis and Short Term Regression', *Environ. Monitor. Assess.* this issue.

Poole, A.F.: 1989, *'Ospreys: A Natural and Unnatural History'* Cambridge University Press.

Postupalsky, S.: 1971, 'Toxic chemicals and declining bald eagles and cormorants in Ontario', Canadian Wildlife Service, Pesticide Section. Report No.20. 45pp.

Postupalsky, S.: 1977a, 'Status of the osprey in Michigan', Pp.153-166 In: Ogden,J.C. (Ed.). *Transactions of the North American Osprey Research Conference.* U.S. Department of the Interior National Park Service, Transactions and Proceedings Series No. 2.

Postupalsky, S.: 1977b, 'Status of the osprey in the Lake Nipigon area'. Pp.193-198 In: Ogden,J.C. (Ed.). *Transactions of the North American Osprey Research Conference.* U.S. Department of the Interior National Park Service, Transactions and Proceedings Series No. 2.

Rattner, B.A., Melancon, M.J., Custer, T.W., Hothem, R.L., King, K.A., LeCaptain, L.J., Spann, J.W., Woodin, B.R., and Stegeman, J.J.: 1993, 'Biomonitoring environmental contamination with pipping black-crowned night heron embryos: induction of cytochrome P450', *Environ. Toxicol. Chem.* **12**,1719-1732.

Reeves,R.R. and Mitchell, E.: 1984, 'Catch history and initial population of white whales (*Delphinapterus leucas*) in the River and Gulf of St. Lawrence, Eastern Canada', *Le Naturaliste canadien* **111**,63-121.

Rogan,W.J. and Gladen,B.C.: 1985, 'Study of human lactation for effects of environmental contaminants: The North Carolina breast milk and formula project and some other ideas', *Environ. Health Perspec.* **60**,215-221.

Rogan,W.J., Gladen,B.C., Hung,K.L., Koong,S.L., Shih,L.Y., Taylor,J.S., Wu,Y.C., Yang,D., Ragan,N.B., and Hsu,C.C.: 1988, 'Congenital poisoning by polychlorinated biphenyls and their contaminants in Taiwan', *Science* **241**,334-336.

Ryder, J.P. and Chamberlain, D.J.: 1972, 'Congenital foot abnormality in the ring-billed gull', *Wilson Bulletin* **84**,342-344.

Scharf, W.C. and Buckingham, B.: 1974, 'Cogenital foot abnormality in a herring gull', *Inland Bird Banding* **46**,30-32.

Scharf, W.C. and Shugart, G.W.: 1984, 'Distribution and phenology of nesting Forster's terns in eastern Lake Huron and Lake St. Clair', *Wilson Bulletin* **96**,309-313.

Sergeant,D.E. and Brodie, P.F.: 1975, 'Identity, abundance, and present status of populations of white whales, *Delphinapterus leucas*, in North America', *J.Fish. Res. Bd. Can.* **32**,1047-1054.

Shugart, G.W.: 1980, 'Frequency and polygyny in Great Lakes herring gulls in 1978', *Condor* **82**,426-429.

Sileo, L., Karstad, L., Frank, R., Holdrinet, M.V.H., Addison, E., and Braun, H.E.: 1977, 'Organochlorine poisoning of ring-billed gulls in southern Ontario', *J. Wildl. Dis.* **13**,313-322.

Spear, P.A., Moon, T.W., and Peakall, D.B.: 1986, 'Liver retinoid concentrations in natural populations of herring gulls (*Larus argentatus*) contaminated by 2,3,7,8-tetrachlorodibenzo-*p*-dioxin and in ring doves (*Streptopelia risoria*) injected with a dioxin analogue', *Can. J. Zool.* **64**,204-208.

Spear, P.A., Bourbonnais, D.H., Norstrom, R.J., and Moon, T.W.: 1990, 'Yolk retinoids (vitamin A) in eggs of the herring gull and correlations with polychlorinated dibenzo-*p*-dioxin and dibenzofurans', *Environ. Toxicol. Chem.* **9**,1053-1061.

Sprunt, A., Robertson, Jr., W.B., Postupalsky, S., Hensel, R.J., Knoder, C.E., and Ligas, F.J.: 1973, 'Comparative productivity of six bald eagle populations', *Trans. N. Am. Wildl. Nat. Res. Conf.* **38**,96-106.

Susser, M.: 1986, 'The logic of Sir Karl Popper and the practice of epidemiology', *American J. Epidemiol.* **12**,711-718.

Swain, W.R.: 1991, 'Effects of organochlorine chemicals on the reproductive outcome of humans who consumed contaminated Great Lakes fish: An epidemiological consideration', *J. Tox. Environ. Health.* **33**,587-639.

Tretjak, Z., Beckmann,S., Tretjak,A., and Gunnerson, C.: 1989, 'Occupational, environmental and public health in Semic: a case study of polychlorinated biphenyl (PCB) pollution', pp. 57-72. In: *Post-Audits of Environmental Programs and Projects.* Proc. Environ. Impact. Anal. Res. Council/ASCE, New Orleans.

Teeple, S.M.: 1977, 'Reproductive success of herring gulls nesting on Brothers Island, Lake Ontario in 1973', *Can. Field-Nat.* **91**,148-157.

Tillitt, D.E., Ankley, G.E., Giesy, J.P., Ludwig, J.P., Kurita-Matsuba, H., Weseloh, D.V., Ross, P.S., Bishop, C.A., Sileo, L., Stromborg, K.L., Larson, J., and Kubiak, T.J.: 1992, 'Polychlorinated biphenyl residues and egg mortality in double-crested cormorants from the Great Lakes', *Environ. Toxicol. Chem.* **11**,1281-1288.

U.S. Department of Health, Education and Welfare.: 1964, '*Smoking and Health*', Report of the Advisory Committee to the Surgeon General. Washington, D.C.: Public Health Service Publication 1103.

U.S. Environmental Protection Agency.: 1995, *'The Effects of Great Lakes Contaminants on Human Health'*, Report to Congress, USEPA-905-R-95-017.

Virtual Elimination Task Force.: 1993, 'A Strategy for Virtual Elimination of Persistent Toxic Substances (Volumes 1)',Final report to the International Joint Commission. Windsor, Ontario, 72 pp.

Weseloh, D.V.: 1984, 'The origins of banded herring gulls recovered in the Great Lakes region', *J. Field Ornithol.* **55**,190-195.

Weseloh, D.V., Teeple, S.M., and Gilbertson, M.: 1983, 'Double-crested cormorants of the Great Lakes: egg-laying parameters, reproductive failure and contaminant residues in eggs, Lake Huron 1972-73', *Can. J. Zool* **61**,427-436.

Weseloh, D.V., Ewins, P.J., Struger, J., Mineau, P., and Norstrom, R.J.: 1994, 'Geographic distribution of organochlorine contaminants and reproductive parameters in herring gulls on Lake Superior in 1983',*Environ. Monit. Assessm.* **29**,229-251.

Weseloh, D.V. and Ewins, P.J.: 1994, 'Characteristics of a rapidly increasing colony of double-crested cormorants *(Phalacrocorax auritus)* in Lake Ontario: population size, reproductive parameters and band recoveries', *J. Great Lakes Res.* **20**,443-456.

Weseloh, D.V., Ewins, P.J., Ewins, J., Mineau, P., Bishop, C.A., Postupalsky, S., and Ludwig, J.P.: 1995, 'Double-crested cormorants of the Great Lakes: changes in population size, breeding distribution and reproductive output, 1913-1991', In: Nettleship,D.N. and D.C.Duffy. (Eds.). *The Double-crested Cormorant: Biology, Conservation and Management. Colonial Waterbirds* **18**,48-59 (Special publication No 1).

Whittle, D.M., Kier, J.J., and Hyatt, W.H.: 1987, 'Contaminant burden and bilateral fin ray asymmetry in forage fish of the Laurentian Great Lakes' Poster Presentation, SETAC 8th Annual Meeting. Nov. 9-12, 1987.

Wiemeyer, S.N., Lamont, T.G., Bunck, C.M., Sindelar, C.R., Gramlich, F.J., Fraser, J.D., and Byrd, M.A.: 1984, 'Organochlorine pesticides, polychlorobiphenyl, and mercury residues in bald eagle eggs-1969-79-and their relationships to shell thinning and reproduction', *Arch. Environ. Contam. Toxicol.* **13**,529-549.

Wiemeyer, S.N., Bunck, C.M., and Stafford, C.J.: 1993, 'Environmental contaminants in bald eagle eggs - 1980-84 - and further interpretations of relationships to productivity and shell thickness', *Arch. Environ. Contam. Toxicol.* **24**,213-227.

Willes, R.F., Nestman, E.R., Miller, P.A., Orr, J.A., and Munro, I.C.: 1993, 'Scientific principles for evaluating the potential for adverse effects from chlorinated organic chemicals in the environment', *Regulat. Toxicol. Pharmacol.* **18**,313-356.

Willford, W.A.: 1984, 'Contaminants', In: Eshenroder, R.L., T.P. Poe, and C.H. Olver. 1984. Strategies for rehabilitation of Lake Trout in the Great Lakes: Proceedings of a conference on lake trout research, August 1983. Great Lakes Fishery Commission. Technical Report No. 40. 63p.

Wren, C.D.: 1991, 'Cause-effect linkages between chemicals and populations of mink *(Mustela vison)* and otter *(Lutra canadensis)* in the Great Lakes basin', *J. Toxicol. Environ. Health* **33**,549-586.

Yamashita, N., Tanabe, S., Ludwig, J.P., Kurita, H., Ludwig, M.E., and Tatsukawa, R.: 1993, 'Embryonic abnormalities and organochlorine contamination in double-crested cormorants (*Phalacrocorax auritus*) and Caspian terns (*Hydroprogne caspia*) from the upper Great Lakes', *Environ. Pollut.* **79**,163-173.

# CURRENT STATUS AND TEMPORAL TRENDS IN CONCENTRATIONS OF PERSISTENT TOXIC SUBSTANCES IN SPORT FISH AND JUVENILE FORAGE FISH IN THE CANADIAN WATERS OF THE GREAT LAKES

W.A. SCHEIDER, C. COX, A. HAYTON, G. HITCHIN, A. VAILLANCOURT

*Environmental Monitoring and Reporting Branch, Ontario Ministry of the Environment, 125 Resources Road, Etobicoke, Ontario, M9W 5L1, Canada.*

**Abstract.** This paper presents a summary of the current status and temporal trends over the past 15 - 20 years in contaminant levels in sport fish and juvenile forage fish in the Canadian waters of the Great Lakes. Fish consumption advisories summarized from the 1995 *Guide to Eating Ontario Sport Fish* showed that 67% of the 1736 consumption advisories in the Canadian waters of the Great Lakes had no restrictions. In the remaining 33% of the advisories, consumption of sport fish was restricted to 4 meals per month or less. Lake Erie had the fewest consumption restrictions (19%) and Lake Ontario the most (45%). PCBs were the principal contaminants of concern responsible for 47% of the consumption restrictions in the Canadian waters of the Great Lakes and caused the most consumption restrictions in each of the Great Lakes except Lake Superior where toxaphene caused 69% of the consumption restrictions. Concentrations of PCBs in sport fish declined in Lake Huron and Lake Ontario over the period 1976-1994. A decline in mirex concentrations in sport fish from Lake Ontario was also observed over the same time period. Concentrations of Hg in sport fish from Lake St. Clair declined over the period 1970-1994, but mercury in sport fish showed no trend over time in Lake Huron or Lake Ontario over the period 1981-1994. Contaminant levels in juvenile forage fish collected in 1993 and 1994 at 44 locations in the lower Great Lakes were assessed against wildlife protection guidelines. Concentrations that exceeded the Forage Fish Contaminant Index were observed at 17 locations with PCBs being the principal contaminant of concern. PCB concentrations in spottail shiners declined at 12 of 16 locations monitored in the lower Great Lakes over the period 1975-1994.

## 1. Introduction

The presence of persistent toxic chemicals in the Great Lakes is recognized as a significant stress on the health of this ecosystem. Persistent toxic substances cause fish consumption advisories in the waters of the Great Lakes, and the evidence strongly suggests that contaminants cause reproductive impairment in fish and wildlife, and increased occurrences of liver tumours in fish (Government of Canada 1991).

To assess trends in the concentrations of persistent toxic substances in the Great Lakes, emission monitoring programs and monitoring programs for air, water, sediment, fish and other wildlife are in place. In this paper we report on the findings of two monitoring programs which measure and assess contaminant concentrations in fish from the Canadian waters of the Great Lakes. We give an overview of the current status and temporal trends of persistent toxic contaminants in sport fish and juvenile forage fish.

A variety of organochlorine contaminants and metals bioaccumulate in fish. Contaminants that are often undetectable in ambient water samples may be detected in fish. Because fish integrate spatial and short term temporal changes in water quality and contaminant availability, tissue residue levels of contaminants in fish provide an excellent basis for assessing trends of persistent toxic chemicals.

Contaminant concentrations in sport fish from the Canadian waters of the Great Lakes have been monitored for over 20 years and the results used to provide advice to the public on fish consumption. Consumption advice has been published since 1977 and is currently provided biennially in the *Guide to Eating Ontario Sport Fish* publication (Ontario Ministry of Environment and Energy, and Ontario Ministry of Natural Resources 1995). Consumption advice is specific to the species of fish, the length of fish and the location where the fish is caught. Consumption advice provided is the recommended maximum number of meals of sport fish per month, with the consumption categories being 8, 4 ,2, 1 and no meals of sport fish per month.

We summarize consumption advisories for all species of fish, for all size ranges and all locations listed in the 1995 *Guide to Eating Ontario Sport Fish* for the Canadian waters of the Great Lakes and report on the principal contaminant of concern causing the consumption advice. We undertake a similar analysis focusing on a specific size range of a cold water and a warm water species. For the principal contaminants of concern causing consumption restrictions, we report examples of trends in contaminant concentrations in sport fish over time.

To complement the sport fish contaminant monitoring program, identify areas with elevated contaminant concentrations and monitor contaminant trends over time, a contaminant surveillance program using juvenile spottail shiners (*Notropis hudsonius*) as biomonitors in the near shore Canadian waters of the Great Lakes was established in 1975. Aspects of this work have been previously reported in the literature (Suns and Rees 1978; Suns *et al.* 1983; Suns *et al.* 1985; Suns *et al.* 1991; and Suns *et al.* 1993).

We report on contaminant concentrations in juvenile spottail shiners collected in 1993 and 1994 from 44 locations in the lower Great Lakes, assessing the significance of the contaminant concentrations using wildlife protection guidelines. We report on temporal trends in contaminant concentrations in spottail shiners at 16 selected sites.

## 2. Materials and Methods

### 2.1 SPORT FISH

The sport fish contaminant monitoring program is co-ordinated by the Ontario Ministry of the Environment and conducted in partnership with the Ontario Ministry of Natural Resources and Health Canada. Most sport fish were collected by the Ministry of Natural Resources using nets or by electro fishing. When possible, 20 fish of each species were caught with lengths and weights representative of the size range of the species in the location being tested. The 1995 *Guide to Eating Ontario Sport Fish* gives a listing of all species of fish caught by location.

The length and weight of each fish were recorded and a skinless, boneless fillet of the dorsal muscle was removed from the fish, packaged and frozen for shipment to Ontario Ministry of Environment and Energy laboratories for analysis. Studies have shown that skinless, boneless dorsal fillets are the most commonly consumed portion type (Cox *et al.* 1993).

All sport fish sampled were analysed for mercury. Depending upon location, analyses were also done for polychlorinated biphenyls (PCBs), pesticides including DDT and toxaphene, mirex, dioxins and furans. Analytical methods are described by Ontario Ministry of Environment and Energy (1994). Analyses were done on the 17 most toxic forms of dioxins and furans and the results reported as toxic equivalents of 2,3,7,8-TCDD (Safe 1990).

Consumption advisories provided to the public in the 1995 *Guide to Eating Ontario Sport Fish* were based on health protection guidelines developed by Health Canada. A consumption advisory is the recommended maximum number of meals of sport fish per month (consumption categories are 8,4,2,1 and no meals per month) and each advisory is specific to the species of fish, the length of fish and the location where the fish was caught. A meal was considered to be 227 gm (8 ounces) for a 60 kg adult. The consumption advice was designed to protect individuals who followed the advice in the 1995 *Guide to Eating Ontario Sport Fish* and who consumed no more than 8 meals of sport fish per month. Health Canada guidelines have been developed to protect the health of the most sensitive individuals, generally considered to be pregnant women and children. However, as an added precaution, the 1995 *Guide to Eating Ontario Sport Fish* recommended that women of childbearing age and children under 15 not consume any fish in the 1 meal per month category as well as the no consumption category.

Contaminant concentrations which triggered consumption restrictions for sport fish in the 1995 *Guide to Eating Ontario Sport Fish* are summarized in Table I.

**Table I.** Concentrations of Contaminants which Triggered Sport Fish Consumption Advisories in the 1995 *Guide to Eating Ontario Sport Fish*

| Compound | 4 Meals/Month | 2 Meals/Month | 1 Meal/Month | No Consumption |
|---|---|---|---|---|
| Mercury | 0.5 ppm | 1.0 ppm | N/A | 1.5 ppm |
| PCBs | 2.0 ppm | 1.0 ppm | | 4.0 ppm |
| Mirex and photomirex | 0.07 ppm | 0.14 ppm | 0.28 ppm | 0.56 ppm |
| DDT | N/A | N/A | N/A | 5.0 ppm |
| Toxaphene | 0.2 ppm | 0.4 ppm | 0.8 ppm | 1.6 ppm |
| Dioxins and furans | 10.1 ppt | 20.1 ppt | 40.2 ppt | 80.5 ppt |

For situations in which several contaminants occurred together in sport fish at concentrations which caused consumption restrictions, the principal contaminant of concern is defined as the contaminant which caused the most restrictive consumption advisory.

For the Canadian waters of the Great Lakes, consumption advisories were provided for blocks or regions of each lake. The boundaries of the blocks are shown in Figures IV and V and are described in more detail in the 1995 *Guide to Eating Ontario Sport Fish.* Contaminant concentrations for all fish of a given species and size were similar throughout a block. The boundaries of the blocks were established in consultation with fisheries biologists of the Ontario Ministry of Natural Resources familiar with local fish populations and after comparing contaminant levels in fish from several adjacent locations. If minor differences were found in contaminant levels in fish of the same species at locations within the same block, then the fish from the more contaminated location were used to set the consumption advisory for the block. Separate consumption advice was provided for localized areas within a block where the fish community or the contaminant levels were different from the remainder of the block.

We also report on consumption advice for 55-65 cm lake trout (*Salvelinus namaycush*). Lake trout were chosen as a useful indicator species for the cold water fishery because of their distribution across all Great Lakes. Additionally, because of their high fat levels, lake trout are particularly useful as monitors of organic

**Figure I.** Spottail Shiner Collection Sites in the Lower Great Lakes in 1993 and 1994.

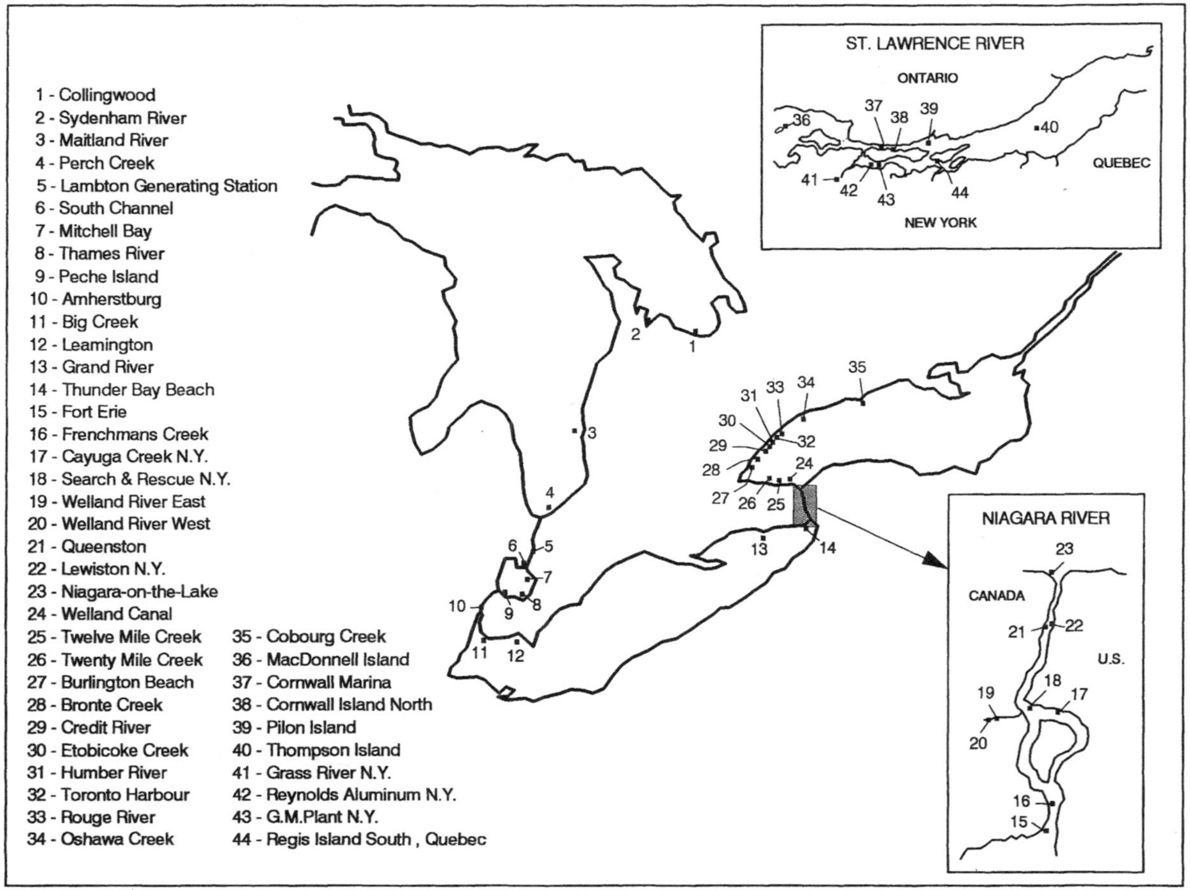

contaminants such as PCBs, mirex and toxaphene. Lake trout from the 55-65 cm size range are commonly kept and consumed by anglers.

No single species of fish is suitable as an indicator of the warm water fishery because none are distributed across all locations in the Great Lakes. Consequently, for the purposes of this paper, we report on consumption advice given for 30-35 cm smallmouth bass (*Micropterus dolomieui*), 35-45 cm walleye (*Stizostedion vitreum*) and 20-25 cm yellow perch (*Perca flavescens*). Fish from these size classes are commonly kept and consumed by anglers. For a given location, we report on the most restrictive consumption advice provided in the 1995 *Guide to Eating Ontario Sport Fish* for these three warm water species and the contaminant causing the restriction.

To assess the changes over time in the concentrations of the principal contaminant of concern in sport fish collected from selected locations in the Canadian waters of the Great Lakes, we focused our analysis on 65 cm lake trout (60 cm rainbow trout (*Onchorynchus mykiss*)where there were no data on lake trout) and 45 cm walleye. For each of these species of fish, for each year and for each location chosen, a power series regression curve $Y=aX^b$, was calculated where Y= predicted contaminant concentration in sport fish and X= fish length in cm. Constants a and b were estimated using an iterative process to solve for the best fit regression (SPSS 1993 ). The best fit regression equation was then solved to calculate the contaminant concentration in a 65 cm lake trout (or a 60 cm rainbow trout) and a 45 cm walleye. This use of standardized fish lengths allowed comparison of contaminant concentrations between collections of fish of different length distributions.

The concentration of the contaminant of concern at the standardized fish length was plotted against the year the fish were collected. Linear, logarithmic, inverse, compound, S, power, growth, exponential and logistic regression curves were fitted to the data (SPSS 1993) and the curve of best fit chosen based on the maximum $r^2$ value.

## 2.2 Juvenile Forage Fish

A common forage fish, the spottail shiner (*Notropis hudsonius*), was selected for monitoring (Suns and Rees 1975). Among the criteria used in the selection of spottail shiners were its limited range in its first year of life, undifferentiated food habits in early life stages, its importance as a forage fish (Scott and Crossman, 1973) and its presence throughout the Great Lakes.

Young of the year spottail shiners were collected at 44 sites throughout the lower Great Lakes in September 1993 or September 1994. Sampling locations are shown in Figure I. Fish were collected from near shore areas using a 0.6 cm. mesh

**Figure II.** Sport Fish Consumption Advisories for the Canadian Waters of the Great Lakes in 1995.

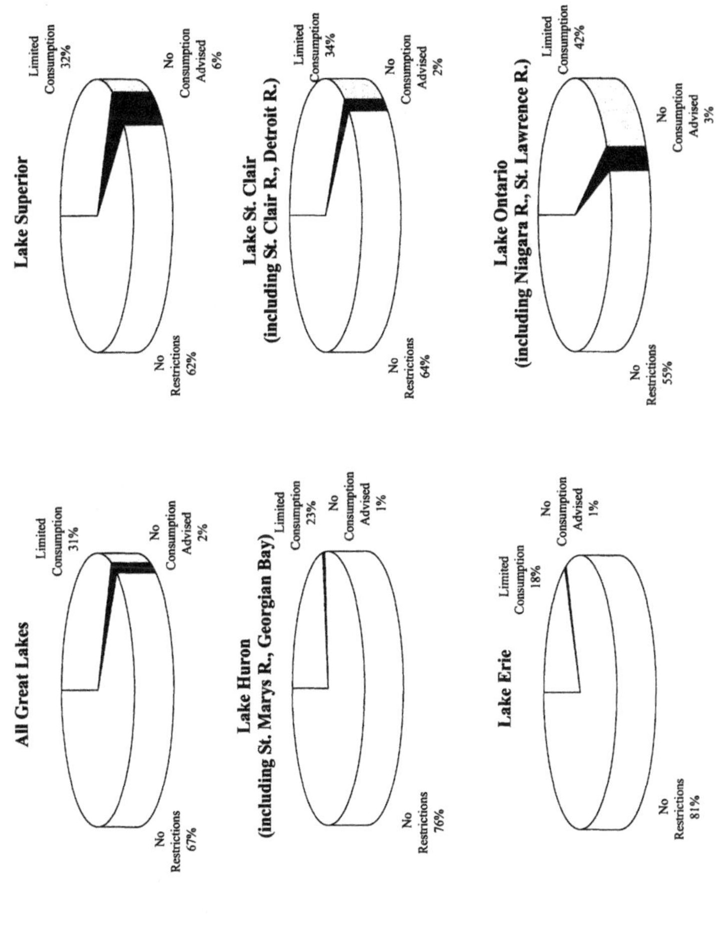

bag seine net. Five to seven 10-fish composites were measured for total length (mm), wrapped in hexane rinsed aluminum foil and frozen at -20 °C until analysed for PCBs and a suite of other organochlorine compounds at the laboratory of the Ontario Ministry of Environment and Energy (1994).

The Forage Fish Contaminant Index (FFCI) developed by Suns *et al.* (1991) was used to assess the significance of the contaminant levels in the forage fish. The FFCI is calculated as the sum of individual contaminant concentrations divided by individual wildlife protection guidelines.

Two sets of guidelines were used in the calculation of the FFCI. These were the aquatic life guidelines published in the Great Lakes Water Quality Agreement (International Joint Commission 1988); and the fish flesh criteria for the protection of piscivorous wildlife (Newell *et al.* 1987) published by the New York State Department of Environmental Conservation. For each contaminant included in the FFCI, the more stringent of these two sets of guidelines was used. Contaminants and guidelines used in the FFCI were PCBs (100 ng/g), DDT (200 ng/g), hexachlorocyclohexane (BHC 100 ng/g), hexachlorobenzene (HCB 330 ng/g), octachlorostyrene (OCS 20 ng/g) and chlordane (500 ng/g). Because the mirex guideline is below detection limits, a value of 1 ng/g was used in calculations.

To assess the changes over time in the concentrations of PCBs in spottail shiners, we chose 16 locations with multiple years of data. For each location and for each year, the mean PCB concentration in the 5-7 composite samples of spottail shiners collected (10 fish in each composite) was calculated together with the 95% confidence limits around the mean. Exponential and linear regression curves were fit to the data (Statistical Graphics Corporation 1986) and the curve of best fit chosen based on the maximum $r^2$ value.

## 3. Results and Discussion

### 3.1 Current Status of Contaminants in Sport Fish

The 1995 *Guide to Eating Ontario Sport Fish* contains 1736 consumption advisories for the Canadian waters of the Great Lakes. A summary of the types of advisories given for the Canadian waters of the Great Lakes as a whole and for the Canadian waters of each Great Lake is given in Figure II. In this figure, the term "limited consumption" means that there was a consumption advisory of either 4, 2 or 1 meal per month in place on the sport fish. "No restrictions" means consumption up to 8 meals per month was advised and "no consumption advised" means that no consumption at all of sport fish was advised.

"No restrictions" was the advice given in 67% of the 1736 consumption

**Figure III.** Principal contaminants of Concern Causing Sport Fish Consumption Advisories in the Canadian Waters of the Great Lakes in 1995.

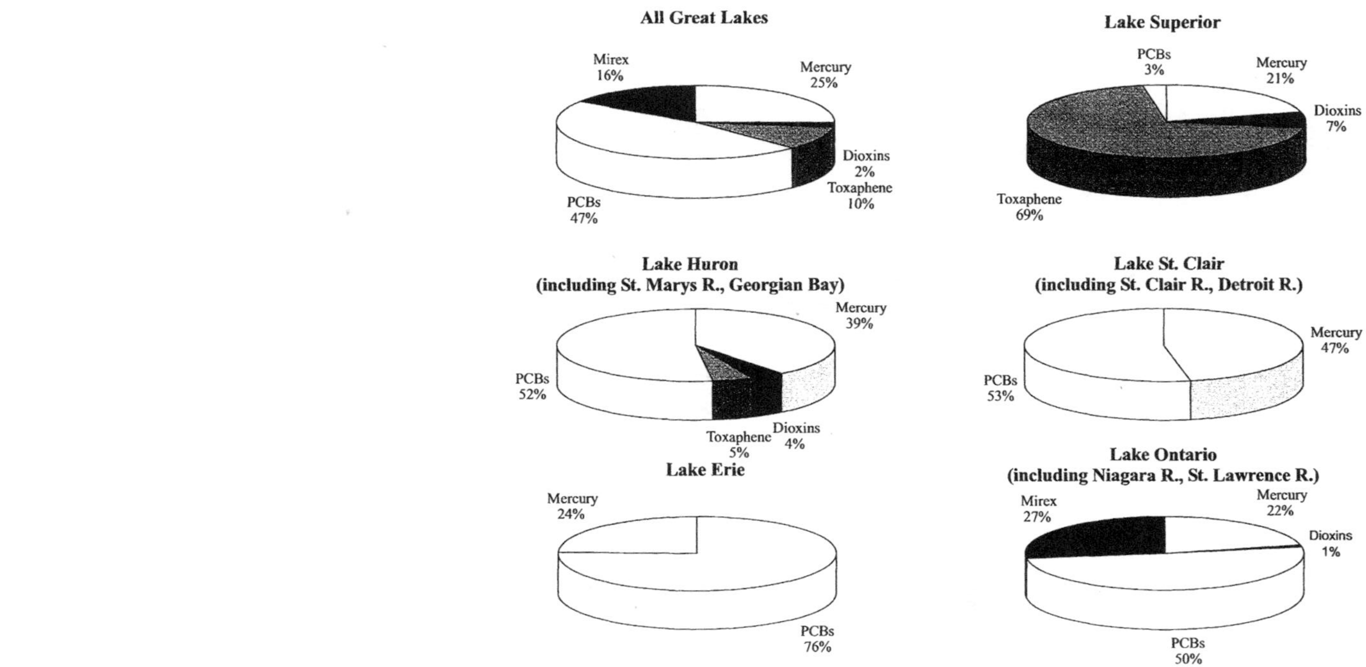

advisories. “Limited consumption” was advised in 31% of the advisories and “no consumption ” was advised in only 2% of the 1736 consumption advisories in place for the Canadian waters of the Great Lakes.

For the individual lakes, the lake with the least consumption restrictions was Lake Erie, with "no restrictions" on sport fish consumption being the advice given in 81% of the consumption advisories on the lake. Lake Ontario (including the Niagara River and the St. Lawrence River) was the lake with the most consumption restrictions (42% of consumption advisories reported as “limited consumption” and 3% as “no consumption”.

The principal contaminant of concern causing each of the fish consumption restrictions in the 1995 *Guide to Eating Ontario Sport Fish* was identified for the Canadian waters of the Great Lakes (Figure III). PCBs were the principal contaminant of concern in 47% of the consumption advisories given where there were some restrictions in effect. In other words, in the 589 individual consumption advisories given where some restrictions were in effect on the consumption of sport fish (out of the total of 1736 consumption advisories given), PCBs were the contaminant causing the restriction in 47% of the advisories. Mercury was the principal contaminant of concern in 25% of the advisories, mirex/photomirex 16%, toxaphene 10% and dioxins/furans 2%.

Figure III also summarizes the principal contaminant of concern for the Canadian waters of each individual Great Lake. In each lake except Lake Superior, PCBs caused the majority of consumption restrictions on sport fish. For Lake Superior, toxaphene was the principal contaminant of concern in 69% of the advisories where consumption restrictions were in effect. Mercury was the principal contaminant of concern in 21% of the advisories, dioxins/furans 7% and PCBs 3%.

A summary of the consumption advisories given in the 1995 *Guide to Eating Ontario Sport Fish* for 55-65 cm lake trout in the Canadian waters of the Great Lakes is given in Figure IV. Consumption categories are the same as those in Figure II. Toxaphene was the principal contaminant of concern in 55-65 cm lake trout in Lake Superior at all locations except Jackfish Bay (Block 8 on Figure II), where dioxins/furans caused the consumption restrictions. PCBs were the principal contaminant of concern in lake trout of this size in Lake Huron and Lake Erie. In Lake Ontario, PCBs caused consumption restrictions in 55-65 cm lake trout in Blocks 1b, 2, 4, 7 and 11; mirex/photomirex in Blocks 4, 6 and 8; and dioxins/furans in Block 4.

A summary of the consumption advisories given in the 1995 *Guide to Eating Ontario Sport Fish* for 30-35 cm smallmouth bass, 35-45 cm walleye and 20-25 cm yellow perch in the Canadian waters of the Great Lakes is given in Figure V. Where there were consumption advisories for more than one of the 3 warm water species, the

**Figure IV.** Boundaries of Blocks with Consumption Advisories for 55-65 cm Lake Trout in the Canadian Waters of the Great Lakes.

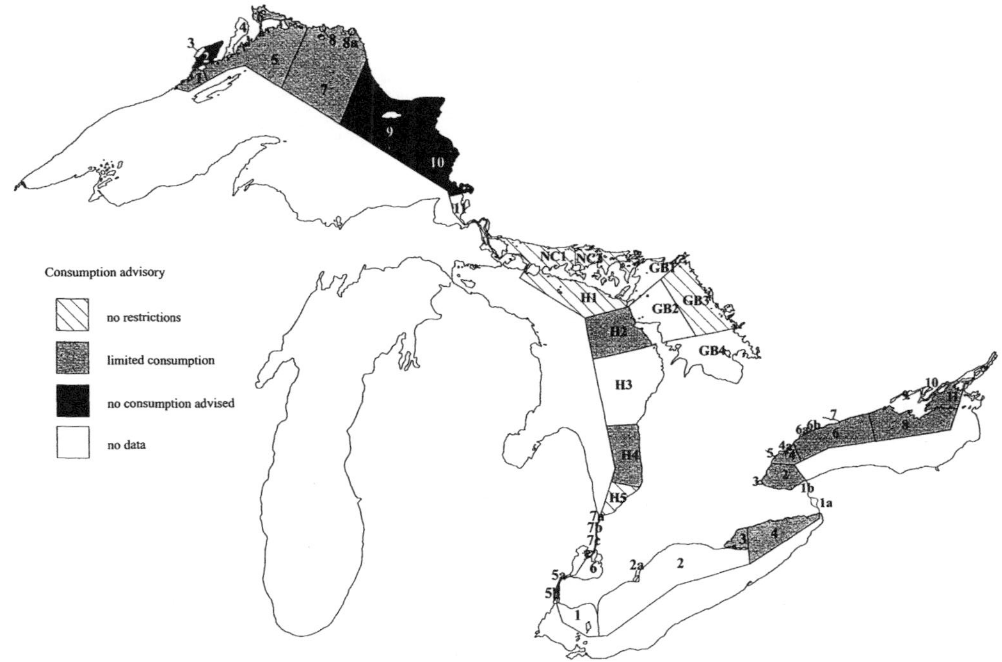

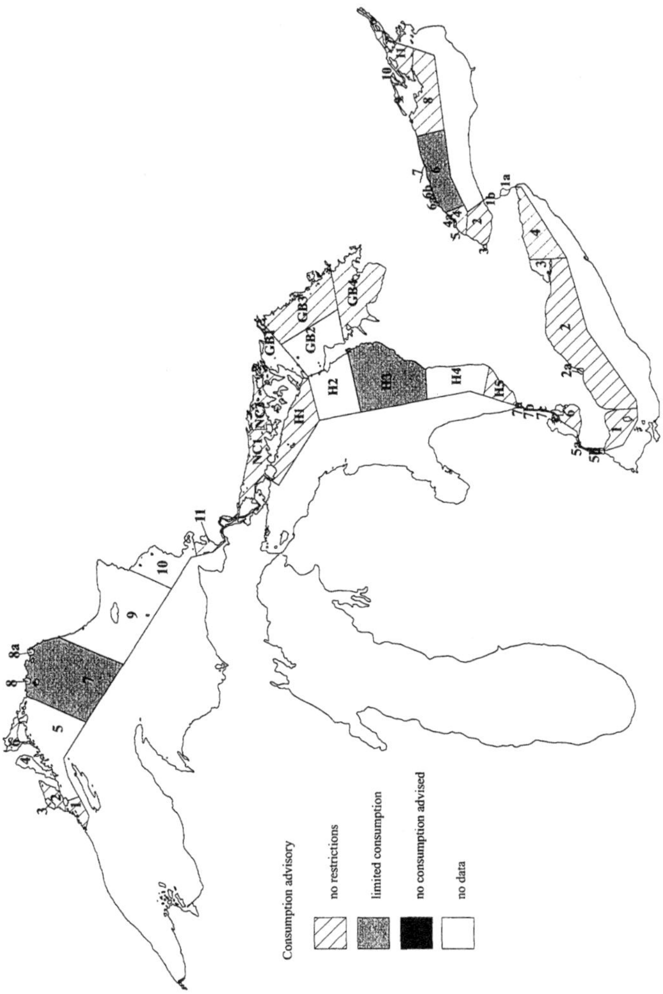


**Figure V.** Boundaries of Blocks with Consumption Advisories for 30-35 cm Smallmouth Bass, 35-45 cm Walleye or 20-25 cm Yellow Perch in the Canadian Waters of the Great Lakes in 1995.

most restrictive consumption advisory is shown. Consumption categories are the same as those in Figure II. Mercury is the principal contaminant of concern at all locations with consumption restrictions with the exception of the Detroit River, where PCBs caused the consumption restriction.

### 3.2 TEMPORAL TRENDS IN CONCENTRATIONS OF CONTAMINANTS IN SPORT FISH

Trends in contaminant concentrations in fish tissue vs time are plotted in Figure VI, focusing in each lake on the principal contaminant of concern in 65 cm lake trout, 60 cm rainbow trout and 45 cm walleye.

Trend information on toxaphene in Lake Superior lake trout is limited, with only 4 collections of sport fish over the period 1986 to 1992. No temporal trend could be identified from this information. No information was available to identify trends in Hg, the principal contaminant of concern in walleye.

Concentrations of PCBs in 65 cm lake trout from southern Lake Huron (Block H5 on Figure IV) have declined from 8.07 ppm in 1976 to 0.47 ppm in 1994. Plotting the concentration of PCB vs time, the line of best fit was an S curve (similar to an exponential decay curve) $[PCB] = e^{(a + b/year)}$. The $r^2$ value of the regression curve was 0.76 ($p<0.001$), indicating a significant decline in PCBs over the period 1976 - 1994..

The concentrations of Hg in 45 cm walleye from southern Lake Huron (Block H5 in Figure V) have varied from 0.41 ppm to 0.25 ppm over the period 1981-1992. The slope of the linear regression curve of concentration vs time was not significantly different than zero ($p>0.05$), indicating no trend in Hg levels over time from 1981-1992.

No information on trends in the concentration of PCBs for either cold water or warm water fish from Lake St. Clair or Lake Erie was available. Mercury concentrations in 45 cm walleye from Lake St. Clair declined from 2.3 ppm in 1970 to 0.62 ppm in 1994. Plotting concentration of Hg vs time over this period, the line of best fit was an S curve with $r^2$ of 0.78 ($p<0.001$). There are indications in the data that the concentrations of Hg in 45 cm walleye may have been increasing from 1990 to 1994, but additional years of data are needed to confirm any trend. The reasons for such an increase are not known.

The concentrations of PCBs in 60 cm rainbow trout from Lake Ontario at the Ganaraska River (Block 7 in Figure IV) have declined from 3.9 ppm in 1976 to 0.97 ppm in 1994. Again, the curve best describing this decline was an S curve with $r^2$ value of 0.74 ($p<0.001$).

Similarly, mirex concentrations in 60 cm rainbow trout at the Ganaraska River declined from 0.26 ppm in 1976 to 0.10 ppm in 1994 with an S decay curve best describing the decline ($r^2$=0.59, $p<0.001$). There are indications in the data that the

**Figure VI.** Temporal Trends in the Concentrations of PCBs, Mirex and Mercury in 65 cm Lake Trout 60 cm Rainbow Trout or 45 cm Walleye from Selected Locations in the Canadian Waters of the Great Lakes.

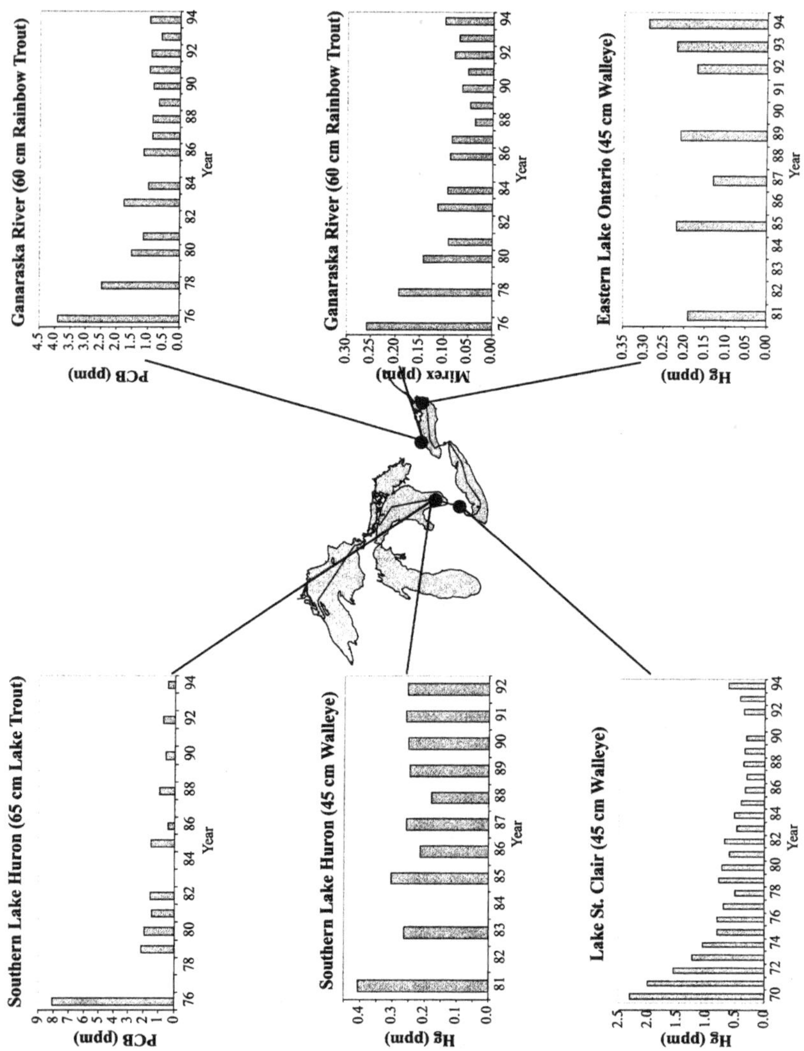

concentrations of mirex in rainbow trout may have increased over the period 1990-1994 but additional years of data are needed to verify whether a trend exists.

Mercury in 45 cm walleye in eastern Lake Ontario (Block 11 in Figure V) have varied between 0.13 ppm and 0.29 ppm over the period 1981 and 1994. The slope of the linear regression curve of concentration vs time was not significantly different from zero ($p$>0.05) indicating no trend in Hg levels over time.

### 3.3 JUVENILE FORAGE FISH

Calculated values of the Forage Fish Contaminant Index (FFCI) together with concentrations of total PCB and DDT in juvenile spottail shiners are shown in Figure VII. An index value of "1" was designated as the Wildlife Risk Level by Suns *et al.* (1991). Higher values represent greater risk for piscivorous wildlife. The FFCI exceeded 1 at 17 of the 44 sites where spotttail shiners were collected. Higher index values were generally more frequent in the lower Great Lakes, with the maximum observed values noted at the Grass River and Reynolds Aluminum sites in the St. Lawrence River and at the Welland Canal.

PCBs were detected in spottail shiners at 31 of 44 sites sampled in 1993 or 1994, and exceeded the IJC Aquatic Life Guideline of 100 ng/g at 13 sites (Figure VII). Concentrations of PCBs were highest in spottail shiners collected on the U.S. side of the St. Lawrence River in the Grass River and at the Reynolds Aluminum site. PCBs generally accounted for the largest fraction of the FFCI at most locations. Exceptions to this occurred at the St. Clair River at Lambton Generating Station (OCS concentrations in shiners were 15 ± 2 ng/g) and locations in Lake St. Clair where OCS was the contaminant accounting for the greatest proportion of the FFCI; at the Welland Canal where mirex was of greater significance than PCBs (concentration of mirex was 5.0 ng/g). This was the only location where mirex was detected in spottail shiners.

Total DDT concentrations in young-of-the-year spottail shiners were below established guidelines (200 ng/g) at all 44 sites sampled. The location with the highest concentration of DDT in juvenile shiners was 12 Mile Creek in Lake Ontario. Concentrations of BHC in spottail shiners were below the guideline of 100 ng/g at all sites, with the highest concentrations measured at Cayuga Creek in 1994 (33 ± 11 ng/g).

Chlordane was detected in spottail shiners at 4 sites, one at Fort Erie and 3 in Lake Ontario. Concentrations in spottail shiners did not exceed 12 ng/g and were below the guideline (500 ng/g) at all locations. HCB residues in spottails were below the aquatic life guideline (330 ng/g) at all 44 sites. The highest concentration of HCB in juvenile shiners was observed at Cayuga Creek on the Niagara River.

**Figure VII.** Values of the Forage Fish Contaminant Index (FFCI), Concentrations of PCBs and DDT for Young of the Year Spottail Shiners Collected at 44 Locations in the Great Lakes in 1993 and 1994

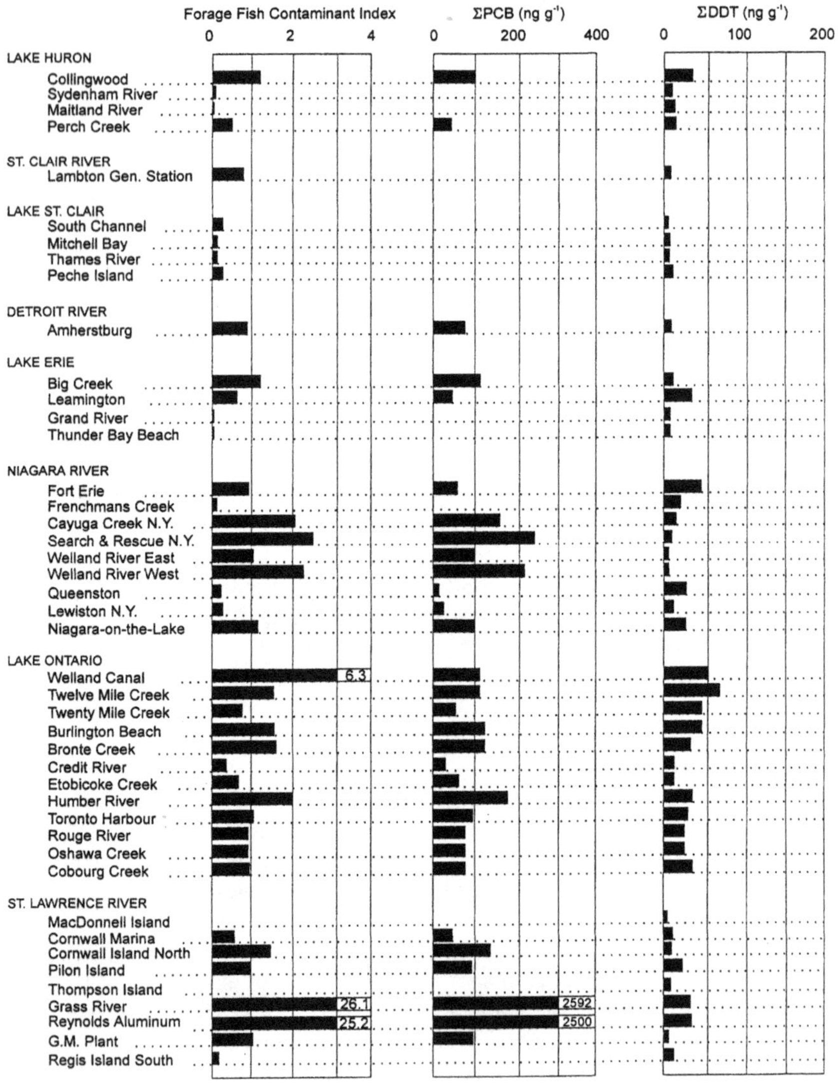

OCS concentrations in spottail shiners were below the guideline (20 ng/g) at all locations and were generally only detected in spottail shiners collected in the St. Clair River, Lake St. Clair and Detroit River. Concentrations of OCS in spottail shiners have declined since the mid-1980s at Lambton Generating Station (as high as 104 ng/g), but still persisted in 1994 (15 ± 2 ng/g).

Temporal trends of PCB concentrations are illustrated in Figure VIII. Values are means ± 95% confidence limits. Lines indicate significant correlations with time ($p < 0.05$). Total PCB concentrations in spottail shiners were negatively correlated with time at 12 of 16 long term sampling sites, indicating that PCB availability in the near shore waters of the Great Lakes decreased at most sites monitored. Further containment of point and non-point sources of PCBs are required to reduce contaminant levels to acceptable levels at all sites.

## 4. Conclusions

Based on the available data for sport fish and juvenile fish, PCBs followed by mercury can be considered as the two principal contaminants of concern in fish in the Canadian waters of the Great Lakes. PCBs were responsible for causing the greatest proportion (47%) of fish consumption restrictions in the Canadian waters of the Great Lakes. They were also the principal contributor to values that exceeded the FFCI for juvenile forage fish.

Exceptions to this generalization occurred in Lake Superior, where toxaphene followed by mercury caused the greatest proportion of sport fish consumption restrictions and in Lake Ontario where PCBs followed by mirex caused the greatest proportion of consumption restrictions in sport fish.

PCBs, being lipophilic, tend to be of greater concern in fish of higher fat content such as salmon and trout whereas mercury is often of greater concern in predators such as walleye, smallmouth bass and yellow perch with lower fat content.

Temporal trends in contaminant concentrations in fish from the Canadian waters of the Great Lakes demonstrated that PCBs and mirex levels have declined over the past 15-20 years, reflecting improved industrial practices, more stringent regulations and restrictions on the use of these chemicals. Mercury levels also fell in areas responding to point source control programs such as in Lake St. Clair. In areas more remote from direct mercury discharges, no temporal trends in tissue concentrations of Hg were observed.

**Figure VIII.** Temporal Trends in the Concentrations of PCBs in Young of the Year Spottail Shiners Collected at 16 Locations in the Great Lakes over the Period 1975-1994. Values are Means ±95% Confidence Limits. Trend Lines Indicate Significant Correlations with Time.

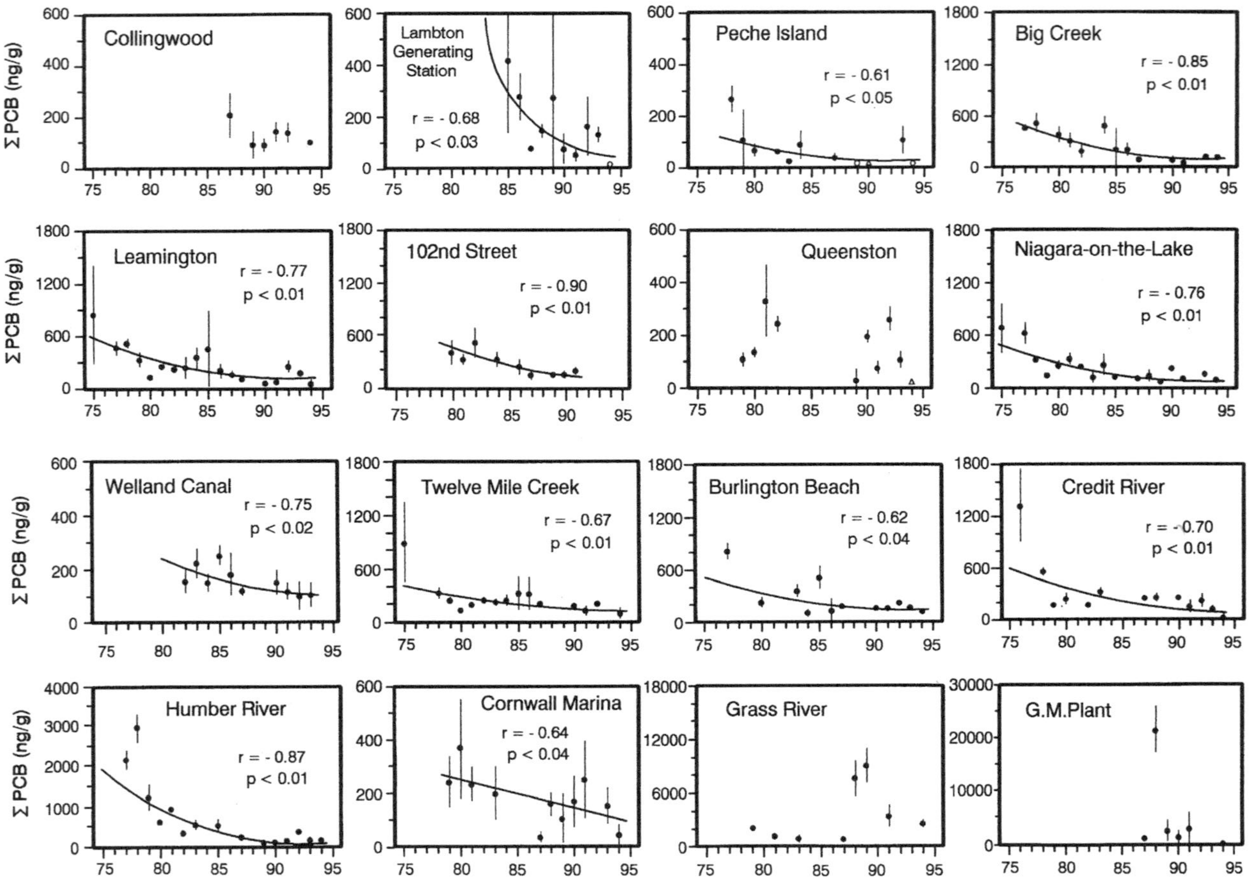

## Acknowledgments

We thank Lynda Nakamoto for assistance with sport fish data tabulation, and Bernie Neary for preparation of sport fish graphics. Comments on the manuscript were given by two anonymous reviewers.

## References

Cox, C., Vaillancourt, A. and Hayton, A.:1993, '*The Results of the 1992 Guide to Eating Ontario Sport Fish Questionnaire*', Ontario Ministry of Environment and Energy, Toronto, Ontario. 26 pp.

Government of Canada: 1991, '*Toxic Chemicals in the Great Lakes and Associated Effects*', Ottawa, Ontario. 755 pp.

International Joint Commission.: 1988, '*Revised Great Lakes Water Quality Agreement of 1978, As amended by Protocol Signed November 18, 1987*'. 130pp.

Newell, A.J., Johnson, D.W., and Allen, L.K.: 1987, '*Niagara River Biota Contamination Project: Fish Flesh Criteria for Piscivorous Wildlife*', New York State Department of Environmental Conservation Tech. Report 87-3.

Ontario Ministry of the Environment and Energy.: 1994, '*The Determination of Polychlorinated Biphenyls (PCBs), Organochlorines (OCs), Chlorobenzenes (CBs) and Chlorophenols (CPs) in Fish, Clams and Mussels by Liquid Chromatography-Electron Capture Detection (GLC-ECD). Method PFAOC-E3136A*'. Toronto, Ontario.

Ontario Ministry of the Environment and Energy and Ontario Ministry of Natural Resources.: 1995, '*The Guide to Eating Ontario Sport Fish*', 173 pp.

Safe, S.: 1990, 'Polychlorinated biphenyls (PCBs), dibenzo-*p*-dioxins (PCDDs), dibenzofurans (PCDFs), and related compounds: environmental and mechanistic considerations which support the development of toxic equivalency factors (TEFs)', *Crit. Rev. Toxicol.* **21**, 51-88.

Scott, W.G. and Crossman, E.J.: 1973, '*Freshwater Fishes of Canada*', Bull. Fish. Res. Board Can. 184, 461 pp.

SPSS.: 1993, '*SPSS for Windows Based System User's Guide Release 6.0*', Chicago, IL. 828 pp.

Statistical Graphics Corporation.: 1986, '*Statgraphics, Statistical Graphics System*'. Rockville, MD.

Suns, K. and Rees, G.: 1975, '*Chlorinated Hydrocarbon Residues from Selected Sites on Lakes Ontario, Erie, and St. Clair*', Ontario Ministry of Environment Technical Report. Toronto, Ontario.

Suns, K. and Rees, G.: 1978, 'Organochlorine contaminant residues in young-of-the-year spottail shiners from Lakes Ontario, Erie, and St. Clair', *J. Great Lakes Res.* **4**, 230-233.

Suns, K., Craig, G., Crawford, G., Rees, G., Tosine, H., and Osborne, J.: 1983, 'Organochlorine contaminant residues in spottail shiners from the Niagara River', *J. Great Lakes Res.* **9**, 335-340.

Suns, K., Crawford, G., and Russel, D.: 1985, 'Organochlorine and mercury residues in young-of-the-year spottail shiners from the Detroit River, Lake St. Clair, and Lake Erie', *J. Great Lakes Res.* **11**,347-352.

Suns, K., Hitchin, G. and Adamek, E.: 1991, 'Present status and temporal trends of organochlorine contaminants in young-of-the-year spottail shiner from Lake Ontario'. *Can. J. Fish. Aquat. Sci.* **48**, 1568 - 1573.

Suns, K., Hitchin, G., and Toner, D.: 1993, 'Spatial and temporal trends of organochlorine contaminants in spottail shiners from selected sites in the Great Lakes (1975-1990)'. *J. Great Lakes Res.* **19**, 703 - 714.

# ORGANOCHLORINE CONTAMINANTS IN HERRING GULL EGGS FROM THE GREAT LAKES, 1974-1995: CHANGE POINT REGRESSION ANALYSIS AND SHORT-TERM REGRESSION

C. Pekarik and D.V. Weseloh

*Canadian Wildlife Service, Environmental Conservation Branch, Ontario Region, Environment Canada, 4905 Dufferin St., Downsview, Ontario. M3H 5T4*

**Abstract.** The temporal trends (1974-1995) of 11 organochlorine contaminants in herring gull eggs from 13 colonies throughout the Great Lakes were statistically analyzed using two regression methods on logarithmically transformed data. Change point analysis was used to determine if there had been significant year to year fluctuations in contaminant levels and/or changes in long-term trends. Short-term regressions were conducted on 6 major compounds for two time periods (early 1990s and early 1980s) to compare the rates of decline. Overall, change point analyses indicated that for most of the comparisons (75%) there had been significant year to year fluctuations in contaminant levels. They also indicated that for most of the comparisons (67%) the rate of decline after the change point was as fast as or faster than before the change point, this pattern was most common for dieldrin and heptachlor epoxide at certain locations. In 19% of the comparisons the rates of decline had slowed or stabilized, this was most common for PCB and pentachlorobenzene. In 14% of the comparisons there were no significant temporal trends, this was most common for photomirex and mirex. Results for short-term regression showed that out of 78 comparisons for each time period, 5 (6%) were declining significantly in the early 1990s and 11 (14%) were declining significantly in the early 1980s. Both types of regression indicated that, for most of the herring gull egg contaminant database, recent logarithmic rates of decline were similar to those seen previously in the sampling period. For PCB 1254:1260, a group of compounds of particular toxicological importance, change point analyses indicated that the logarithmic rates of decline in herring gull eggs from western Lake Ontario were slower from 1987-1995 than they were from 1974-1986. At both Lake Superior colonies and the Niagara River colony PCB 1254:1260 concentrations ceased to decline in the mid-1980s. At the colony in Green Bay, Lake Michigan, PCB 1254:1260 levels have shown no significant temporal trend since 1976. At the remaining 8 colonies, PCB 1254:1260 levels continue a logarithmic decline in recent years at the same rate as or faster than previously.

## 1. Introduction

During this century, organochlorine chemical manufacturing and use in the Great Lakes basin have resulted in widespread environmental pollution and contamination of fish and wildlife resources, resulting in population declines and even local extirpations (Environment Canada 1991). The herring gull (*Larus argentatus*) has been a useful biomonitor of the concentrations of toxic chemicals in the Great Lakes basin for almost 25 years. Since 1974, eggs have been collected in a yearly monitoring program from up to 13 nesting colonies throughout the Great Lakes and connecting channels.

Herring gull eggs were chosen to be monitored because adult individuals are year-round residents on the Great Lakes and they are top predators that integrate temporal and geographic variations in contaminant concentrations. They are able to bioaccumulate high levels of toxicants which are usually at detectable concentrations. The ability to measure higher concentrations gives the opportunity for more precise data analysis and allows for the early detection of potential problems, for example the discovery of mirex, and the investigation of chemically-induced epizootics such as chick-edema disease caused by PCBs and TCDD in the Great Lakes ecosystem (Hallett *et al.* 1976, Norstrom *et al.* 1982, Environment Canada 1991).

Recent statistical analyses of fish and herring gull egg data suggest that toxic substances are no longer declining as rapidly as they did following restrictive regulation (Borgmann and Whittle 1991, 1992). Particular attention has been given to the statistical analysis of the data sets for PCBs (Stow *et al.* 1994, Stow *et al.* 1995, Stow 1995) and in fact, there have been claims that concentrations have reached a steady state. Smith (1995b) found that the short-term deviations from the regression lines describing the long-term trends of PCBs in herring gull eggs were synchronized throughout the Great Lakes. He also found that these short-term deviations from long-term trends were correlated to the deviations of spring temperatures (degree days). Smith (1995b) hypothesized that the fluctuations in weather were affecting springtime productivity of the food chain, resulting in the observed deviations from the calculated long-term trends of contaminants in herring gull eggs, and that the recent period of slow decline was the result of these deviations. There has been an impetus to determine whether toxic compounds in the biota of the Great Lakes have in fact reached stable equilibria or continue to decline. If there have been long term changes in the rates of decline of Great Lakes contaminants, special statistical techniques such as change point or piecewise regression models (Draper and Smith 1981, Gujarati 1988) would aid in detecting when the change occurred and the statistical significance of these changes. We feel that by pinpointing the specific year when changes in long term trends occurred and establishing the nature of these changes we provide an aid in determining the significance of factors such as contaminant loadings, weather, and food web dynamics in influencing changes in the temporal trends of contaminants in herring gull eggs.

Our objectives for this paper were to statistically analyze the Great Lakes herring gull egg contaminants dataset, 1974-1995 and determine the extent to which the data fit change point or piecewise regression models (Draper and Smith 1981, Gujarati 1988). We wanted to determine the model of best fit and identify the change point year (if applicable) for each compound at each colony. We also wanted to compare the short-term temporal trends during 2 different time periods as a second means of assessing recent changes in trends. The purpose was to determine if the rate of change (decline) has changed.

## 2. Methods

### 2.1. Collections and Chemical Analysis

Egg contents were chosen as the tissue for sampling because their lipid content is less variable than other tissues and collection from large ground-level colonies is easy (Weseloh *et al.* 1979, Mineau *et al.* 1984, Bishop and Weseloh 1990, Ewins *et al.*1992). Fresh herring gull eggs (10-13) were collected annually, with some exceptions (Appendix I-A), from 13 sites throughout the Great Lakes from 1974 to 1995 (Mineau *et al.* 1984) (Figure I). The study is ongoing. Prior to 1985 each egg was analyzed individually. Since 1986, eggs from four of the five Great Lakes each year were pooled and analyzed by site. To maintain an assessment of within-site variability, eggs from the fifth lake were still analyzed individually on a rotating basis (Bishop *et al.* 1992). Pooling samples was adopted as a measure to reduce analytical costs. All statistical analyses were conducted using one (n=1) data point per year per site based either on the arithmetic average or on the pooled estimate. Turle and Collins (1992) indicated that pooled samples were adequate estimates of arithmetic means. This applies to all compounds with the possible exception of mirex for which levels may be overestimated by pooled analysis (Turle and Collins 1992).

The compounds discussed in this study and their corresponding abbreviations are given in Table I (Pettit *et al.* 1994). The structure of the herring gull egg dataset, including yearly egg collections by colony (Appendix 1-A) and the chemical analysis of each compound by year (Appendix 1-B).

**Table I.** Compounds Analyzed, their Corresponding Abbreviations, the First Year of Analysis (and Range in Number of Years Sampled).

| Compounds | Abbreviation | First year of Analysis |
|---|---|---|
| Oxy-chlordane | o-CHL | 1977 (15-19) |
| Pentachlorobenzene | QCB | 1979 (15-17) |
| Hexachlorobenzene | HCB | 1974 (16-21) |
| *p,p'*-DDE | DDE | 1974 (16-21) |
| Dieldrin | DIEL | 1974 (15-21) |
| Heptachlor Epoxide | HE | 1974 (15-21) |
| Alpha-hexachlorocyclohexane | α-HCH | 1982 (12-14) |
| Beta-hexachlorocyclohexane | ß-HCH | 1977 (15-19) |
| Mirex | MIR | 1974 (16-21) |
| Photomirex | P-MIR | 1977 (10-13) |
| PCB: 1254-1260 | PCB | 1974 (16-21) |

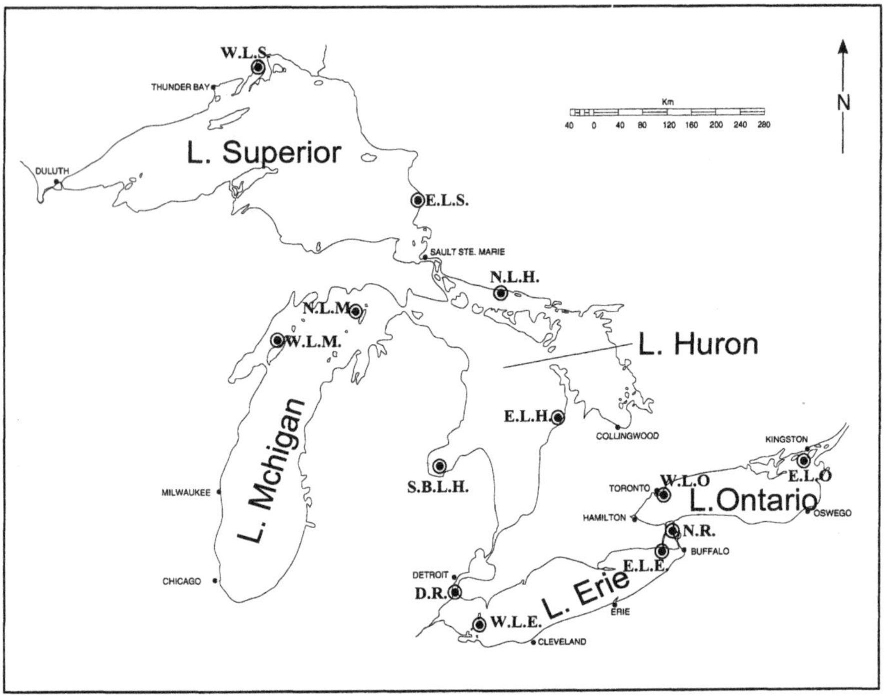


**Figure I.** The location of the Herring Gull colonies examined in this study: **E.L.O)** Snake Island, Eastern Lake Ontario; **W.L.O)** Muggs Island, Western Lake Ontario; **N.R.)** Unnamed Island, Niagara River; **E.L.E)** Port Colborne Lighthouse, Eastern Lake Erie; **W.L.E.)** Middle Island, Western Lake Erie; **D.R.)** Fighting Island, Detroit River; **E.L.H)** Chantry Island, Eastern Lake Huron; **N.L.H)** Double Island, Northern Lake Huron; **S.B.L.H)** Channel Shelter Island, Saginaw Bay, Lake Huron; **N.L.M)** Gull Island, Northern Lake Michigan; **W.L.M)** Big Sister Island, Western Lake Michigan; **E.L.S)** Agawa Rock, Eastern Lake Superior; **W.L.S)** Granite Island, Western Lake Superior.

From 1974 to 1985 chemical analyses were conducted at the Ontario Research Foundation (also named ORTECH) in Mississauga, Ontario; since 1986 samples have been analyzed at the Canadian Wildlife Service (CWS) National Wildlife Research Centre (Bishop *et al.* 1992). Changes in laboratory methodology may have affected the change point analyses of tetrachlorobenzenes (1,2,3,4-tetrachlorobenzene and 1,2,3,5-/1,2,4,5-tetrachlorobenzene), thus they were omitted from the results presented here.

Eggs were stored at 4 °C within 48 hours of collection from the nesting colonies. Within 2 weeks they were cut open around the girth with hexane-rinsed scissors, contents were placed in hexane-rinsed jars with aluminum foil lids and kept at -20 °C (Ewins et al. 1992). Aliquots from pooled and whole egg samples were analyzed by gas chromatography as specified by Reynolds and Cooper (1975) and modified by Norstrom *et al.* (1988). A weighed amount of ground homogenate was combined with anhydrous sodium sulfate to form a free flowing blend, the mixture was placed in a chromatographic column and moistened with 50% (v/v) dichloromethane in hexane. After 30 minutes the lipid in the sample was eluted from the column with approximately 250 ml 50% (v/v) dichloromethane in hexane. After the solvent was evaporated, a portion was used to determine lipid content using gravimetric analysis. A second portion was processed to separate organochlorine compounds from the lipid. The lipid soluble compounds were separated and fractionated on Florisil (deactivated with 1.2% water). The first fraction was eluted with hexane. It contained chlorobenzenes, DDE, MIR, P-MIR and PCB Aroclors. The second fraction contained DDT, DDD, α-HCH, o-CHL, and ß-HCH. It was eluted with 15% (v/v) dichloromethane in hexane. The third elution used 50% (v/v) dichloromethane in hexane. It contained DIEL and HE. Once concentrated, each fraction was analyzed utilizing gas chromatography with an electron capture detector. Calibration was achieved with multi-component standards. PCB 1254:1260 was estimated by a 1:1 mixture of Aroclors 1254 and 1260 corresponding to the single peak of PCB 138 (Turle *et al.* 1991).

All data have been calculated on a wet weight basis and reported in g/g (ppm). Detection limits varied with the laboratory and methods used but the following can be used as a guide (Pettit *et al.* 1994):-

| | |
|---|---|
| Chlorinated benzenes | 0.001 g/g |
| Organochloride pesticides | 0.005 g/g |
| Polychlorinated biphenyls | 0.01 g/g |

A value of half the detection limit was used in the statistical analyses when compounds were not detected.

## 2.2. Statistical Analysis

All data analyses were conducted with the statistical program SAS, versions 6.10 and 6.12 for Windows and Windows 95, respectively (SAS Institute Inc. 1988a, b). Results were determined at a 5% level of significance for all tests ( =0.05).

## 2.3. Long-Term Temporal Trends

Diagnostic tests were conducted to determine the suitability of regression analysis. The Durbin-Watson test (Draper and Smith 1981) established that autocorrelation among the errors was not a problem for the data set. Visual examination of the plots of studentized residuals by year indicated heterogeneity of the variance (Draper and Smith 1981). Examination of stem and leaf plots of the studentized residuals and the Shapiro-Wilk statistic indicated that the errors were not normally distributed (SAS Institute Inc. 1988a). The Box-Cox method (Draper and Smith 1981) was used to choose the natural log transformation for the data. Subsequent diagnostic tests revealed homogeneous variance and normally distributed errors for the transformed data. Percent lipid of the egg was not used as a covariable because it was found to be uncorrelated to the compounds in 92% of the tests (Hebert and Keenleyside 1995).

The data span a long interval of time and on a non-transformed scale appeared to demonstrate a period of rapid decline followed by a period of more gradual change or stability. To examine whether there was a real change in the rate of decline a change point regression model on the logarithmically transformed data was used rather than the usual single trend regression (Draper and Smith 1981). Change point regression is an application of piecewise linear regression (Gujarati 1988). The dataset was divided on the x axis (year). Separate trends could then be determined for data points before and after the change point year, or split in the time period. The usual method of discerning the possible year in which a change point occurred is from plotting the data. This method, however, can introduce bias and was not used. Instead all data points within 3 years from the start or finish of sampling were tested as the best possible change point. The likelihood ratio test (Draper and Smith 1981) was used to determine the change point year and the corresponding model that minimized the mean square error or the squared distance from the trend lines to the data points.

From the change point model we established:

1) **The change point year**. This corresponds to the most significant year to year change in contaminant concentration and/or the year in which there was a

significant change in the slope. The change point year was determined as the year after the most significant year to year change, and pertains to the later regression line. When there were missing data, the change point was determined by comparing contaminant values to the previous year with data. For example, if there were no data for 1976, data for 1977 were tested as a change point by comparing 1977 values to 1975 values. Due to differences in the years when data collection was initiated (Table I), most years did not have an equal probability of being significant change points. For example, there were 48 colony-contaminant comparisons whose change point could have occurred in 1977 and 143 comparisons whose change point could have occurred in 1990. Results for change point years have been indexed to compensate for this difference: the number of comparisons whose change point occurred in a given year was divided by the total number of comparisons for which that year was a possible change point. For example, if 1977 and 1990 had one significant change point each, the indices would be calculated as 1/48 =0.02 and 1/143=0.007, respectively.

2) **The slopes before and after the change point year.** Establishing two slopes allows greater flexibility and interpretation as well as taking full advantage of the temporal extent of the data. The reader is cautioned that when change points occurred in recent years (1990-1993), trends after the change point were based on relatively few data points (n= 5 years or less). For example, when a change point occurred in 1993, only two years (1994 and 1995) were available to determine post change point trends. The change point analysis should not be used to extrapolate future trends of contaminants in herring gull eggs.

There were four general models that the data could fit:

1) A model without a change point where a single slope indicates the rate of change (or no change) over the entire time period (single trend model) (Figure II).
2) A change point model where the slopes before and after do not differ and there is a significant change in contaminant concentration at the change point year (Figure III).
3) A change point model where slopes are different from one another before and after the change point and there is no significant change in contaminant concentration at the change point year (Figure IV).
4.) A change point model where slopes are different from one another before and after the change point and there is a significant change in contaminant concentration at the change point year (Figure V).

A change point model was established initially. The slopes for the regression lines before and after the change point were tested for equality. If the slopes were found to be equal, the model was tested for no significant change in the contaminant concentration at the change point year. If this was the case, a single trend model was established for the entire time period (Figure II).

If the slopes were equal but there was a significant change in the contaminant concentration at the change point year, the data were best represented by two parallel lines about a significant change in contaminant concentration at the change point year (Figure III). If the slopes were found to be unequal the model was tested for no significant change in the contaminant concentration at the change point year. If this was the case, the data were best represented by two significantly different trends before and after a smooth transition at the change point year (Figure IV). If the slopes were unequal and there was a significant change in the contaminant concentration at the change point year, the data were best represented by two significantly different trends before and after the significant change in contaminant concentration at the change point year (Figure V). The four possible models are described algebraically as follows:

### *2.3.1. Single trend model (Figure II)*

$$\textbf{ln (contaminant)} = \beta_0 + \beta_1 t$$

where:

ln represents the natural logarithm of the contaminant concentration

$\beta_0$ represents the intercept of the line on the y axis. It is set at one year prior to data collection. For example, if data collection was initiated in 1974, the intercept was set as the concentration corresponding to 1973.

$\beta_1$ represents the slope of the line at time t (the entire time period).

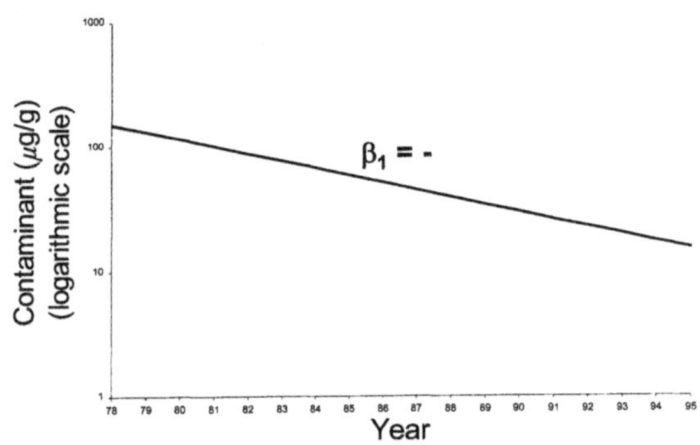


**Figure II.** Constant Rate of Decline, No Significant Change Point in the Regression Model

### 2.3.2. *Change point model with equal slopes and a significant change in the contaminant concentration at the change point year (Figure III)*

**ln (contaminant) =** $\beta_0 + \beta_1 t_1 + S + \beta_1 t_2$

where:

ln represents the natural logarithm of the contaminant concentration.

$\beta_0$ represents the intercept of the line before the change point. It is set at one year prior to data collection.

$\beta_1$ represents the common slope at time $t_1$ and $t_2$ (before and after the change point).

S represents the change in the contaminant concentration at the change point year. A positive value indicates that contaminant concentrations rose significantly at the change point year, a negative value indicates that contaminant concentrations dropped significantly at the change point year.

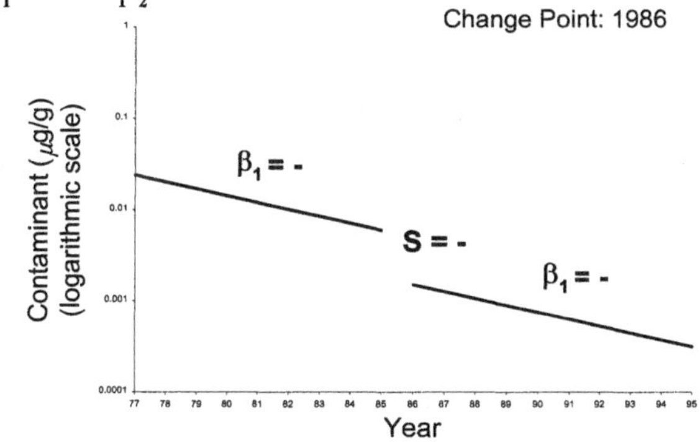


**Figure III**. Compound Levels Changing (Declining) at the Same Rate before and after the Change Point.

### 2.3.3. *Change point model with unequal slopes and no significant change in the intercept (Figure IV)*

**ln (contaminant) =** $\beta_0 + \beta_1 t_1 + \beta_2 t_2$

where:

ln and $\beta_0$ represent the natural logarithm of the contaminant concentration and the intercept of the line on the y axis before the change point, respectively (see above)

$\beta_1$ represents the slope of the line at time $t_1$ (before the change point).

$\beta_2$ represents the slope of the line at time $t_2$ (after the change point).

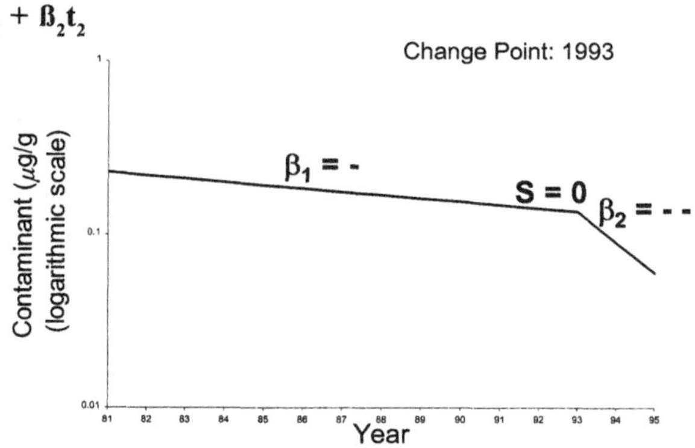


**Figure IV**. Compound Declining Faster after the Change Point Year.

*2.3.4. Change point model with unequal slopes and a significant change in the intercept (Figure V)*

**ln (contaminant) =** $\mathbf{ß_0 + ß_1t_1 + S + ß_2t_2}$

where:

ln and $ß_0$ represent the natural logarithm of the contaminant concentration and the intercept of the line on the y axis before the change point, respectively (see above).

$ß_1$ represents the slope of the line at time $t_1$ (before the change point).

S represents the change in the intercept. A positive value indicates that contaminant concentrations rose significantly at the change point year, a negative value indicates that contaminant concentrations dropped significantly at the change point year.

$ß_2$ represents the slope of the line at time $t_2$ (after the change point).

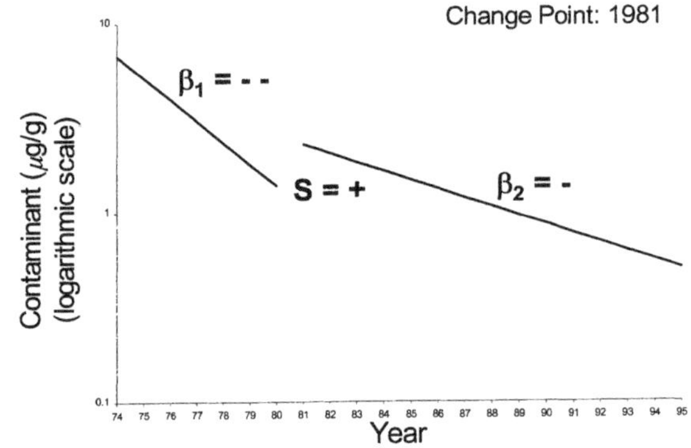


**Figure V.** Compound Declining Slower after the Change Point Year.

The extra sum of squares principle was used to choose among the four models (Draper and Smith 1981). The change point model with two distinct slopes and a change in concentration at the change point year had the most parameters to estimate, namely four ($ß_0$, $ß_1$, S, $ß_2$). The model with different slopes before and after the change point, but no change in the intercept had three parameters ($ß_0$, $ß_1$, $ß_2$) as did the model with a common slope before and after the change point ($ß_0$, $ß_1$, S). The single trend model had two parameters to estimate ($ß_0$, $ß_1$).

The test statistic (F) was calculated and compared as such (Draper and Smith 1981):

$$\frac{\left(S_f - S_r\right) / \left(q_f - q_r\right)}{S^2_f} \approx F_{(q_f - q_r),\,(n - q_f)}$$

where:

$S_f$ represents the (model) sum of squares for the "full" regression model (more parameters).

$S_r$ represents the (model) sum of squares for the "reduced" model (fewer parameters).

$q_f$ represents the number of parameters in the "full" model.

$q_r$ represents the number of parameters in the "reduced" model.

$S^2_f$ represents the variance of the errors, estimated by the mean square error of the "full" model.

$n$ represents the sample size.

The chi-square test was used to determine if there were significant differences in the number of comparisons with significant decreases, increases or no change at the change point year (Zar 1996).

## 2.4. Short-Term Temporal Trends

To further elucidate how recent rates of change in chemical concentrations compared to earlier rates of change, we considered six major compounds, PCB 1254:1260, DDE, DIEL, HCB, HE and MIR and conducted regression analyses during two different 5 year periods. These time periods were chosen arbitrarily and may include short-term temporal fluctuations in contaminant levels. The objective was to select the most recent years of data and to compare them with an earlier time period. For the earlier time period, data from 1981-1985 (inclusive) were used for all colonies except northern Lake Michigan, western Lake Michigan and western Lake Superior which had one year of missing data in that range, thus for those colonies data from 1980-1985 were used (Appendix I). For the later time period, data from 1991-1995 (inclusive) were used at all colonies except for eastern Lake Erie where data from 1990-1995 were used, since no data were available at that site in 1994 (Appendix I). For all regressions the sample size was equal to five. The natural log was chosen as the appropriate transformation for the data using the Box-Cox method (Draper and Smith 1981). The model fit was the following:

$$\ln(\text{contaminant}) = \beta_0 + \beta_1 t$$

where:

$\beta_0$ represents the intercept of the line on the y axis (set one year prior to the first year of analysis).

$\beta_1$ represents the slope of the line at time t.

## 3. Results

### 3.1. Long-Term Temporal Trends

Of 143 analyses (colony-contaminant comparisons) conducted on the logarithmically transformed data, 36 (25%) did not have significant change points in the regression models and 107 (75%) did. The models without change points had a constant rate of change (33 analyses), or no change (3 analyses) over the study period. Of the models that had change points, there were significant increases or decreases in contaminant levels at the change point year for most (95) of the comparisons. For 45 analyses the slope after the change point was significantly different than before the change point. Of 107 comparisons with significant change points, 39 (36%) had change points in 1990 or later. For those analyses post change point trends were based onrelatively few data points (n= 5 or less). Compounds where a high proportion of comparisons had recent (1990-1993) change points included: QCB (100% of the comparisons with change points in the model), DIEL (64%), DDE (46%), P-MIR (43%), PCB 1254:1260 (36%) and HE (33%) (Table II). The following compounds had a high proportion of early (1977-1982) change points: MIR (70% of the comparisons with change points in the model) and o-CHL (37%).

#### *3.1.1. No change point; constant rate of change*

Thirty-six analyses of 143 (25%) did not have significant change points in the regression models, indicating that the rate of change since the first sampling was constant. Of these 36 analyses, the majority (92%) indicated that compounds were declining significantly over the entire time period (Table II) (Figure II).

The pattern of steady decline was most evident for QCB (where it occurred in 46% (6/13) of the analyses), ß-HCH (46%), α-HCH (38%) and P-MIR (31%).

The remaining 8% (3/143) of the analyses without a change point had slopes which were not significantly different from zero, indicating no change had taken place in concentration over the entire sampling period. These were MIR and P-MIR at Saginaw Bay, Lake Huron and P-MIR at western Lake Erie (Table II).

*3.1.2. Change point; constant rate of change*

Sixty-two analyses of 143 (43%) had significant change points in the models and slopes which were not significantly different from each other before and after the change point year (Figure III).

In these analyses the rate of change in the contaminant level was the same before and after a significant shift in contaminant concentration at the change point year. Of these 62 analyses, the majority (45) were declining significantly before and after the change point year (Figure III). For most (58%) of these analyses there was a significant increase in contaminant concentration at the change point. This pattern was characteristic of: DDE (where it occurred in 54% (7/13) of the analyses), HCB (31%) and PCB 1254:1260 (31%) (Table II). For the remaining 42% of the analyses with declining levels before and after the change point, there was a significant decrease in contaminant concentrations at the change point year. This pattern was characteristic of ß-HCH (where it occurred in 62% (8/13) of the analyses) (Table II). The remaining 17 analyses with the same rate of change before and after the change point had slopes not significantly different from zero and significant decreases in contaminant levels at the change point year. This model was characteristic of o-CHL (where it occurred in 23% (3/13) of the analyses), HE (23%), α-HCH (15%), MIR (15%) and P-MIR (15%) (Table II).

*3.1.3. Change point; different rate of change*

Forty-five analyses of 143 (32%) required a change point in the model and showed a change in the slope (Figures IV and V). Based on the assumption of declining contaminant levels, slopes before and after the change point year were classified into two further categories:-

**A.) Faster rate of decline after the change point** (Figure IV): This category pertains to cases where contaminant concentrations were:

1.) Declining before and after the change point year, but the rate of decline was faster (steeper slope) after the change point (Figure IV).
2.) Not changing significantly or increasing before the change point and declining subsequent to the change point year.
3.) Increasing significantly before the change point and not changing (slope = 0) significantly subsequent to the change point year.

**Table II.** Results of Change Point Regression Analyses on Organochlorine

| Compound | PCB 1254:1260 | | | | | DDE | | | | | DIEL | | | | | HE | | | | | HCB | | | | |
|---|---|---|---|---|---|---|---|---|---|---|---|---|---|---|---|---|---|---|---|---|---|---|---|---|---|
| **Site** | **Yr.** | **B** | **S** | **A** | **$R^2$** | **Yr.** | **B** | **S** | **A** | **$R^2$** | **Yr.** | **B** | **S** | **A** | **$R^2$** | **Yr.** | **B** | **S** | **A** | **$R^2$** | **Yr.** | **B** | **S** | **A** | **$R^2$** |
| ***E.L.O.*** | 92 | - | 0 | - | 0.92 | 92 | - | 0 | - | 0.84 | 93 | - | 0 | -- | 0.82 | 81 | - | 0 | - | 0.61 | | - | | | 0.86 |
| ***W.L.O.*** | 87 | -- | | - | 0.92 | 89 | - | 0 | - | 0.88 | 89 | - | 0 | -- | 0.78 | 77 | -- | 0 | - | 0.64 | 83 | - | - | - | 0.95 |
| ***N.R.*** | 87 | - | - | 0 | 0.94 | 87 | 0 | - | 0 | 0.89 | 87 | 0 | | - | 0.79 | 84 | 0 | - | - | 0.91 | 92 | - | 0 | - | 0.93 |
| ***E.L.E.*** | 81 | - | + | - | 0.91 | 81 | - | + | - | 0.91 | 90 | - | - | - | 0.79 | 90 | 0 | - | 0 | 0.81 | 91 | - | - | 0 | 0.93 |
| ***W.L.E.*** | | - | | | 0.72 | 78 | - | - | - | 0.85 | 87 | 0 | - | 0 | 0.82 | 87 | 0 | - | 0 | 0.75 | 78 | - | - | - | 0.93 |
| ***D.R.*** | | - | | | 0.88 | 92 | - | 0 | - | 0.89 | 85 | 0 | - | - | 0.82 | 89 | 0 | - | 0 | 0.84 | | - | | | 0.72 |
| ***E.L.H.*** | 90 | - | + | - | 0.88 | 90 | - | + | - | 0.84 | | - | | | 0.63 | | - | | | 0.32 | 87 | - | | 0 | 0.92 |
| ***N.L.H.*** | 78 | - | - | - | 0.85 | 79 | 0 | - | 0 | 0.83 | | - | | | 0.61 | 93 | - | + | -- | 0.43 | 78 | - | - | - | 0.86 |
| ***S.B.L.H.*** | 93 | - | 0 | -- | 0.71 | 93 | 0 | 0 | - | 0.59 | 93 | - | | -- | 0.75 | 86 | - | + | - | 0.73 | 88 | - | 0 | - | 0.92 |
| ***N.L.M.*** | 90 | - | + | - | 0.80 | 90 | - | + | - | 0.71 | 90 | - | + | -- | 0.71 | 90 | 0 | + | - | 0.38 | 90 | - | + | - | 0.72 |
| ***W.L.M.*** | 83 | 0 | - | 0 | 0.83 | 90 | - | 0 | - | 0.75 | 91 | 0 | - | 0 | 0.72 | 91 | 0 | - | + | 0.64 | 84 | -- | 0 | - | 0.85 |
| ***E.L.S.*** | 84 | - | - | 0 | 0.92 | 83 | - | | 0 | 0.89 | 93 | - | - | - | 0.68 | 88 | 0 | | - | 0.44 | | - | | | 0.89 |
| ***W.L.S.*** | 86 | - | | 0 | 0.92 | 80 | -- | | - | 0.86 | 92 | - | + | -- | 0.84 | 86 | 0 | + | - | 0.51 | 88 | - | + | - | 0.87 |

**Site:** Herring gull colony, arranged east to west: ***E.L.O.*** Eastern Lake Ontario ***W.L.O.*** Western Lake Ontario ***N.R.*** Niagara River ***E.L.E.*** Eastern Lake Erie ***W.L.E.*** Western Lake Erie ***D.R.*** Detroit River ***E.L.H.*** Eastern Lake Huron ***N.L.H*** Northern Lake Huron ***S.B.L.H.*** Saginaw Bay, Lake Huron ***N.L.M.*** Northern Lake Michigan ***W.L.M.*** Western Lake Michigan ***E.L.S*** Eastern Lake Superior ***W.L.S.*** Western Lake Superior.

# Contaminants in Herring Gull Eggs

| QCB | | | | | MIR | | | | | P-MIR | | | | | o-CHL | | | | | α-HCH | | | | | ß-HCH | | | | |
|---|---|---|---|---|---|---|---|---|---|---|---|---|---|---|---|---|---|---|---|---|---|---|---|---|---|---|---|---|---|
| Yr. | B | S | A | $R^2$ | Yr. | B | S | A | $R^2$ | Yr. | B | S | A | $R^2$ | Yr. | B | S | A | $R^2$ | Yr. | B | S | A | $R^2$ | Yr. | B | S | A | $R^2$ |
| | - | | | 0.75 | 81 | -- | + | - | 0.90 | | - | | | 0.77 | 81 | - | + | - | 0.80 | | - | | | 0.61 | | - | | | 0.87 |
| 91 | - | - | 0 | 0.98 | 79 | -- | + | - | 0.79 | | - | | | 0.68 | 83 | + | - | - | 0.87 | | - | | | 0.60 | 91 | - | - | - | 0.97 |
| | - | | | 0.81 | 87 | 0 | - | 0 | 0.78 | 87 | 0 | - | 0 | 0.76 | 88 | 0 | - | 0 | 0.38 | | - | | | 0.78 | 91 | - | - | - | 0.94 |
| 91 | - | - | - | 0.87 | 81 | - | + | - | 0.83 | 90 | 0 | - | 0 | 0.84 | 82 | - | + | - | 0.83 | 91 | - | + | - | 0.84 | 86 | - | - | - | 0.88 |
| 91 | - | - | 0 | 0.92 | 78 | 0 | - | 0 | 0.72 | | 0 | | | 0.02 | 82 | - | + | - | 0.58 | 87 | - | - | - | 0.90 | | - | | | 0.77 |
| 91 | - | | 0 | 0.93 | | - | | | 0.35 | 92 | - | + | - | 0.71 | 85 | 0 | - | - | 0.88 | | - | | | 0.64 | 86 | 0 | - | 0 | 0.92 |
| | - | | | 0.65 | 81 | -- | + | - | 0.78 | | - | | | 0.55 | 87 | 0 | - | 0 | 0.51 | 88 | 0 | - | 0 | 0.88 | 86 | - | - | - | 0.90 |
| | - | | | 0.57 | 81 | -- | + | - | 0.68 | | - | | | 0.33 | 81 | - | + | 0 | 0.51 | 88 | 0 | - | 0 | 0.89 | | - | | | 0.84 |
| 91 | - | - | - | 0.87 | | 0 | | | 0.08 | | 0 | | | 0.04 | 83 | + | - | 0 | 0.40 | 88 | - | - | - | 0.98 | | - | | | 0.79 |
| | - | | | 0.70 | 90 | - | + | - | 0.57 | 87 | - | + | + | 0.78 | 84 | 0 | - | 0 | 0.58 | | - | | | 0.88 | | - | | | 0.83 |
| 93 | - | + | -- | 0.85 | | - | | | 0.39 | 87 | - | + | - | 0.55 | 88 | - | - | + | 0.80 | 90 | - | | 0 | 0.81 | 86 | - | - | - | 0.83 |
| | - | | | 0.70 | 77 | - | - | - | 0.81 | 93 | - | + | - | 0.75 | | - | | | 0.83 | 89 | -- | - | - | 0.99 | 86 | - | - | - | 0.89 |
| 91 | - | - | + | 0.78 | 89 | - | | + | 0.81 | 89 | - | + | 0 | 0.81 | | - | | | 0.29 | 88 | - | - | - | 0.93 | | - | | | 0.83 |

**Yr: Last two digits of the change point year** (if applicable).
**B: Slope before the change point. Key:** -: declining significantly, --: declining significantly faster than after the change point year, +: increasing significantly, 0: no significant trend. **Note:** If the model has no change point this column indicates the trend over the entire time period.
**S: Direction of shift in contaminant concentration at the change point year. Key:** -: decreased significantly, +: increased significantly. **Note:** If this column is blank and there is a change point, the parameter representing the shift in contaminant concentration was non-significant. The data is best modeled by two regression lines (slopes shown in columns B and A), joined by a smooth transition at the change point year.
**A: Slope after the change point. Key:** -: declining significantly, --: declining significantly faster than before the change point year +: increasing significantly, 0: no significant trend.
**$R^2$: $R^2$ value for the regression model.**

A total of 18 analyses was classified as having faster rates of decline (or a steeper negative slope) after the change point than before. Of these, 55% had significant increases, 28% had significant decreases and 17% showed no change in contaminant concentrations at the change point year. In eight analyses where contaminants were declining before and after the change point, the rate was faster after the change point. This was most characteristic of DIEL (where it occurred in 38% (5/13) of the analyses). In eight analyses compounds did not begin to decline significantly (after a period of no change) until the change point year. This model was characteristic of HE (where it occurred in 31% (4/13) of the analyses) and DIEL (15%). O-CHL levels at western Lake Ontario were increasing significantly before the change point, and declining significantly after the change point (Table II). O- CHL levels at Saginaw Bay, Lake Huron were increasing significantly before the change point and not changing subsequently (Table II). Overall, the models classified as declining faster after the change point were most characteristic of DIEL (where they occurred in 54% (7/13) of the analyses) and HE (31%)

**B.) Slower rate of decline after the change point** (Figure V): This category pertains to cases where contaminant concentrations were:

1.) Declining before and after the change point year, but the rate of decline was slower after the change point (Figure V).
2.) Declining significantly before the change point, and not changing significantly (slope=0) subsequent to the change point year.
3.) Declining significantly or not changing significantly before the change point and increasing significantly after the change point year.

A total of 27 (19%) analyses was classified as having slower rates of decline (or a more positive slope) after the change point than before. These were equally divided (9 each) among those showing significant increases, decreases or no change at the change point. In comparing trends before and after the change point, there were 9 analyses where the rate of decline was significantly slower after the change point year. This model was characteristic of MIR (where it occurred in 31% (4/13) of the analyses). There were 12 analyses where levels were declining significantly before and not changing significantly after the change point. This trend was evident for the following compounds: PCB 1254:1260 (where it occurred in 23% (3/13) of the analyses), QCB (23%) and HCB (15%) (Table II). There were 6 analyses where concentrations were declining or not changing significantly before the change point year and increasing significantly after the change point, these were DIEL, HE and o-CHL at western Lake Michigan, QCB and MIR at western Lake Superior and P-MIR at northern Lake Michigan (Table II).

Overall, results from change point regression based on significant slopes show that in 98 of 143 analyses (69%) long-term contaminant trends up to 1995 continue to change at similar rates as seen previously in the sampling period. This was comprised of 55% of comparisons where contaminants were declining and 14% where contaminant levels were not changing significantly. In the remaining 45 comparisons (31%) slopes changed subsequent to the change point year. Of these, 12% had greater rates of decline after the change point and 19% had slower rates of decline after the change point. In other words, in 67% of the comparisons the rate of decline, to 1995, was the same as or faster than seen previously in the sampling period, in 19% of the comparisons the rate of decline was slower after the change point and in 14% of the comparisons results indicated no significant trend in contaminant values.

*3.1.4. Change Point Years*

The distribution of the significant change point years (N=107), indexed by the probability of each year being a change point, and the direction of change in contaminant concentration for each change point is presented in Figure VI. The four most common change point years (comprising 37% of the total number of change point years), based on the index of probability, were 1981 (index=0.11), 1978 (index=0.09), 1987 (index=0.09) and 1990 (index=0.09) (Figure VI). Change point years where contaminants tended to decline in concentration included the following: 1978 (100%), 1984 (75%), 1986 (63%), 1987 (62%), 1988 (67%) and 1991(83%) (Figure VI). Change point years characterized by significant increases in concentration were: 1981 (100%), 1982 (100%), 1990 (69%), 1992 (100%), and 1993 (75%) (Figure VI). Results of the chi-square tests indicated that more change points were associated with significant increases or decreases at the change point year than were associated with no change in contaminant concentration. They also indicated that there was no significant difference in the number of change points that had significant increases (45 comparisons) or decreases (50 comparisons).

For the herring gull dataset as a whole, there was not a single change point or model which predominated. Nonetheless there were some evident patterns. For the Niagara River, 56% (5/9) of the analyses (with change points) occurred in 1987. This was seen for PCB 1254:1260, DDE, DIEL, MIR and P-MIR (Table II). Most (4/5) of the change points at the Niagara River in 1987 were associated with significant decreases in contaminant levels (Table II). At northern Lake Michigan most (75%) of the change points occurred in 1990 and were associated with significant increases (Table II). Change point years for PCB 1254:1260 and DDE appeared to be synchronized at some locations. Of the comparisons with significant change

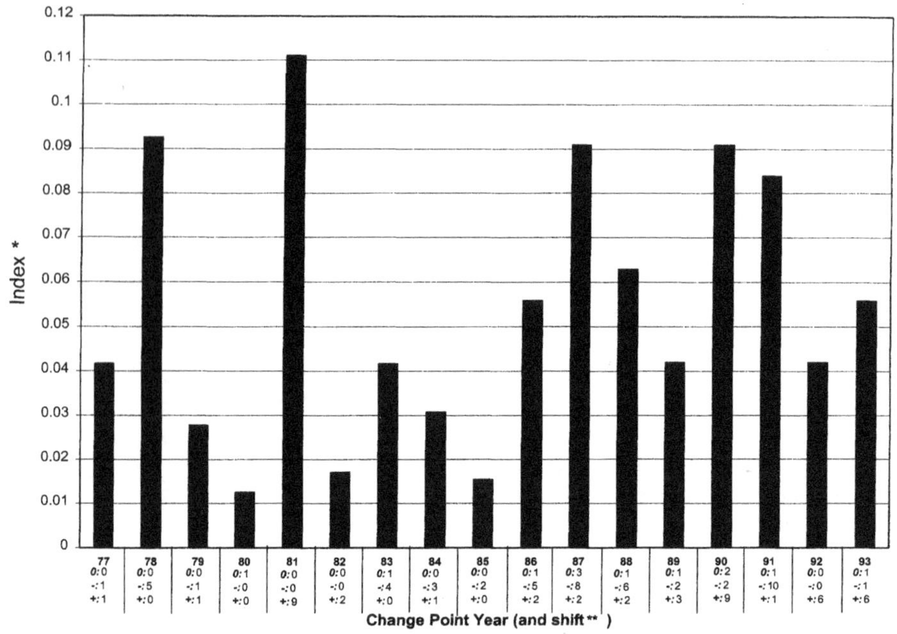


**Figure VI.** Distribution of Significant Change Point Years (N=107).

Key:

**Index** *: Refers to the number of change points which occurred in that year divided by the total number of comparisons for which that year was a possible change point.

**shift** **:
- ***0:*** Number of significant change points where contaminant concentrations did not change significantly that year.
- ***-:*** Number of significant change points where contaminant concentrations decreased significantly that year.
- ***+:*** Number of significant change points where contaminant concentrations increased significantly that year.

points for both compounds, 56% (6/11) occurred in the same year at the same location. This was the case at eastern Lake Ontario, the Niagara River, eastern Lake Erie, eastern Lake Huron, Saginaw Bay, Lake Huron and northern Lake Michigan (Table II). For QCB 86% (6/7) of the analyses with change points occurred in 1991 and were associated with significant decreases (Table II). MIR levels rose significantly in 1979 or 1981 for 50% (5/10) of the comparisons with change points. These include analyses for both colonies on Lake Ontario and two colonies on Lake Huron. At all four sampling locations on Lake Ontario and Lake Huron MIR levels were declining significantly faster before the change points (Table II). P-MIR levels rose significantly in 1987 at both Lake Michigan colonies, levels continued to increase since then at northern Lake Michigan (Table II). P-MIR levels also rose significantly between 1989 and 1993 in Lake Superior and the Detroit River (Table II). For α-HCH and ß-HCH there was some synchrony in the occurrence of change point years. For α-HCH 44% (4/9) were significant declines in 1988 and for ß-HCH 100% (7/7) were significant declines in 1986 or 1991 (Table II).

*3.1.5. Examples of long-term temporal trends: PCB 1254:1260 and DDE*

There were 9 different patterns for PCB 1254:1260 and DDE (Figure VII). A pattern of steady decline was seen for PCB 1254:1260 at the Detroit River and western Lake Erie (Figure VII-a). Patterns that had significant change points and contaminant concentrations that were declining equally before and after the change point occurred for PCB 1254:1260 at 5 of 13 colonies (38%) and for DDE at 8 of 13 colonies (62%) (Figures VII-b and VII-c). These patterns tended to have recent change points (1981-1992) when contaminant concentrations increased significantly at the change point year, and had an earlier change point (1978) when concentrations decreased significantly at the change point year (Figures VII-b and VII-c). For PCB 1254:1260 at western Lake Michigan there were no significant temporal trends before or after a significant decrease in levels in 1983 (Figure VII-d). This pattern also fit DDE data from the Niagara River and northern Lake Huron with change points in 1987 and 1978, respectively. At Saginaw Bay, Lake Huron, PCB 1254:1260 levels have been declining more rapidly, and DDE levels began to decline after significant increases in 1993 (Figures VII-e and VII-f). Levels of PCB 1254:1260 at western Lake Ontario and DDE at western Lake Superior have been declining at a slower rate in recent years than previously seen (Figure VII-g). PCB 1254:1260 and DDE concentrations ceased to decline in the mid 1980s at 3 colonies and 1 colony, respectively (Figures VII-h and VII-i). At two of these, stabilizing trends were associated with significant decreases at the change point year, while 2 were associated with no change in contaminant levels at the change point year. (Figure VII-g).

**Figure VII.** Examples Showing the 9 Representative Change Point Patterns for PCB 1254:1260 and DDE.

Key: **Change point:** Change point year

**Model:** Indicates the patterns of the long term change point model: **B/S/A**, where:

**B** indicates the direction of the slope before the change point

(key: -: declining significantly, --: declining significantly faster than after the change point year, +: increasing significantly, 0: no significant trend.) **NOTE:** If the model has no change point this indicates the trend over the entire time period.

**S** indicates the shift in contaminant concentration at the change point year

(key: -: significant decrease, +: significant increase, 0: no significant change)

**A** indicates the direction of the slope after the change point

(key: -: declining significantly, --: declining significantly faster than before the change point year, +: increasing significantly, 0: no significant trend)

**$R^2$:** $R^2$ value for the comparison shown in each figure

■ : Observed Values

—— : Predicted Values

Colony abbreviations mentioned in the captions (in ***bold italic*** text) pertain to those indicted in Figure I. The last two digits of change point years for the analyses mentioned in the captions are indicated in parentheses after each colony abbreviation.

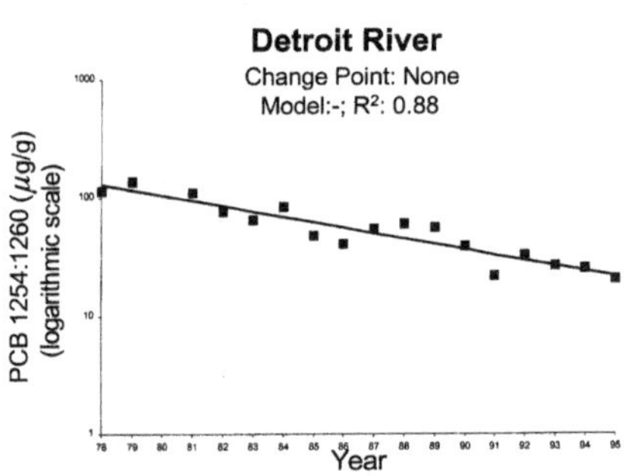


**Figure VII-a.** Model with no significant change point. Contaminant levels declining significantly over the entire time period. This model also occurred for PCB 1254:1260 at *W.L.E.*

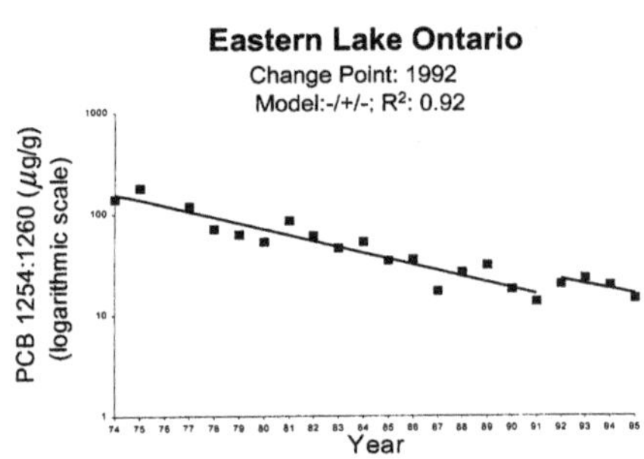


**Figure VII-b.** Model showing the same rate of decline before and after a significant increase in contaminant levels at the change point. This model also occurred for PCB 1254:1260 at *E.L.E.* (81), *E.L.H.* (90), and *N.L.M.* (90) and for DDE at *E.L.O.* (92), *W.L.O.* (89), *E.L.E.* (81), *D.R.* (92), *E.L.H.* (90), *N.L.M.* (90) and *W.L.M.* (90).

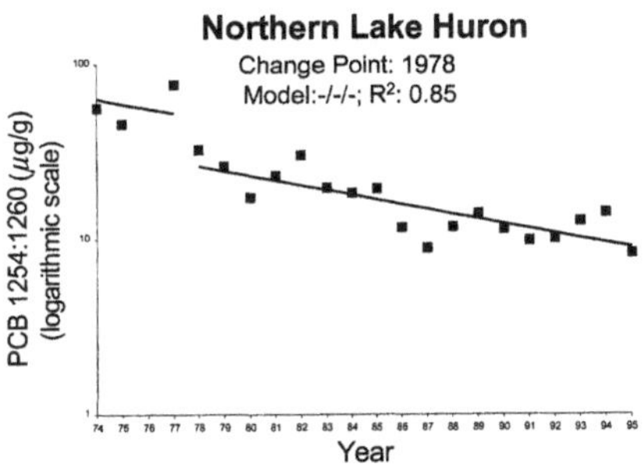


**Figure VII-c.** Model showing the same rate of decline before and after a significant decrease in contaminant levels at the change point. This also occurred for DDE at *W.L.E.* (78).

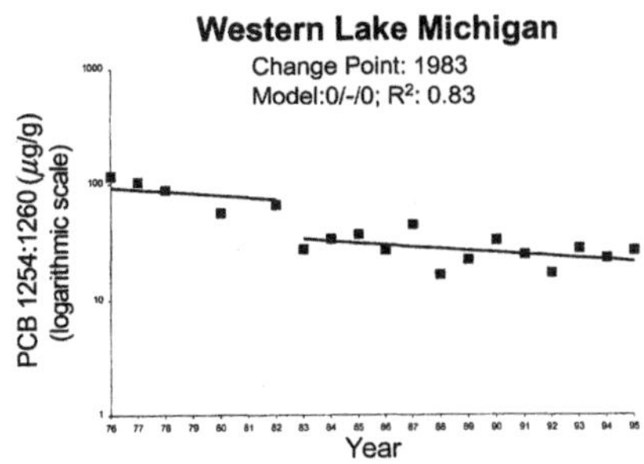


**Figure VII-d.** Model showing no significant trend before or after a significant decrease at the change point. This model also occurred for DDE at *N.R.* (87) and *N.L.H.* (79).

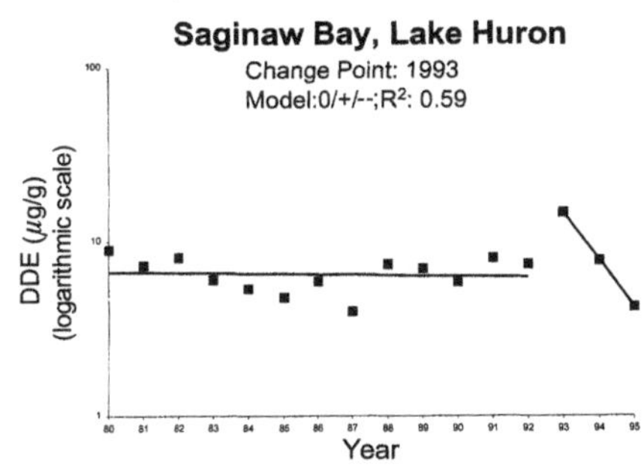


**Figure VII-e.** Model showing levels declining faster after a significant increase in levels at the change point. This model did not occur for PCB 1254:1260 or DDE at any other site.

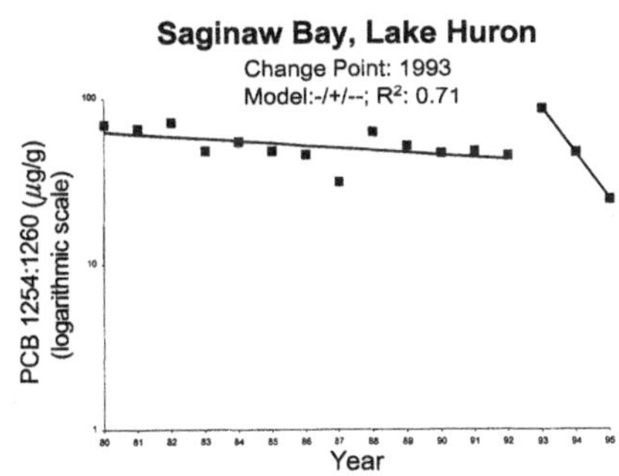


**Figure VII-f.** Model showing no significant trend before, and declining levels after a significant increase at the change point. This model did not occur for PCB 1254:1260 or DDE at any other site.

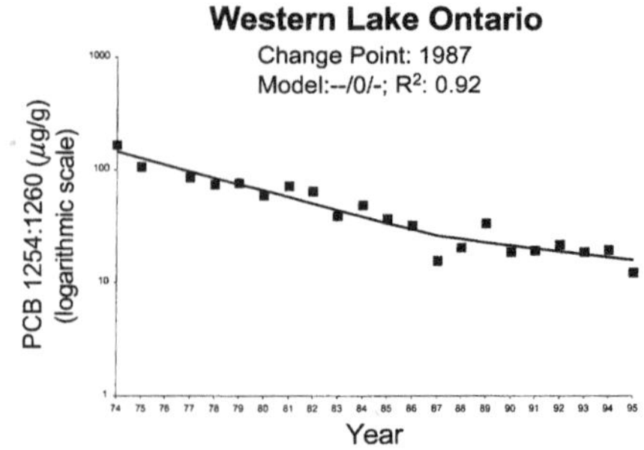


**Figure VII-g.** Model showing levels declining faster before the change point than after. There was no significant change in contaminant levels at the change point. This also occurred for DDE at *W.L.S.* (80).

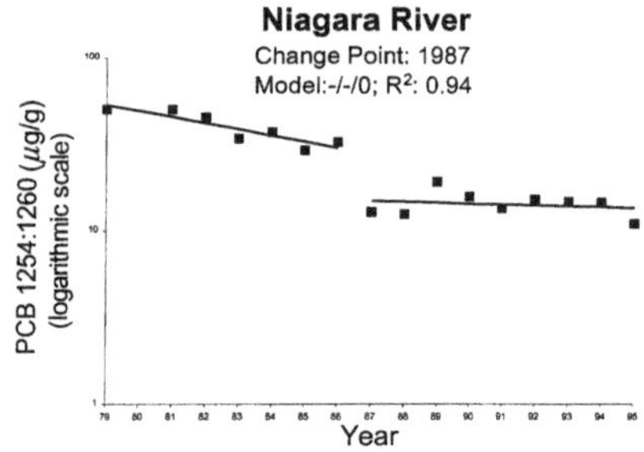


**Figure VII-h.** Model showing a significant declining trend before the change point, a significant decrease at the change point and no significant trend after. This model also occurred for PCB 1254:1260 at *E.L.S.* (84).

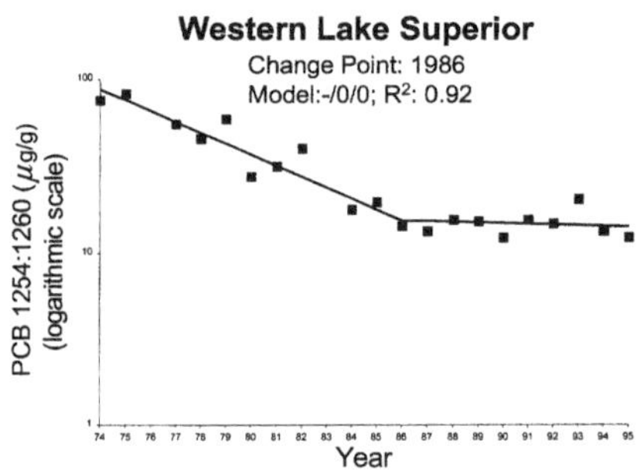


**Figure VII-i.** Model showing levels declining before the change point and no significant trend after. There was no significant change in contaminant levels at the change point. This also occurred for DDE at *E.L.S.* (83).

### 3.2. Short-Term Temporal Trends

For the 6 major compounds analyzed (PCB 1254:1260, DDE, DIEL, HCB, HE and MIR), results indicated that the majority of the analyses had non-significant temporal trends (93% during 1991-1995 and 85% during 1981-1985). Of the analyses with significant results 5 comparisons (6%) were declining significantly during 1991-1995 and 11 (14%) were declining significantly during 1981-1985 (Table III). During both time periods MIR was increasing significantly at one site, at Eastern Lake Superior from 1991-1995 and the Detroit River from 1981-1985. Overall, for 57% of the comparisons where contaminants were declining significantly in the 1980s, the rate of decline had slowed by the 1990s. Nonetheless for most of the analyses, significant temporal trends could not be determined using five years of data.

## 4. Discussion

The structure of the herring gull egg contaminant dataset contains some anomalies that could affect the results of the change point regression analyses. There were isolated years when eggs were not collected at selected sites or years when all compounds were not analyzed for. This could potentially affect the results of change point analyses since missing year(s) of data would increase the probability that contaminant levels would differ between the years when data were collected. When comparing the change point regression analyses to the gaps in data we found only one analysis (HE at western Lake Ontario) where the change point occurred the year after one in which no data were collected. Thus, we did not find any indication that the years when data were missed were influential in determining change points. Turle and Collins (1992) indicated that pooling samples for contaminant analysis could overestimate the results for mirex. This did not appear to influence the change point analyses since no change points for MIR occurred in years when pooling was initiated at any given site. Differences in the years when data collection/chemical analysis was initialized could also influence the results of change point analysis. Data collection for o-CHL, QCB, α-HCH, ß-HCH and P-MIR was initiated after 1974. The shorter length of the dataset affected the results of the change point analyses for QCB, α-HCH, ß-HCH and P-MIR. These compounds had a greater proportion of comparisons with a steady rate of change over the entire time period or more recent change points. The shorter length of the dataset did not appear to influence the results for o-CHL.

For the dataset as a whole, the change point regression indicated continuing

**Table III.** Temporal Trends for Six Major Compounds during the Early 1980s and Early 1990s.

| TIME PERIOD | EARLY 1980s (~81-85) | | | EARLY 1990s (~91-95) | | |
|---|---|---|---|---|---|---|
| **COMPOUND** | **Declining Significantly** | **Increasing Significantly** | **No significant trend** | **Declining Significantly** | **Increasing Significantly** | **No significant trend** |
| PCB 1254:1260 | 23% | 0% | 77% [+: 0%, -: 100%] | 15% | 0% | 85% [+: 36, -: 54%] |
| DDE | 15% | 0% | 85% [+: 0%, -:100%] | 8% | 0% | 92% [+: 17, -: 83%] |
| DIEL | 8% | 0% | 92% [+:33%, -: 67%] | 15% | 0% | 85% [+: 9%, -: 91%] |
| HCB | 23% | 0% | 77% [+: 0%, -: 100%] | 0% | 0% | 100% [+: 15%, -: 85%] |
| HE | 8% | 0% | 92% [+: 17%, -: 83%] | 0% | 0% | 100% [+: 15%, -: 85%] |
| MIR | 8% | 8% | 84% [+: 18%, -: 82%] | 0% | 8% | 92% [+: 50%, -: 50%] |
| **TOTAL** | **14%** | **1%** | **85% [+: 12%, -: 88%]** | **6%** | **1%** | **93% [+: 24%, -: 76%]** |

Key: For the comparisons with no significant trends:

-: Percentage of comparisons where contaminant concentrations were declining non-significantly.

+: Percentage of comparisons where contaminant concentrations were increasing non-significantly.

declining trends (equal or faster than earlier) in contaminant concentrations in herring gull eggs for 67% of the total colony-contaminant comparisons performed. These declining patterns were detected in more than half of the comparisons for all compounds except MIR. In an additional 14% of the total comparisons there was no significant temporal trend over the 22 year period of the study. All compounds (except QCB and HCB) showed this pattern at at least one site. Finally, only 19% of the comparisons were declining more slowly than earlier in the sample period. One- third of these were continuing to decline to 1995 (most notably mirex) while two- thirds have stabilized (slope does not differ significantly from zero) or show increasing trends since their change point.

Results from short-term regression analyses for the six major compounds indicated that for most of the comparisons significant trends could not be determined using five years of data. Out of the few where significant declines were determined there was not a large percentage (57%) where the rate of decline had slowed in the early 1990s compared to the early 1980s. These results correspond with those found using the long-term change point analyses which indicated that for most (67%) of the analyses the rates of decline to 1995 were the same or faster than those seen previously in the sampling period. The results of both long-term change point analyses and the short-term regression analyses point to the need for continuous long-term monitoring. Short-term regression analyses indicated that the proportion of results for which significant trends could be determined was minimal (15% for 1981-1985 and 7% for 1991-1995). Results of the change point analyses, using many more years of data and accounting for one significant year to year change in the time period indicated that for most (86%) of the analyses significant trends were evident and that for the majority of the analyses levels were declining at the same or similar rates as seen previously in the time period. The lack of significant results found by short-term regression and the many significant trends found using change point regression (which indicated many different change points related to significant increases or decreases in contaminant levels) suggest that long-term continuous data collection can give more insight into the temporal trends and associated changes than can be determined by the comparison of data analyzed during two short (arbitrarily chosen) time periods.

For the Niagara River the most common change point was 1987, 56% of all the change points occurred that year. Most of the change points for the Niagara River in 1987 were associated with significant decreases in contaminant levels. Change point models for DDE, MIR and P-MIR at the Niagara River appear to indicate that 1987 was a year when an environmental change was responsible for the decrease in contaminants in herring gull eggs. For all three compounds the rate of change remained constant before and after the significant decrease in 1987.

O'Gorman *et al.* (1991) found that the water in Lake Ontario was much warmer in 1987 than it was in 1986 or 1988. When this was the case, they found that in the spring, alewife tended to spawn earlier in the shallow waters near the shoreline (O'Gorman *et al.* 1991). Hebert *et al.* (1997) found that PCB levels in herring gull eggs were higher than expected when alewife condition was low. At present, it appears that the significant decreases in contaminants in herring gull eggs from the Niagara River in 1987 may be related to changes in their pattern of food consumption. For example, if the waters of the Niagara River were warmer than usual, alewives would experience less overwinter mortality. Thus there would be fewer fish in poor condition to be consumed by herring gulls in the early spring. It is interesting that significant decreases were not seen in 1987 for any contaminants in herring gull eggs from Lake Ontario. This indicates that other year to year factors had more significant effects on the long-term trends of contaminants in herring gull eggs from Lake Ontario than those from the Niagara River.

The synchrony of change points for PCB 1254:1260 and DDE appears to be an indication of the common chemical characteristics of these contaminants. They are both highly insoluble in water but highly soluble in fat (WHO/EURO 1987, WHO 1989b). Most of the exposure to both contaminants probably took place early in the sampling period. Subsequent trends and change points are most likely indications of environmental factors which could affect the rates of decline and uptake of both contaminants (for example, year to year changes in the diet of herring gulls).

DIEL and HE had synchronous change points within colonies in Lake Erie (1987 for DIEL and 1990 for HE) and Lake Michigan (1990 for DIEL and 1991 for HE). At all these colonies, except northern Lake Michigan, the contaminants decreased significantly at the change point year. Both contaminants are epoxides of parent compounds which were used as soil insecticides and which have a strong affinity to soil particles (WHO 1984, 1989a). The significant decreases could be associated with increased sedimentation rates resulting in lower quantities of the compounds being available for uptake in the aquatic system. They could also indicate years in which herring gulls foraged less in fields (Fox *et al.* 1990). Likewise the significant increases in 1990 at northern Lake Michigan could be the result of increased turbulence of the sediments or increased foraging in fields (Fox *et al.* 1990). Though the use of both pesticides was banned in the 1970s, limited use for the subterranean control of termites was allowed until recent years (WHO 1984, 1989a). The prevalent pattern, of faster rates of decline to 1995 than earlier in the sampling period, may be indicative that herring gulls were still being exposed to new sources of the compounds until their use ceased entirely.

The results from the change point regression analyses are highly variable. The degree of variability is reflected in the numerous different change point years and models that fit the data. At this point we do not have explanations for the variation in years when change points occurred. Change point years tended to be more synchronous within compounds or within locations for contaminants with similar chemical properties (and/or uses) than within geographic locations. This would appear to indicate that the long-term trends of contaminants in herring gull eggs have been affected by changes in environmental parameters, and that the results of these effects are correlated to the chemical properties of the contaminants. Reasons for year to year variation in contaminant trends may include factors such as changes in predator-prey dynamics, the invasion of dreissenid mussels, changes in temperature, rainfall and wind currents (Baumann and Whittle 1988, Borgmann and Whittle 1991, 1992, O'Gorman *et al.* 1991, Fontaine and Stewart 1992, Lesht and Brander 1992, Rowan and Rasmussen 1992, Bruner *et al.* 1994a, b, Chan *et al.* 1994, Madenjian *et al.* 1995, Moll *et al.* 1995, Smith 1995b, Hebert *et al.* 1997).

### 4.1. Previous Herring Gull Egg Data Analyses (Stow 1995)

The change point model is one of many that has been used recently to describe the temporal trends of contaminants in herring gull eggs (Smith 1995b, Stow *et al.* 1995, Stow 1995, Hebert *et al.* 1997). Stow (1995) used a portion of the CWS herring gull egg contaminant database (1978-1992) to determine the long-term trends of lakewide values of PCB 1254:1260 (Bishop *et al.* 1992, Pettit *et al.* 1994). Based on the results of extra sums of squares tests, he found that data from Saginaw Bay, Lake Huron had undergone a different statistical trend than at the other sites on that lake, thus trends were determined separately for that site. For all the colonies on a given lake, including sites other than Saginaw Bay, data within a given year were pooled. Since not all of the colonies he used in his analyses were sampled each year, the same database was not maintained (within each lake) from year to year (Bishop *et al.* 1992, Pettit *et al.* 1994). Applying one of three possible models, he found that:

1.) PCB 1254:1260 concentrations had ceased to decline in herring gull eggs from Lake Ontario and Lake Michigan.
2.) PCB 1254:1260 concentrations had started to increase in Lake Huron and Lake Superior
3.) PCB 1254:1260 concentrations were declining at the same rate over the entire time period in Lake Erie.

Using change point regression analyses on data from the same colonies throughout the analytical period, we found that PCB 1254:1260 levels in herring gull eggs continued to decline significantly at both colonies in Lake Ontario. Nonetheless at the western Lake Ontario site the rate of decline has been slower since 1987 and at the eastern Lake Ontario site the (steady) rate of decline had been decelerated by a significant increase in 1992. For Lake Michigan the change point analyses found that at the western colony PCB 1254:1260 levels dropped significantly in 1983 though levels were not changing significantly before or after. At northern Lake Michigan, levels were declining at the same rate as earlier in the time period, though they were decelerated by a significant increase in 1990. Change point analyses found that PCB 1254:1260 concentrations were not increasing significantly at any colonies on Lake Huron or Lake Superior. At the eastern and northern Lake Huron sites, the rate of decline had not changed significantly, though the trends had been affected by a significant increase in 1990 at eastern Lake Huron and a significant decrease in 1978 at northern Lake Huron. Similar to Stow's (1995b) result, PCB 1254:1260 levels continued to decline significantly at similar rates during the entire time period at both colonies in Lake Erie, though the change point analyses found that at western Lake Erie the rate had been decelerated by a significant increase in 1990. Overall, the results of the change point analyses do not concur with the conclusions drawn by Stow (1995) for Lake Ontario, Lake Michigan and Lake Huron. The results of the change point analyses do concur with the conclusions drawn by Stow (1995) for Lake Erie and Lake Superior, both studies found that PCB 1254:1260 levels were declining at steady rates in Lake Erie and they had stopped declining in Lake Superior.

### 4.2. Previous Herring Gull Egg Data Analyses (Smith 1995b)

Smith (1995b) found that for 5 contaminants in herring gull eggs, the short-term deviations from the long-term trends were synchronized within and among the Great Lakes. He found that these synchronous deviations were correlated with short-term weather patterns. Smith (1995b) proposed several hypotheses to explain the mechanism by which weather would affect contaminant levels in herring gulls. He concluded that the data were most consistently (though perhaps not exclusively) explained by the following: warm spring temperatures lead to increased phytoplankton growth which, through growth dilution, leads to less contaminated phytoplankton that spring. This effect is transferred up through the food chain to herring gulls and contributes to lower organochlorine levels in herring gull eggs laid that year. The opposite effect would be seen for cold spring temperatures. Smith (1995b) found that in 1980, 1983 and 1987 contaminant values were below expected and in 1981, 1982, 1991 and 1992 they were above expected.

Change point regression analyses indicated that 1980 was not a significant change point year for many analyses. Only one significant change point occurred in 1980 and it was not associated with a significant change in contaminant levels. Five change points occurred in 1983 and 80% of them were associated with significant decreases. Likewise, the majority (62%) of the regression models with change points in 1987 were associated with significant decreases in contaminant levels. In summary, of the three years when Smith (1995b) found that contaminant levels were lower than expected, 63% of the change points were associated with significant decreases in contaminant concentrations. The most common change point was 1981, and all the analyses that year were associated with significant increases in contaminant concentrations. Smith (1995b) also found that contaminant levels in 1981 were higher than expected. Change point regression found significant increases in contaminant levels in 1982, though only two change points occurred that year. Change point analyses did not find that 1991 was associated with significant increases in contaminant levels. On the contrary 83% of the analyses with change points in 1991 decreased significantly that year. Nine of the change points associated with significant decreases in levels occurred for compounds not analyzed by Smith (1995b). Nonetheless for the remaining contaminants which were included in both studies, change point models were all associated with significant decreases in 1991. For 1992, 100% (6/6) of the change point analyses were associated with significant increases. In summary, of the four years when Smith (1995b) found that contaminant levels were higher than expected 62% of the change points were associated with significant increases in contaminant concentrations. For the most part the results of the change point analyses concur with the expected deviations from the long-term trends found by Smith (1995b). The one exception is 1991 where Smith's (1995b) analysis found that concentrations were relatively higher than expected, yet most of the change points in that year were associated with significant decreases in contaminant levels.

### 4.3. Previous Herring Gull Egg Data Analyses (Hebert *et al.* 1997)

Hebert *et al.* (1997) refined the relationship between PCB 1254:1260 accumulation in herring gull eggs and weather at two Lake Ontario colonies. They analyzed the statistical influence of winter/spring degree days, autumn/winter degree days, spring alewife biomass, alewife condition and rainbow smelt biomass relative to the residuals from an exponential model with a non-zero asymptote. The significant amount of variation explained by these factors differed between the two colonies. For example, at Snake Island in eastern Lake Ontario, spring alewife biomass explained 30% of the variation in the dataset while it explained 7% at Muggs Island in western Lake Ontario (Hebert *et al.* 1997). Hebert *et al.* (1997) also measured the stable isotope ($[^{15}N]/[^{14}N]$) values in herring gull eggs to determine the year to year differences in the trophic level at which herring gulls were feeding and to determine if this was correlated to deviations in the long-term trends of PCB 1254:1260 . The value of the stable isotope ($[^{15}N]/[^{14}N]$) would be higher in years when herring gulls consumed more carnivorous fish (such as alewife). Their results indicated that ($[^{15}N]/[^{14}N]$) levels were correlated to the deviations of PBC 1254:1260 from the long-term trends at eastern Lake Ontario but not at western Lake Ontario. Hebert *et al.* (1997) concluded that the consumption of alewives by herring gulls (which is affected by the metabolism of the gulls, alewife abundance and condition and alewife overwinter mortality) and alewife population dynamics could be important factors in understanding the deviations of PCB 1254:1260 levels from the long-term trends. The differences in the results between the two colonies indicated that the weather mediated mechanisms which affect the year to year changes in contaminants differ between colonies and probably between lakes (Hebert *et al.* 1997). Herring gulls are opportunistic feeders whose diet may vary between years, seasons and sites. For example, Ewins *et al.* (1994) discovered that the species of fish found in regurgitated pellets were significantly different between two sampling years (1982 and 1983) at Snake Island, Lake Ontario. They also found that for a colony in Hamilton Harbour, Lake Ontario, there was a significant decrease in the amount of fish found in pellets between December and February of the same winter. This decrease in fish remains was concomitant with an increase in the amount of bird and mammal remains. This points to the need for careful interpretation of contaminant data accounting for behavioral differences in herring gulls due to environmental conditions. These differences may be affected by the geographic location of the colony, proximity to shore, and the abundance of alternative food sources; such factors vary among regional populations of Great Lakes herring gulls (Fox *et al.* 1990). The results of the change point regression analyses appear to corroborate the conclusions drawn by Hebert *et al.* (1997). For the colony in eastern Lake Ontario, 1992 was a significant change point when PCB 1254:1260 levels increased significantly. Hebert *et al.*(1997) indicated that alewife dieoffs were thought to have occurred in the winters of 1991/92 and 1992/93 in Lake Ontario (Ontario Ministry of Natural Resources 1993, National Biological Service 1994, cited in Hebert *et al.* 1997). They also found that alewife biomass was more of a significant factor for the changes in the temporal trends of PCB 1254:1260 at eastern Lake Ontario than at western Lake Ontario (where the same change point was not observed).

### 4.4. Contaminant Trends in Fish

There has been much recent discussion about stabilizing rates of decline in contaminant levels in fish and herring gull eggs. Non-declining trends have been reported from Lake Michigan for PCBs in lake trout (*Salvelinus namaycush*), rainbow trout (*Oncorhynchus mykiss*), brown trout (*Salmo trutta*), alewife (*Alosa pseudoharengus*), bloater chub (*Coregonus hoyi*), chinook salmon (*Oncorhynchus tshawytscha*) and coho salmon (*Oncorhynchus kisutch*) and for DDE in bloater chub (Baumann and Whittle 1988, Hesselberg *et al.* 1990, Miller *et al.* 1992, Stow 1994, Stow *et al.* 1995). In herring gull eggs, PCB 1254:1260 levels continued to decline significantly, to 1995, at northern Lake Michigan but they were not declining significantly at the colony in Green Bay (western Lake Michigan); DDE levels continued to decline significantly, to 1995, at both Lake Michigan colonies. Slower rates of decline in the 1980s (as compared to the 1970s) have been reported for DDE in spottail shiners (*Notropis hudsonius*) from the Niagara River and for PCBs in spottail shiners from Lake Ontario (Suns *et al.* 1991, 1993). In herring gull eggs, DDE levels decreased significantly in 1987 but they were not declining before or after the change point. PCB 1254:1260 levels have been declining more slowly at western Lake Ontario since 1987; at eastern Lake Ontario they continue to decline at the same rate as before a significant increase in 1992. Change point regression indicated that PCB 1254:1260 concentrations were no longer declining in herring gull eggs from Lake Superior, similar results were reported for spottail shiners (Suns *et al.* 1993). Overall, the results of the change point analyses of PCB 1254:1260 and DDE in herring gull eggs only concur with the results found for PCBs in fish from Lake Superior. For the most part DDE and PCB 1254:1260 in herring gull eggs from the other Great Lakes continue to decline at similar rates throughout the entire time period, nonetheless, many of these declines have been decelerated by significant increases in concentrations at the change point.

De Vault *et al.* (1996) analyzed the long-term temporal trends of PCBs, DDT, DIEL and chlordane in lake trout from Lakes Ontario, Huron, Michigan and Superior and in walleye from Lake Erie. Data collection was initiated in the late 1970s, in all lakes except Lake Ontario where it began in 1982, and continued until 1992, in all lakes except Superior where it ended in 1990. For most analyses they used a first order loss rate model. Overall, they found that the contaminants in lake trout fit the single trend models with significantly declining trends over the entire time period. The analyses of PCB and DDT in walleye from Lake Erie was divided into two time sections. Results indicated that levels declined until 1982, and afterwards the slope was not significantly different from zero. Change point analyses on herring gull eggs indicated that PCB 1254:1260 levels continued to decline significantly to 1995 at most colonies except those on Lake Superior where levels stabilized in the mid 1980s and DDE levels continued to decline significantly to 1995 at most colonies. Unlike the results of De Vault *et al.* (1996), change point analyses on herring gull eggs from western Lake Erie (which was closest to the area sampled by De Vault *et al.* 1996) indicated that both PCB 1254:1260 and DDE levels continued to decline to 1995. De Vault *et al.* (1996) reported that DIEL levels in lake trout from Lakes Michigan, Superior, Huron and Ontario had peaks in 1979 and 1984, these were not found to be significant change points for DIEL in herring gull eggs. Overall, they found that DIEL trends in lake trout and walleye were declining at a constant rate. This is unlike DIEL levels in herring gull eggs, which have been declining at faster rates in recent years. De Vault *et al.* (1996) also found that o-CHL levels were declining at a constant rate in most Great Lakes. In herring gull eggs the predominant patterns indicated that levels were not changing significantly before or after significant decreases or that levels began to decrease in the 1980s. De Vault *et al.* (1996) used a change point model on the data for PCBs in lake trout from Lake Michigan to evaluate the hypothesis that a recent lack of decline in the contaminant is the result of changes in the food chain. They found that no individual change point could be determined. In herring gull eggs, change points (and significant changes in PCB 1254:1260 levels) were determined at both Lake Michigan sites, though similar to the results in Lake Trout, there were no significant changes in the trends before and after the change point year. Overall, the results of the analyses of long-term contaminant trends in lake trout and herring gull eggs only concur for o-CHL in Lake Superior and DIEL in Lake Huron. For these sites the contaminants in herring gull eggs tended to decline at a steady rate over the entire time period. For the remaining lake trout analyses, the trends in herring gull eggs are not similar . Trends in herring gull eggs have undergone changes in the rate of decline and/or significant changes in concentration at the change point year, rather than a single trend model. The non- significant loss rates seen for PCB and DDT from 1982-1992 in walleye from Lake Erie were not observed for PCB 1254:1260 and DDE in herring gull eggs where the rates of decline were the same as those seen previously in the sampling period.

## 5. Conclusions

The change point regression for organochlorine contaminants in herring gull eggs allowed us to summarize the trends in contaminant levels to 1995 from 13 sites throughout the Great Lakes. The lack of significant trends found by short-term regression conducted on two (arbitrarily) chosen time periods indicates the importance of the continuous data collection and does not confirm or deny the results of the change point analysis. Using the longer dataset (and incorporating a change point model) provided the statistical power for a greater proportion of significant results. For the majority of the comparisons, long-term trends continue to decline but they have been affected by year to year changes within the dataset. There were many different change point years and models which applied to the data. The reasons for the many different patterns of changes in long-term trends are not all apparent. Nonetheless for the most part, they appeared to be more correlated with the chemical properties of the compounds. Overall, the long-term trends do not lead to the model of gradual phasing out or a reversal of declining trends. Some contaminants in herring gull eggs have ceased to decline at some locations, but this was not the case for the majority of the contaminants sampled.

## Acknowledgments

The authors wish to express their gratitude to D. Ryan for computer programming and insightful lecture notes. Thanks to C. Hebert, B. Wakeford and R. Norstrom for their invaluable assistance provided during discussions and manuscript reviews. Thanks also to D. Ryckman T. Heyland, S. Huestis, B. Collins, K. Hughes, J. Huang, K. Pettit and K. Timm for logistic assistance. Two anonymous reviewers, L. Shutt and D. Stewart provided valuable advice on improving the manuscript.

## References

Baumann, P.C. and Whittle, D.M.: 1988, 'The status of selected organics in the Laurentian Great Lakes: an overview of DDT, PCBs, dioxins, furans, and aromatic hydrocarbons', *Aquatic Toxicology* **11**, 241-257.

Bishop, C. and Weseloh, D. V.: 1990, *Contaminants in Herring Gull eggs from the Great Lakes*, A State of the Environment Fact Sheet No. 90-2, Environment Canada, 12 pp.

Bishop, C. A., Weseloh, D. V., Burgess, N. M., Struger, J., Norstrom, R. J., Turle, R., Logan, K. A.: 1992, *An atlas of contaminants in eggs of fish-eating colonial birds of the Great Lakes (1970-1988). Vol. I.*, Can. Wildl. Serv. Tech. Rep. Ser. No. 152, Canadian Wildlife Service, Ontario Region, 313 pp.

Borgmann, U. and Whittle, D. M.: 1991, 'Contaminant concentration trends in Lake Ontario Lake Trout (*Salvelinus namaycush*): 1977 to 1988', *J. Great Lakes Res.* **17(3)**, 368-381.

Borgmann, U. and Whittle, D. M.: 1992, 'DDE, PCB and mercury concentration trends in Lake Ontario Rainbow Smelt (*Osmerus mordax*) and Slimy Sculpin (*Cottus cognatus*): 1977 to 1988', *J. Great Lakes Res.* **18(2)**, 298-308.

Bruner, K. A., Fisher, S. W., Landrum, P. F.: 1994a, 'The Role of the Zebra Mussel, *Dreissena polymorpha*, in Contaminant Cycling: I. The Effect of Body Size and Lipid Content on the Bioconcentration of PCBs and PAHs', *J. Great Lakes Res.* **20(4)**, 725-734.

Bruner, K. A., Fisher, S. W., Landrum, P. F.: 1994b, 'The Role of the Zebra Mussel, *Dreissena polymorpha*, in Contaminant Cycling: II. Zebra Mussel Contaminant Accumulation from Algae and Suspended Particles, and Transfer to the Benthic Invertebrate, *Gammarus fasciatus*', *J. Great Lakes Res.* **20(4)**, 735-750.

Chan, C. H., Bruce, G., Harrison, B.: 1994, 'Wet Deposition of Organochlorine Pesticides and Polychlorinated Biphenyls to the Great Lakes', *J. Great Lakes Res.* **20(3)**, 546-560.

De Vault, D. S., Hesselberg, R.H., Rodgers, P.W. and Feist. T.J.: 1996, 'Contaminnat trends in Lake Trout and Walleye from the Laurentian Great Lakes', *J. Great Lakes Res.* **22(4)**, 884-895.

Draper, N. R. and Smith, H.: 1981, *Applied Regression Analysis*, John Wiley and Sons. New York, U.S.A, 709pp.

Environment Canada: 1991, *Toxic Chemicals in the Great Lakes and Associated Effects- Synopsis*, Ministry of Supply and Services Canada, 51pp.

Ewins. P. J., Weseloh, D. V., Mineau, P.: 1992, 'Geographical distribution of contaminants and productivity measures of Herring Gulls in the Great Lakes: Lake Huron 1980', *J. Great Lakes Res.* **18(2)**, 316-330.

Ewins. P. J., Weseloh, D. V., Groom J. H. Dobos, R. Z., Minueau, P.: 1994, 'The diet of Herring Gulls (*Larus argentatus)* during winter and early spring on the lower Great Lakes', *Hydrobiologia* **279/280,** 39-55.

Fontaine, T. D., and Stewart, D. J.: 1992, 'Exploring the Effects of Multiple Management Objectives and Exotic Species on Great lakes Food Webs and Contaminant Dynamics', *Environmental Management* **16(2)**, 225-229.

Fox, G. A., Allan, L. J., Weseloh, D. V., Minueau, P.: 1990, 'The diet of herring gulls during the nesting period in the Canadian waters of the Great Lakes', *Can J. Zool.* **68,** 1075-7085.

Gujarati, D. N.:1988, *Basic Econometrics*, McGraw Hill. New York, U.S.A.

Hallett, D. J., Norstrom, R. J., Onuska, F .I., Comba, M. E., Sampson, R.: 1976, 'Mass spectral confirmation and analysis by the Hall detector of mirex and photmirex in herring gulls from Lake Ontario' *J Agric. Food Chem.* **24**, 1189-1193.

Hebert, C. E. and Keenleyside, K. A.: 1995, 'To normalize or not to normalize: fat is the question', *Environ. Toxicol. Chem.* **14(5)**, 801-807.

Hebert, C. E., Shutt, J. L., Norstrom, R. J.: 1997, 'Dietary Changes Cause Temporal Fluctuations in Polychlorinated Biphenyl Levels in Herring Gull Eggs from Lake Ontario', *Environ. Sci. Technol.* **31**,1012-1017.

Hesselberg, R. J., Hickey, J. P., Nortrup, D. A., Willford, W. A.: 1990, 'Contaminant residues in the Bloater (*Coregonus hoyi*) of Lake Michigan, 1969-1986', *J. Great Lakes Res.* **16(1)**, 121-129.

Lesht, B. M. and Brander, D. J.: 1992, 'Functional Representation of Great Lakes Surface Temperatures', *J. Great Lakes Res.* **18(1)**, 98-107.

Madenjian, C. P., Whittle, D. M., Elrod, J. H., O'Gorman, R., Owens, R. W.: 1995, 'Use of a Simulation Model to Reconstruct PCB Concentrations in Prey of Lake Ontario Lake Trout, *Environ. Sci. Technol.* **29**, 2610-2615.

Miller, M. A., Madenjian, C. P., Masnado, R. G.: 1992, 'Patterns of organochlorine contamination in Lake Trout from Wisconsin waters of the Great Lakes', *J. Great Lakes Res.* **18(4)**, 742-754.

Mineau, P., Fox, G. A., Norstrom, R. J., Weseloh, D. V., Hallet, D. J., Ellenton, J. A.: 1984, 'Using the Herring Gull to monitor levels and effects on organochlorine contamination in the Canadian Great Lakes', In Nriagu, J. O. and Simmons, M. S. eds., *Toxic Contaminants in the Great Lakes*, John Wiley & Sons, New York, N. Y., pp. 425-452.

Moll, R. A., Jude, D., Rossmann, R., Kantak, G. V., Barres, J., DeBoe, S., Giesy, J., Tuchman, M.: 1995, 'Movement and Loadings of Inorganic Contaminants Through the Lower Saginaw River', *J. Great Lakes Res.* **21(1)**, 17-34.

National Biological Service: 1994, In *Status of prey fishes in the U.S. waters of Lake Ontario, 1993*, Department of the Interior, Lake Ontario Biological Station, Oswego, NY.

Norstom, R. J., Hallet, D. J., Simon, M., Mulvihill, M. J.: 1982, 'Analysis of Great Lakes herring gull eggs for tetrachlorodibenzo-*p*-dioxins', In Hutzinger *et al.* eds., *Chlorinated dioxins and related compounds; Impacts on the environment,* Pergamon Press, New York, N. Y. pp. 173-181.

Norstom, R. J., Simon, M., Muir, D. C. G., Schweinsburg, R. E.: 1988, 'Organochlorine Contaminants in Arctic Marine Food Chains: Identification, Geographical Distribution, and Temporal Trends in Polar Bears', *Environ. Sci. Technol.* **22(9)**, 1063-1071.

O'Gorman, R., Mills, E. L., DeGisi, J.: 1991, 'Use of Zooplankton to Assess the Movement and Distribution of Alewife (*Alosa pseudoharengus)* in South-Central Lake Ontario in Spring', *Can. J. Fish. Aquat. Sci.* **48**, 2250-2257.

Ontario Ministry of Natural Resources (OMNR): 1993, In *Annual Report, Lake Ontario Fisheries Research Station*, OMNR, Glenora, Ontario.

Pettit, K. E., Bishop, C. A., Weseloh, D. V., Norstrom, R. J.: 1994, *An atlas of contaminants in eggs of fish-eating colonial birds of the Great Lakes (1989-1992). Vol. I.*, Can. Wild. Serv. Tech. Rep. Ser. No. 193, Canadian Wildlife Service, Ontario Region, 319 pp.

Reynolds, L. M. and Cooper, T.: 1975, 'Analyses of organochlorine residues in fish', *Water Qual. Param. ASTM STP.* **573**, 196-205.

Rowan, D. J. and Rasmussen, J. B.: 1992, 'Why Don't Great Lakes Fish Reflect Environmental Concentrations of Organic Contaminants?- An Analysis of Between-Lake Variability in the Ecological Partitioning of PCBs and DDT', *J. Great Lakes Res.* **18 (4)**, 724-741.

SAS Institute Inc.: 1988a, *SAS® Procedures Guide, Release 6.03 Edition*, Cary, NC, SAS Institute Inc., 441 pp.

SAS Institute Inc.: 1988b, *SAS/STAT® User's Guide, Release 6.03 Edition*, Cary, NC, SAS Institute Inc., 1028 pp.

Smith, D. W.: 1995a, 'Are PCBs in the Great Lakes approaching a "new equilibrium"?', *Environ. Sci. Technol.* **29**, 42A-46A.

Smith, D.: 1995b, 'Synchronous response of hydrophobic chemicals in Herring Gull eggs from the Great Lakes', *Environ. Sci. Technol.* **29**, 740-750.

Stow, C. A., Carpenter, S. R., Amrhein, J. F.: 1994, 'PCB concentration trends in Lake Michigan Coho (*Oncorhynchus kisutch*) and Chinook Salmon' *Can. J. Fish. Aquat. Sci.* **51**, 1384-1390.

Stow, C. A.: 1995, 'Great Lakes Herring Gull Egg PCB Concentrations Indicate Approximate Steady-State Conditions', *Environ. Sci. Technol.* **29**, 2893-2897.

Stow, C. A., Carpenter, S. R., Eby, L. A., Hesselberg, R. J.: 1995, 'Evidence that PCBs are approaching stable concentrations in Lake Michigan fishes', *Ecological Applications* **5(1)**, 248-260.

Suns, K., Hitchin, G., Toner, D.: 1991, *Spatial and temporal trends of organochlorine contaminants in Spottail Shiners (Notropis hudsonius) from the Great Lakes and their connecting channels (1975-1988)*,Water Resources Branch - Ontario Ministry of the Environment, 97pp.

Suns, K. R., Hitchin, G. G., Toner, D.: 1993, 'Spatial and temporal trends of organochlorine contaminants in Spottail Shiners from selected sites in the Great Lakes (1975-1990)' *J. Great Lakes Res.* **19(4)**, 703-714.

Turle, R., Norstrom, R. J., Collins, B.: 1991, 'Comparison of PCB quantitation methods: re-analysis of archived specimens of Herring Gull eggs from the Great Lakes', *Chemosphere* **22**, 201-213.

Turle, R. and Collins, B.: 1992, 'Validation of the use of pooled samples for monitoring of contaminants in wildlife', *Chemosphere* **25(4)**, 463-469.

Weseloh, D. V., Mineau, P., Hallett, D. J.: 1979, 'Organochlorine contaminants and trends in reproduction in Great Lakes herring gulls, 1974-1978', In *Transactions*

*of the 44th North American Wildlife and Natural Resources Conference*, Wildlife Mgmt. Inst., pp. 543-557.
World Health Organization (WHO).: 1984, *Heptachlor*, Finland, Vammala, 81 pp.
WHO.:1989a, *Aldrin and Dieldrin*, Finland, Vammala, 335 pp.
WHO.:1989b, *DDT and it derivatives – environmental aspects*, Finland, Vammala, 98 pp.
WHO/EURO.: 1987, *PCBs, PCDDs and PCDFs: prevention and control of accidental and environmental exposures*, Copenhagen, World Health Organization, Regional office for Europe (Environmental Health Series 23).
Zar, J. H.: 1996, *Biostatistical Analysis*, Prentice Hall Inc. New Jersey, U.S.A., 899 pp.

**Appendix I.** Structure of the Great Lakes Herring Gull Egg Contaminant Dataset (1974-1995).

**Appendix I-A.** Yearly Egg Collections from Great Lakes Colonies.

**YEAR** **COLONY** (Abbreviations Pertain to those Shown in Figure I)

| | *E.L.O.* | *W.L.O.* | *N.R.* | *E.L.E.* | *W.L.E.* | *D.R.* | *E.L.H.* | *N.L.H.* | *S.B.L.H.* | *N.L.M.* | *W.L.M.* | *E.L.S.* | *W.L.S.* |
|---|---|---|---|---|---|---|---|---|---|---|---|---|---|
| 1974 | + | + | 0 | + | + | 0 | + | + | 0 | 0 | 0 | + | + |
| 1975 | + | + | 0 | + | + | 0 | + | + | 0 | 0 | 0 | + | + |
| 1976 | 0 | 0 | 0 | 0 | 0 | 0 | 0 | 0 | 0 | 0 | + | 0 | 0 |
| 1977 | + | + | 0 | + | + | 0 | + | + | 0 | + | + | + | + |
| 1978 | + | + | 0 | + | + | + | + | + | 0 | + | + | + | + |
| 1979 | + | + | + | + | + | + | + | + | 0 | 0 | 0 | + | + |
| 1980 | + | + | 0 | + | + | 0 | + | + | + | + | + | + | + |
| 1981 | + | + | + | + | + | + | + | + | + | 0 | 0 | + | + |
| 1982 | + | + | + | + | + | + | + | + | + | + | + | + | + |
| 1983 | + | + | + | + | + | + | + | + | + | + | + | + | 0 |
| 1984 | + | + | + | + | + | + | + | + | + | + | + | + | + |
| 1985 | + | + | + | + | + | + | + | + | + | + | + | + | + |
| 1986 | + | + | + | + | + | + | + | + | + | + | + | + | + |
| 1987 | + | + | + | + | + | + | + | + | + | + | + | + | + |
| 1988 | + | + | + | + | + | + | + | + | + | + | + | + | + |
| 1989 | + | + | + | + | + | + | + | + | + | + | + | + | + |
| 1990 | + | + | + | + | + | + | + | + | + | + | + | + | + |
| 1991 | + | + | + | + | + | + | + | + | + | + | + | + | + |
| 1992 | + | + | + | + | + | + | + | + | + | + | + | + | + |
| 1993 | + | + | + | + | + | + | + | + | + | + | + | + | + |
| 1994 | + | + | + | 0 | + | + | + | + | + | + | + | + | + |
| 1995 | + | + | + | + | + | + | + | + | + | + | + | + | + |

Key: +: indicates that eggs were collected at that colony that year.
0: indicates that eggs were not collected at that colony that year.

**Appendix I-B.** Chemical Analysis of Compounds, by Year (Assuming Availability of Eggs).

**YEAR** **COMPOUND** (Compound Abbreviations Pertain to those Listed in Table I).

| | *o-CHL* | *QCB* | *HCB* | *DDE* | *DIEL* | *HE* | *a-HCH* | *b-HCH* | *MIR* | *P-MIR* | *PCB 1254:1260* |
|---|---|---|---|---|---|---|---|---|---|---|---|
| 1974 | 0 | 0 | + | + | + | + | 0 | 0 | + | 0 | + |
| 1975 | 0 | 0 | + | + | + | + | 0 | 0 | + | 0 | + |
| 1976 | 0 | 0 | + | + | + | + | 0 | 0 | + | 0 | + |
| 1977 | + | 0 | + | + | + | + | 0 | + | + | + | + |
| 1978 | + | 0 | + | + | + | + | 0 | + | + | + | + |
| 1979 | + | + | + | + | + | + | 0 | + | + | + | + |
| 1980 | + | + | + | + | + | + | 0 | + | + | 0 | + |
| 1981 | + | + | + | + | + | + | 0 | + | + | 0 | + |
| 1982 | + | + | + | + | + | + | + | + | + | 0 | + |
| 1983 | + | + | + | + | + | + | + | + | + | 0 | + |
| 1984 | + | + | + | + | + | + | + | + | + | 0 | + |
| 1985 | + | + | + | + | + | + | + | + | + | 0 | + |
| 1986 | + | + | + | + | + | + | + | + | + | + | + |
| 1987 | + | + | + | + | + | + | + | + | + | + | + |
| 1988 | + | + | + | + | + | + | + | + | + | + | + |
| 1989 | + | + | + | + | + | + | + | + | + | + | + |
| 1990 | + | + | + | + | + | + | + | + | + | + | + |
| 1991 | + | + | + | + | + | + | + | + | + | + | + |
| 1992 | + | + | + | + | + | + | + | + | + | + | + |
| 1993 | + | + | + | + | + | + | + | + | + | + | + |
| 1994 | + | + | + | + | + | + | + | + | + | + | + |
| 1995 | + | + | + | + | + | + | + | + | + | + | + |

Key: +: indicates that eggs were analyzed for that compound that year.

# REPRODUCTIVE AND PHYSIOLOGICAL EFFECTS OF ENVIRONMENTAL CONTAMINANTS IN FISH-EATING BIRDS OF THE GREAT LAKES: A REVIEW OF HISTORICAL TRENDS

KEITH A. GRASMAN,[1] PATRICK F. SCANLON,[2] and GLEN A. FOX[3]
[1]*Department of Biological Sciences, Wright State University, Dayton, Ohio, USA, 45435;* [2]*Department of Fisheries and Wildlife Sciences, Virginia Polytechnic Institute and State University, Blacksburg, Virginia, USA, 24061;* [3]*National Wildlife Research Centre, Canadian Wildlife Service, Environment Canada, Hull, Quebec, Canada, K1A 0H3.*

**Abstract.** During the 1950s and 1960s, reproductive failures and population declines were observed in fish-eating birds such as gulls, terns, cormorants, herons, and eagles in the Great Lakes. DDE-induced eggshell thinning contributed to these declines, but other factors such as embryo toxicity also were implicated. With reduced releases of many pollutants, reproduction recovered in some species. However, biomonitoring during the 1980s and 1990s indicates continuing effects at highly contaminated sites. Improved analytical techniques and bioassays have allowed the assessment of the total dioxin-like toxicity of complex mixtures of organochlorines (TCDD-equivalents). Developmental defects such as embryo mortality, deformities, and edema have been associated with dioxin-like PCBs in several avian species. Improved biochemical techniques have allowed the measurement of biomarkers that detect physiological alterations associated with contaminants. Specific biomarkers studied in Great Lakes birds include cytochrome P-450 monooxygenases, highly carboxylated porphyrins, thyroxine, vitamin A, and immune function. Reproductive and physiological alterations are associated with population-level effects in Caspian terns and bald eagles that feed on highly contaminated fish. Biomonitoring using biomarkers and population-level measures in fish-eating birds will continue to be important for assessing the effects of contaminants on the Great Lakes ecosystem.

## 1. Introduction

During the late 1950s, several studies suggested that environmental contaminants were disrupting reproduction in fish-eating birds in the Great Lakes of North America. Subsequent research confirmed these effects and contributed to restrictions on many of the more persistent chemicals. Over the past several decades, identification of point sources, improved pollution control technologies, and more stringent anti-pollution laws have reduced the releases of many toxic chemicals into the Great Lakes. Despite these measures, a variety of contaminants continue to impact the Great Lakes ecosystem. The Great Lakes Water Quality Board has identified 11 critical pollutants that still present risks to human health and the aquatic ecosystem: 2,3,7,8-tetrachlorodibenzo-*p*-dioxin (TCDD), 2,3,7,8-tetrachlorodibenzofuran (TCDF), benzo[a]pyrene (B[a]P), 1,1,1-trichloro-2,2-bis(4-chlorophenyl)ethane (DDT), dieldrin, hexachlorobenzene (HCB), alkylated lead, mirex, mercury, polychlorinated biphenyls (PCBs), and toxaphene (International Joint Commission, 1991).

Some of these pollutants continue to affect a number of fish, reptiles, birds, and mammals, including humans, that eat contaminated fish. At highly contaminated sites, biomagnification of persistent, lipophilic chemicals results in high concentrations in the tissues of upper trophic level species. These contaminants cause abnormalities that are evident at biochemical, physiological, organismal, and population levels in fish-eating birds of the Great Lakes. Numerous studies have demonstrated that halogenated aromatic hydrocarbons (HAHs), which include PCBs, TCDD, and TCDF, presently are associated with many of these biological effects in gulls, terns, cormorants, herons, and eagles. This paper reviews the past and present reproductive and population-level effects of contaminants on fish-eating birds in the Great Lakes. It also describes the more recent advances in analytical chemistry, biochemistry, and physiology that have allowed scientists to examine more closely cause-effect associations between contaminants and biological effects.

## 2. Early Indications of Reproductive Problems

The bald eagle *(Haliaeetus leucocephalus)* was probably the first North American species that was widely affected by the introduction of DDT, demonstrating effects parallel to those seen in three raptor species in Great Britain (Ratcliffe, 1967). Within one year of the widespread introduction of DDT in 1946, C.L. Broley noted nesting failures in the Florida subpopulation that he had studied extensively for many years (Broley, 1952). These nesting failures were characterized by failure of the adults to return to traditional nests, failure of those adults that did return to lay eggs, and failure of many eggs to hatch. The quick onset of these reproductive effects in Florida eagles probably was caused by the widespread application and high application rate of DDT for mosquito control. On his Canadian study area immediately north of Lake Ontario, Broley observed a marked decrease in nesting success of bald eagles in 1951, when only one of 12 pairs under observation produced young (Broley, 1951). The last year that bald eagles nested successfully on Lake Ontario (Figure I) was 1955 (Quilliam, 1965). A 1996 survey showed that no pair of eagles had yet reestablished a territory on the shorelines of this lake. In the 1950s, the bald eagle population on the shores of

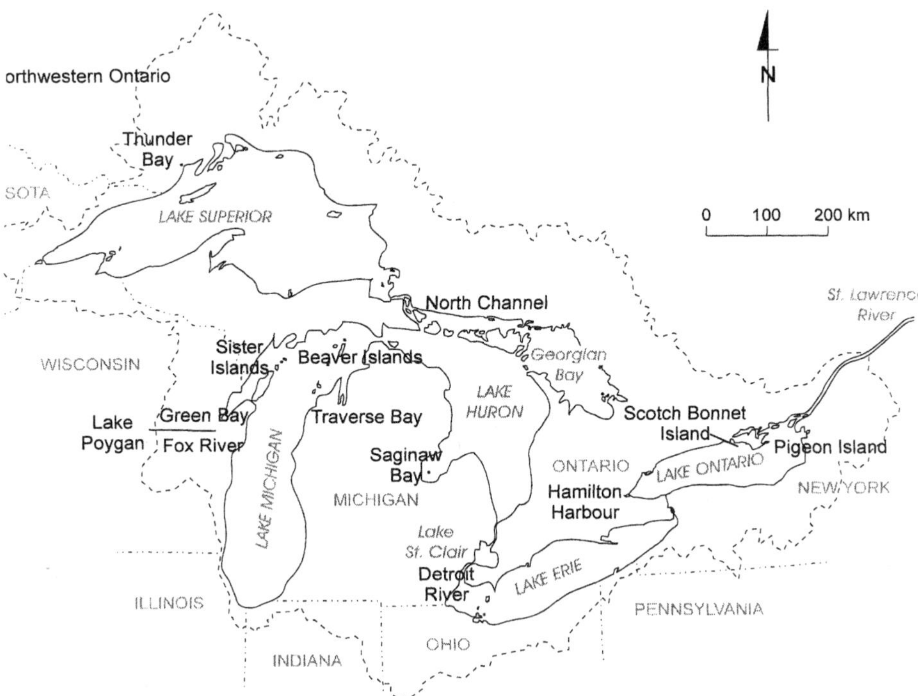


**Figure I** Fish-Eating Bird Study Locations in the Great Lakes Basin

Lake Superior crashed, and few young were produced. In the late 1960s, eggshells of eagles in this subpopulation had thinned by more than 20% in association with DDE (1,1-dichloro-2,2-bis(4-chlorophenyl)ethylene, a breakdown product of DDT) levels of more than 50 $\mu$g/g, the highest found in any subpopulation (Postupalsky, 1971). PCBs in those same eggs were found at 28 $\mu$g/g. Between 1968 and 1970, bald eagles on Lakes Superior, Michigan, Huron, and Erie (Figure I) averaged only 0.13 young/occupied nest. Production of 0.7 and 1.0 young/ occupied nest are associated with maintaining stable and healthy populations, respectively.

The Kingston Field Naturalists visited Pigeon Island (Figure I) in eastern Lake Ontario every year between 1961 and 1973 and observed a marked decrease in the size and reproductive success of the black-crowned night heron *(Nycticorax nycticorax)* colony in 1965, culminating with complete reproductive failure in 1969 (Edwards, 1970). On their June 1970 visit, the Kingston Field Naturalists first observed a number of herring gull *(Larus argentatus)* nests in which the eggs were dented or collapsed, apparently under the weight of the adult birds (Edwards, 1970). No double-crested cormorants *(Phalacrocorax auritus)* are known to have fledged from colonies in the Canadian waters of Lake Ontario between 1954 and 1977 (Price and Weseloh, 1986).

The population of common terns *(Sterna hirundo)* nesting in the lower Great Lakes began to decrease in 1965, and the marked decline continued through the 1970s. Observations of common terns nesting on two man-made islands in Hamilton Harbour (Figure I) on Lake Ontario in 1970 and 1971 revealed very low breeding success with a high proportion of the eggs failing to hatch and deformities in some chicks (Gilbertson and Reynolds, 1972; Gilbertson, 1975). Banders who had worked in the largest colony on Lake Ontario for many years also observed many chicks with deformed bills and feet. Residues in 13 failed eggs from the Hamilton Harbour colony in 1970 contained mean levels of DDE, HCB, and PCBs of 16 g/g, 3.5 $\mu$g/g, and 90 $\mu$g/g, respectively (Gilbertson and Reynolds, 1972). In 1971, eggs collected over a 60 day period revealed that the terns arrived at the colony with negligible burdens of PCBs and DDE and subsequently became grossly contaminated with PCBs, DDE, HCB and dieldrin from the fish they captured near the colony (Gilbertson, 1974b). PCBs were accumulated most rapidly.

During 1972, five herring gull colonies in Lake Ontario produced only 0.06 to 0.21 fledglings/nest, approximately 1/10 of normal. Eggshell thickness was correlated with DDE concentrations, and DDE-induced eggshell thinning produced eggshell breakage and flaking in the most contaminated colonies

(Gilbertson, 1974a). In eggs with flaking shells, significant numbers of embryos died just before hatching. On Scotch Bonnet Island (Figure I) in Lake Ontario during 1973, 27% of herring gull nests had no eggs, 20% of eggs exhibited early embryonic mortality, hatching success was only 17%, chick mortality was 74% during the first 9 days and 88% before fledging, and fledging success was only 0.06 fledglings/pair (Gilbertson and Hale, 1974a,b). These problems were associated with high contaminant residues in the eggs: approximately 35 $\mu$g/g (wet wt.) DDE and 138 $\mu$g/g PCBs (converted from 140 $\mu$g/g DDE and 550 $\mu$g/g PCBs dry wt. assuming 75% water content; Gilbertson, 1988). Dieldrin, heptachlor epoxide, and HCB also were detected. In Lake Huron's Georgian Bay (Figure I) in 1972, 95% of the double-crested cormorant eggs broke or disappeared from the nest before they were incubated to term (Weseloh *et al.*, 1983). Eggshells of the Lake Huron cormorant population were 24% thinner than normal and contained high concentrations of DDE (14.5 $\mu$g/g) and PCBs (23.8 $\mu$g/g; Weseloh *et al.*, 1983).

In 1964, Keith (1966) found exceptionally high embryo mortality and low chick survival in herring gulls nesting on the Sister Islands (Figure I) in upper Green Bay in Lake Michigan. Many of the dead eggs had large portions of the shell that had been fractured into small sections and chipped off. Keith suggested that there were two components to the embryo mortality (1) the effect of the residues upon the embryo, and (2) the effect of the adult's residue burden on its incubation behavior. Hickey *et al.* (1966) confirmed that the eggs, tissues, and food of these gulls were heavily contaminated with DDT and its metabolites, and Hickey and Anderson (1968) confirmed that their eggshells were significantly thinner than the pre-DDT mean for this population. Cormorants ceased to breed in Lake Michigan in 1963 (Anderson and Hamerstrom, 1967).

### 3. Further Documentation of Decreased Reproduction and Survival

These initial reports documented associations between reproductive problems and high concentrations of organochlorine contaminants in piscivorous birds of the Great Lakes. Reproductive processes are very sensitive to toxic chemicals and other environmental stressors. Although gross reproductive abnormalities are easily and frequently detected, the contribution of the various stages of the reproductive process to these abnormalities has been assessed less frequently. Reductions in eggshell thickness and strength were certainly major components in the reproductive failures observed in the 1950s and 1960s. However, there is little doubt that eggshell breakage often masked the occurrence of developmental toxicity. The manifestations of developmental toxicity most frequently observed in Great Lakes fish-eating birds are growth retardation, malformation, and embryo death, often associated with edema, hepatomegaly, liver necrosis, hemorrhaging, hepatic porphyria, and gastroschisis (failure to resorb the yolk sac and close the abdominal wall). Gilbertson *et al.* (1991) named this suite of abnormalities Great Lakes embryo mortality, edema, and deformity syndrome (GLEMEDS). There is considerable similarity between the pathobiology of GLEMEDS and a condition, seen in domestic poultry fed HAH-contaminated feed, that is known as chick edema disease (Gilbertson *et al.*, 1991).

GLEMEDS contributed to the early reproductive failures noted in the previous section, and its importance was confirmed by later studies. Excessive embryo mortality was associated with the poor reproductive success of bald eagles (1950s through 1970s), herring gulls in Lakes Michigan (1960s) and Ontario (1970s), and common terns in Lake Ontario (1970s; see previous section). Edema is the excessive accumulation of fluid in tissues leading to swelling. Subcutaneous, pericardial, and abdominal edema were observed in herring gull embryos from Lake Ontario in the mid-1970s but not in the late 1970s (Gilbertson and Fox, 1977; Gilbertson, 1983). Since then this syndrome has not been looked for in Lake Ontario gulls. The most common contaminant-associated deformities in Great Lakes fish-eating birds include crossed bills, other craniofacial abnormalities, club feet, hip dysplasia, and gastroschisis (Ludwig *et al.*, 1996). Deformities were observed in black-crowned night herons, ring-billed gulls *(Larus delawarensis)*, common terns, and Caspian terns *(Sterna caspia)* in the lower Great Lakes during the early 1970s (Gilbertson *et al.*, 1976). This suite of effects led Gilbertson to hypothesize that chick edema-active compounds such as TCDD and TCDF were present in the Lake Ontario food web. However, contemporary analytical methods were incapable of detecting these chemicals at environmentally relevant concentrations. Once technologies improved, retrospective analysis of herring gull eggs from the lower Great Lakes showed that TCDD concentrations averaged 500 pg/g during 1974 but were as high as 1200 pg/g in earlier years (Gilbertson, 1988). Excessively high rates of deformities, especially crossed bills, also were found in Forster's terns *(Sterna forsteri)*, common terns, ring-billed gulls, herring gulls, Virginia rails *(Rallus limicola)* and double-crested cormorants in Green Bay, WI, starting in about 1973 (Gilbertson, 1989; Gilbertson *et al.*, 1991; Kubiak *et al.*, 1989).

Several important studies confirmed earlier suggestions (Keith, 1966) that poor reproduction was caused not only by factors intrinsic to the egg (i.e., developmental toxicity of contaminants) but also by extrinsic factors (i.e., contaminant-induced abnormalities in parental behavior). Peakall *et al.* (1980) switched eggs between a contaminated herring gull colony on Lake Ontario and an uncontaminated colony in New Brunswick. During 1975, 86% of "clean" eggs incubated by "clean" adults hatched. "Dirty" eggs incubated by "clean" adults had low hatchability (only 10%), demonstrating intrinsic factors. But "clean" eggs incubated by "dirty" adults also had low hatchability (only 7%), demonstrating extrinsic effects. "Dirty" eggs incubated by "dirty" adults had a hatchability of only 2%. Fox *et al.* (1978) used wax-filled, telemetered eggs (temperature and light sensors) in Lake Ontario and New Brunswick colonies to investigate the nature of these extrinsic effects. Lake Ontario herring gulls exhibited poor incubation behavior compared to New Brunswick controls. At the Lake Ontario colony, parents of unsuccessful nests left their nests unattended for three times longer than parents of successful nests. The average egg temperature was 1 °C less for unsuccessful nests compared to successful nests. The length of time that a nest was left unattended was positively correlated with organochlorine concentrations in the egg. Abnormal parental behavior allowed greater intraspecific predation and possibly contributed to chilling or overheating of the embryos.

Two studies suggested that organochlorine contaminants were affecting survival throughout the year as well as causing the decreases in reproduction noted above. Sileo *et al.* (1977) measured organochlorine contaminants in the brains of ring-billed gulls from two major die-offs that occurred in southern Ontario during the summers of 1969 and 1973. Ring-billed gulls found dying with signs of neurologic poisoning and ring-billed gulls found dead with no apparent cause of death had significantly higher (30-90X) residues of DDE, dieldrin, and PCBs than healthy ring-billed gulls shot for comparison. Dieldrin, an organochlorine insecticide, was implicated as an important factor. The dead and dying gulls were severely emaciated, and the sequence of events that led to their death was not clear, but it was apparent that some ring-billed gulls were, at that time, carrying lethal burdens of organochlorine compounds. Ludwig and Ludwig (1969) captured 15 nesting adult herring gulls from Traverse Bay, MI (Figure I). Upon starvation, six gulls died within 10 days. In previous experiments, relatively uncontaminated herring gulls from Lake Huron withstood a 30% loss of body mass in 17 days without any mortality or loss of activity. Traverse Bay herring gulls that died during starvation had brain concentrations of 12 $\mu$g/g DDT, 6 $\mu$g/g DDD (1,1-dichloro-2,2-bis(*p*-chlorophenyl)ethane; a metabolite of DDT), 180 $\mu$g/g DDE, and 2 $\mu$g/g dieldrin, suggesting the lethal mobilization of organochlorine pesticides from body fat during starvation.

## 4. Restrictions on Organochlorine Compounds: Improvements in Reproduction and Population Recoveries in Some Species

During the 1970s and early 1980s, the use of most organochlorine compounds was greatly restricted in the U.S. and Canada. In the U.S., crop and non-health use of DDT was suspended in 1972. Most uses of PCBs were banned in the U.S. in 1979, with the exception of totally closed systems such as electrical transformers and capacitors and in hydraulic fluids, or by special exemptions. With growing knowledge about the dangers of HCB, releases into the environment decreased. Identification of pollution point sources helped reduce inputs of TCDD and mirex into Lake Ontario, TCDD and PCBs into Saginaw Bay (Figure I), and PCBs into Green Bay. Decreases in organochlorine residues in wildlife through the late 1970s were associated with increased reproduction and population recovery in some affected species, including herring gulls and double-crested cormorants.

Significant decreases in DDT, DDE, and mirex concentrations occurred in herring gull eggs from Lakes Ontario, Erie, Huron, and Superior between 1974 and 1979 (Mineau *et al.*, 1984). PCB and dieldrin concentrations decreased in herring gull eggs from Lakes Ontario, Erie, and Huron but not Superior during this same time period. Weseloh *et al.* (1989) found that DDE and PCB concentrations in the eggs of common terns declined 80-90% in four colonies on Lakes Huron, Erie, and Ontario between 1969-73 and 1981.

Decreased contaminant concentrations were associated with significant increases in the reproductive success of some species. In egg switching experiments during 1975, Peakall *et al.* (1980) had demonstrated severe intrinsic and extrinsic effects that were limiting reproduction in Lake Ontario herring gulls. During 1976, these researchers found only intrinsic effects. Neither intrinsic nor extrinsic effects were observed during 1977. By 1978, the number of fledged herring gull young/breeding pair had increased to over 1.0 for Lakes Ontario and Erie, both of which had shown severe reproductive impairment during the early 1970s (Weseloh *et al.*, 1979; Mineau *et al.*, 1984; Gilbertson, 1988). By 1979-81 the Great Lakes double-crested cormorant population had begun to recover. Although 5-14% eggshell thinning was still observed, reproductive output had increased to a mean of 1.5-2.0 young/nest (Peakall 1988). In Lake Ontario, the double-crested cormorant population increased at an annual rate of 56% between 1974 and 1982 (Price and Weseloh, 1986). Cormorant populations have continued to grow exponentially on all five Great Lakes through the early 1990s (Weseloh *et al.*, 1995).

Bald eagles, which feed one trophic level higher in the food web than gulls, terns, and cormorants, continued to experience reproductive problems during the middle and late 1970s, presumably because of their greater exposure to organochlorine compounds. Low reproductive success was observed on Lakes Superior, Michigan, Huron, and Erie; no adults bred on Lake Ontario (Colborn, 1991). At one breeding site on Lake Superior, a pair of bald eagles attempted breeding in most years between 1960 and 1977, but were unsuccessful until 1977. In 1970, this pair produced an egg with a shell 40% thinner than normal pre-DDT eagle eggs (Postupalsky, 1978). In 1977, five nests on Lake Superior's south shore produced six young. Using data on productivity and residues in eggs for a single subpopulation of bald eagles breeding in Northwestern Ontario (outside the Great Lakes ecosystem), over the period 1966 to 1981, Grier (1982) has confirmed the negative association between DDE exposure, reproduction, and population size in this species.

## 5. Continued Physiological and Population-level Problems during the 1980s and 1990s

Despite the improvements observed during the late 1970s, a number of problems, especially at highly contaminated sites, have persisted into the 1990s. Although concentrations of many contaminants declined markedly during the late 1970s, concentrations have declined slowly, leveled off, or increased during the 1980s and 1990s (Hebert *et al.*, 1994; Stow, 1995). As biological and chemical methods have improved, the research emphasis for Great Lakes toxics has shifted to examining problems in finer detail. Improvements in analytical chemistry and bioassays have made possible congener-specific chemical analyses and have become important tools for assessing the effects of complex mixtures of contaminants. Although expensive, more detailed chemical analyses have helped to separate the effects of the various PCB and dioxin congeners, which vary widely in their toxicity and environmental persistence. Much emphasis also has been placed on the development and use of biomarkers, which are biochemical, physiological, or histological changes that measure effects of, or exposure to, toxic chemicals (Fox, 1993). Beyond these physiological effects, limited population studies in several species have suggested continued population-level problems such as reduced recruitment at highly contaminated sites.

## 5.1. Improvements in Measuring Environmental Contaminants

Observations of embryonic mortality, edema, and deformities in fish-eating birds suggested to Gilbertson the presence of chick edema-active compounds such as TCDD and TCDF (Gilbertson 1983). However, initial efforts to detect such chemicals in herring gull tissues were unsuccessful. But as analytical techniques improved in the early 1980s, R. Norstrom successfully measured TCDD concentrations greater than 1000 pg/g in Great Lakes herring gull eggs archived during the early 1970s. Toxicity studies suggested that some PCB congeners have TCCD-like effects and could have contributed to GLEMEDS during the 1960s and 1970s.

Assessing the toxicity of HAHs is complicated by the fact that multiple congeners occur in environmental samples. For example, there are 209 PCB congeners, although only 90 typically are found in environmental samples. The planar (non *ortho*- and mono *ortho*-chlorinated) PCB congeners bind to the Ah receptor and exert toxic effects that are similar to TCDD (Safe, 1990). After binding to the Ah receptor in the cytoplasm, the contaminant-receptor complex travels to the nucleus and initiates gene transcription and protein synthesis, including genes for cytochrome P450 1A. TCDD has the strongest affinity for the Ah receptor and has the highest toxicity by this mechanism.

An additive model can be used to sum the dioxin-like activity of a complex environmental mixture. The additive model does not account for interactive (synergistic or antagonistic) effects, but it is an improvement over simply measuring total PCB concentrations based on comparisons to commercial PCB mixtures. The concentrations of specific congeners must first be multiplied by toxic equivalence factors to correct for differences in toxicity. These toxicity equivalency factors are determined by comparing the potency ($LC_{50}$ or $EC_{50}$) of each congener to that of TCDD (the most toxic Ah-active compound) determined in laboratory experiments (Safe, 1990; Metcalfe and Haffner, 1995). The total TCDD-like toxicity (TCDD-equivalents or TCCD-EQs) can then be calculated by adding the individual toxicities of each congener. Most toxic equivalency factors have been determined using laboratory mammals, but Kennedy *et al.* (1994) determined factors using liver cells from various avian species, including the herring gull.

As alternatives to expensive congener-specific chemistry and additive assumptions, *in vitro* bioassays have been developed to measure the TCDD-EQs of complex mixtures of HAHs in biological and environmental samples. Bioassays for TCDD-EQs use induction of cytochrome P-450 activity in rat hepatoma cells (Tillitt *et al.*, 1991) and chicken embryo hepatocytes (Kennedy *et al.*, 1993, 1996) cultured in extracts containing HAHs. The enzymatic activity induced by the extract from the environmental sample is compared to that induced by TCDD. These bioassays presumably respond to the additive, synergistic and antagonistic interactions of the substances in the mixtures, likely giving a more biologically relevant measure than the additive model from the analytical chemistry approach.

These advancements in measuring PCB and TCDD contamination have led to new insights about the current characteristics of organochlorine contamination in the Great Lakes. Tillitt *et al.* (1992) noted that total PCB concentrations have declined 10 to 100 times from the concentrations present in the mid-1960s, but now these concentrations appear to have reached a steady state. Point source discharges, leachates from chemical landfill sites, long range atmospheric transport and deposition and internal cycling are balancing degradation. However, the PCB congeners found in fish-eating birds today appear to be the more toxic forms with TCDD-like activity. Tillitt *et al.* (1992) found that the relative potencies of extracts from Great Lakes double-crested cormorant eggs were 3-4 times greater than the potencies of any PCB technical standards, which represent the mixtures originally released into the environment. This study suggested the relative enrichment of toxicity in PCB mixtures in fish-eating birds of the Great Lakes, a phenomenon subsequently reinforced by other studies (Jones *et al.*, 1993a,b; Yamashita *et al.*, 1993). These more toxic congeners are more resistant to physicochemical degradation and biological metabolism and (or) may be sequestered by binding tightly to the Ah receptor inside cells, leading to preferential accumulation up the food web. It is important to note that while TCDD-EQs have clarified many reproductive effects of HAHs thought to be mediated through the Ah-receptor, HAHs may exert toxic effects by other mechanisms such as disruption of hormone-receptor systems. Therefore, TCDD-EQs may not completely assess the toxicity of HAH mixtures (Ludwig *et al.*, 1996).

## 5.2 Continued Embryonic Mortality and Deformities

Improved chemistry and bioassays have helped clarify cause-effect relationships between different organochlorines and reproductive effects in Great Lakes birds. Reproductive problems, including GLEMEDS, have persisted for birds in lower Green Bay near the outflow of the Fox River (Figure I). During 1983, congenital abnormalities, including crossed bills, were observed in herring gulls, common terns, double-crested cormorants, and Virginia rails in Green Bay (Gilbertson, 1989). Edema was observed in embryos of Forster's terns from Green Bay in 1983 (Kubiak *et al.*, 1989). During the same year, Kubiak *et al.* (1989) performed egg-switching experiments between Forster's tern colonies on lower Green Bay and Lake Poygan (Figure I), an inland reference site. PCBs were the only contaminants present at concentrations high enough to cause the intrinsic (developmental effects) and extrinsic effects (parental inattentiveness) that were demonstrated by the egg-switching. "Dirty" eggs incubated by "dirty" adults had a hatching success that was 75% lower than that of "clean" eggs incubated by "clean" adults. For nests that were not manipulated, eggs from nests in lower Green Bay had a hatching success that was 52% lower than eggs from the inland lake nearby. The incubation period in the Green Bay colony was extended by eight days compared to the Lake Poygan colony. In this first study to analyze PCBs in Great Lakes birds on a congener-specific basis, two PCB congeners, 2,3,3′,4,4′- and 3,3′,4,4′,5-pentachlorobiphenyl (PCB numbers 105 and 126), appeared to occur at particularly high concentrations, although later one of the authors suggested that 3,3′,4,4′,5-tetrachlorobiphenyl (PCB number 77) may have contributed significantly to the embryonic mortality (Hoffman *et al.*, 1996).

In a followup study 1988, there was little embryo mortality, no edema was reported, and the incubation period was normal compared to the inland lake (Harris *et al.*, 1993). In 1988, PCB and TCDD-EQs were 67% and 42% lower than in 1983. The lower contaminant concentrations presumably were the result of lower river flow in the years preceding 1988. Contaminant concentrations in lower Green Bay are strongly influenced by the downstream movement of highly contaminated sediments, a process driven by flow rates in the Fox River (Velleux and Endicott, 1994). Despite improvements in embryonic mortality and incubation period, in 1988 total reproductive success was still lower than normal because of reduced chick survival associated with body mass loss beginning approximately 14-20 days after hatch. Fledging success was significantly lower than at a reference colony. This body mass loss in Green Bay Forster's terns was similar to TCDD-induced wasting syndrome in laboratory animals. It was also similar to the severe body mass loss and high chick mortality observed during some years in Saginaw Bay in herring gull and Caspian tern chicks despite the apparent presence of abundant food (Grasman, 1995; Grasman *et al.*, 1996).

Another cluster of GLEMEDS has been observed in double-crested cormorants and Caspian terns at contaminated sites in the upper Great Lakes, with particularly high frequencies of abnormalities in upper Green Bay, Lake Michigan, and Saginaw Bay, Lake Huron (Gilbertson *et al.*, 1991; Ecological

Research Services, 1991; Fox *et al.*, 1991; Yamashita *et al.*, 1993; Ludwig *et al.*, 1995, 1996). Although these deformities first were documented in 1986, they may have dated back to before 1967, which was the last year during which Caspian terns bred in Saginaw Bay until the early 1980s. Despite exponential growth of double-crested cormorant populations throughout much of the Great Lakes basin during the past decade, cormorants have been experiencing elevated frequencies of abnormalities in upper Green Bay and did not successfully breed in Saginaw Bay until 1991. By using the H4IIE bioassay with extracts of eggs from double-crested cormorants and Caspian terns, Tillitt *et al.* (1989, 1991) found that colonies with greater reproductive problems (Green Bay, WI, and Saginaw Bay, MI) had the highest concentrations of TCDD-equivalents. Other studies showed that the incidence of congenital abnormalities and other pathologies in double-crested cormorant embryos and chicks was strongly correlated with TCDD-equivalents (Ecological Research Services, 1991; Giesy *et al.*, 1994). During 1986-1988, Tillitt *et al.* (1992) collected eggs from 10 double-crested cormorant colonies in the Great Lakes basin and from one control colony in Manitoba, Canada. They also monitored hatching success at these colonies. They used the H4IIE bioassay for TCDD-equivalents. They found a statistically significant correlation between total PCB concentrations and egg mortality ($r^2$=0.32). However, the correlation between TCDD-equivalents and egg mortality was much stronger ($r^2$=0.70). Likewise, in Caspian terns, egg mortality was strongly correlated with TCDD-EQs ($r^2$=0.68) but less strongly with total PCBs ($r^2$=0.29). Death rates in eggs from colonies in the upper Great Lakes in 1987-1991 ranged from 25-42% in Caspian terns and from 22-33% in cormorants (Ludwig *et al.*, 1996). The hatching rate of Caspian tern eggs from Saginaw Bay dropped from 82% in 1986 to 28% in 1987 after a 100-year flood event and returned gradually to 75% by 1991 (Ludwig *et al.*, 1993). This flood mobilized highly contaminated sediments, causing an increase in PCB concentrations and reproductive and developmental problems. No Caspian tern chicks fledged in 1987, and fledging success remained greatly suppressed through at least 1991.

Ludwig *et al.* (1996) conducted a comprehensive analysis of deformity rates in dead eggs, live eggs, and chicks of double-crested cormorants and Caspian terns in the upper Great Lakes during 1986-91. Subcutaneous and abdominal edema were observed in dead embryos of cormorants and Caspian terns collected from colonies in the upper Lakes 1986-1991 and were most prevalent in Green Bay, (Figure I), and Saginaw Bay (Ludwig *et al.*, 1996). Using the percentage of deformed embryos in dead eggs and the rate of egg death, these authors calculated that the percentage of deformed embryos in all eggs laid in upper Great Lakes colonies was 5.2% for cormorants and 7.7% for Caspian terns. For cormorants, these rates ranged from a low of 2.6% in southeastern Lake Superior to highs of 7.0 and 8.2% in Thunder Bay (Lake Huron) and Green Bay (Lake Michigan), respectively. For Caspian terns, deformity rates ranged from a low of 3.9% in the North Channel (Figure I) of Lake Huron to a high of 9.6% in Saginaw Bay, Lake Huron. In Caspian terns, gastroschisis was the most common deformity compared to bill defects in cormorants. Deformity rates based on live eggs (sampled at fewer sites and fewer years) for both species yielded rates twice as high as those based on dead eggs. Autolysis of embryos in some dead eggs obscured deformities so that they were conservatively classified as "normal." Analyses based on live or dead eggs both underestimate true deformity rates because deformities cannot be detected in embryos that have died before 10 days. Ludwig *et al.* (1996) reported that deformity rates in cormorant eggs were strongly correlated with TCCD-EQs ($r^2$=0.70 for dead eggs and $r^2$=0.86 for live eggs) but less so with total PCBs ($r^2$=0.26 for dead eggs and $r^2$=0.67 for live eggs). Correlations between deformities and these contaminants were generally lower for Caspian tern eggs.

Fox *et al.* (1991) found that the frequency of bill defects in double-crested cormorant chicks was 52.1 crossed bills/10,000 chicks in Green Bay and 12.3/10,000 in the Beaver Islands (Lake Michigan) (Figure I) as compared to only 0.6/10,000 at reference colonies in the Canadian prairies. Considering total abnormalities associated with GLEMEDS (not just crossed-bills), Ludwig *et al.* (1996) reported 24.8 deformities/10,000 chicks for all Great Lakes sites, ranging from 6.9/10,000 in the North Channel of Lake Huron to 61.4/10,000 in northern Green Bay. The rate of deformities in hatched cormorant chicks was correlated with total PCBs ($r^2$=0.67) and TCDD-EQs ($r^2$=0.47). In Caspian tern chicks, deformity rates ranged from 0/1000 in Georgian Bay, a low contamination area, to 7.5/1000 in Saginaw Bay, a highly contaminated site. Statistical correlations between deformity rates and contaminants in terns were generally low.

Overall, these studies in the 1980s and 1990s have shown continuing associations between GLEMEDS and organochlorine contaminants, especially TCDD-EQs. In these studies, coplanar PCBs, rather than TCDD or TCDF, have contributed the most to the dioxin-like contamination.

## 5.3. BIOCHEMICAL AND PHYSIOLOGICAL BIOMARKERS OF CONTAMINANT EXPOSURE AND EFFECTS

Many biomarkers have been proposed for use in monitoring exposure to, and effects of, persistent toxic chemicals in Great Lakes wildlife (Table 1; Fox, 1993). Biomarkers have several important uses in ecotoxicological investigations. First, they may help strengthen cause-effect linkages between a contaminant and organismal or population-level effects by demonstrating that the contaminant is not simply present but is biologically active on the biochemical or physiological level. Second, biomarkers can be used to monitor improvements or deterioration in physiological function as contaminant exposure changes following site remediation or changes in pollutant emissions. Third, the observations of effects on the physiological level may provide an early warning of potential effects on the organismal or population levels. Such warning may be particularly important for the conservation of threatened, endangered, and pollution-sensitive species.

5.3.1. *Cytochrome P-450 Monooxygenase Enzymes*

Cytochrome P-450 monooxygenases belong to a class of liver detoxification enzymes known as mixed function oxidases. When an animal is exposed to certain contaminants or drugs, the expression of cytochrome P-4501A is induced via the Ah receptor. In 1981, herring gull embryos collected from Saginaw Bay and Lake Ontario had higher liver microsomal aryl hydroxylase activity (AHH; a cytochrome P-4501A-mediated reaction) than embryos collected from Kent Island, a reference site on the Atlantic coast (Ellenton *et al.*, 1985). In 1983, AHH activity in embryos from Lake Superior, open Lake Huron, and the Detroit River (Figure I) did not differ from the reference site. The activity of ethoxyresorufin O-deethylase (EROD), another cytochrome P-4501A-mediated reaction, was elevated in herring gull embryos from one Lake Ontario colony in 1982 (Boersma *et al.*, 1986). Hoffman *et al.* (1987) found that liver AHH activity was three times greater in Forster's tern hatchlings from Green Bay, WI, as compared to those from Lake Poygan, the nearby reference site. In 1984 and 1985, AHH activity in livers was two to three fold higher in pipping common tern embryos from Saginaw Bay and lower Green Bay (Hoffman *et al.*, 1993). Liver AHH activity also was induced in pipping black-crowned night heron embryos from these sites. Several cytochrome P-450 monooxygenase activities were induced in pipping night herons from Green Bay during 1989, including four-fold increase in AHH and a 20-fold increase in EROD (Rattner *et al.*, 1993). EROD activity and cytochrome P-4501A and 2B proteins were correlated with total PCBs and TCDD-EQs determined using additive models and the H4IIE bioassay (Rattner *et al.*,1994).

**Table I.** Summary of Physiological and Biochemical Markers Shown to be Affected by Environmental Contaminants in Colonial Waterbirds of the Great Lakes.

| Biomarker | Species | Main Finding(s) at Sites with High Organochlorine Contamination | Reference(s) |
|---|---|---|---|
| Cytochrome P450 Enzymes | Herring Gull<br>Forster's Tern<br>Common Tern<br>Black-Crowned Night Heron | Increased activity of AHH and EROD activity in embryos and hatchlings. | Ellenton *et al.*, 1985<br>Boersma *et al.*, 1986<br>Hoffman *et al.*, 1987, 1993<br>Rattner *et al.*, 1993, 1994 |
| Porphyrins | Herring Gull | Increased total and highly carboxylated porphyrins in the liver of adults at sites | Gilbertson and Fox, 1977<br>Fox *et al.*, 1988<br>Fox *et al.*, This Issue |
| Vitamin A | Herring Gull<br>Caspian Tern | Reduced concentrations of vitamin A in liver, blood plasma, and (or) egg yolk | Spear *et al.*, 1985, 1990<br>Fox, 1993<br>Grasman *et al.*, 1996<br>Fox *et al.*, This Issue |
| Thyroid | Herring Gull | Increased thyroid mass and microfollicular hyperplasia of thyroid tissue | Moccia *et al.*, 1986<br>Fox *et al.*, This Issue |
| Immune Function | Herring Gull<br>Caspian Tern | No effect on antibody-mediated immunity in chicks<br>Suppression of T lymphocyte-mediated immunity in chicks of both species and thymic atrophy in gull chicks | Grasman, 1995<br>Grasman *et al.*, 1996 |

5.3.2. *Highly Carboxylated Porphyrins*

Some HAHs, including TCDD, TCDF, HCB, and PCBs, can inhibit heme synthesis and produce a buildup of highly carboxylated porphyrins in liver tissue. This inhibition of enzymes presumably occurs independent of the Ah receptor. Total liver porphyrins were first measured in Lake Ontario herring gull embryos in 1974 (Gilbertson and Fox, 1977). Fox *et al.* (1988) examined the concentrations of highly carboxylated porphyrins in the livers of adult herring gulls collected throughout the Great Lakes during the early 1980s. Herring gulls also were collected from a reference site on the Atlantic coast. The highest concentrations of porphyrins were found in herring gulls from the sites most heavily contaminated with HAHs (Green Bay, WI; Saginaw Bay, MI; and Lake Ontario). Highly carboxylated porphyrins hold much promise as an early warning indicator of HAH toxicity since elevated porphyrin concentrations usually are observed before other toxic effects of HAHs in laboratory studies (Fox *et al.*, 1988), and changes over time have been used to monitor physiological status (Fox *et al.*, this issue).

5.3.3. *Vitamin A Homeostasis*

Retinol, or vitamin A, is essential for proper development, immune function, and vision. Environmental contaminants such as HAHs can disrupt vitamin A homeostasis (Zile, 1992), and such effects have been observed in fish-eating birds of the Great Lakes in the early 1980s (Spear *et al.*, 1985, 1990; Fox *et al.*, this issue). Retinoids stored in the liver were lower for adult herring gulls from the Great Lakes as compared to those from the Atlantic coast. Although the greater amount of vitamin A in the marine food web contributed to this difference, environmental contaminants also appeared to be important. Within the Great Lakes region, liver retinoid stores varied significantly among sites and between years. During 1992 and 1993, Grasman *et al.* (1996) found a strong association between reduced plasma retinol concentrations in three week old Caspian tern chicks and increasing exposure to PCBs and DDE at five Great Lakes sites. A weaker but statistically significant association was found in herring gull chicks. Although vitamin A is known to be important for immune function, few correlations were found between retinol concentrations in individual birds and measures of T-cell and antibody-mediated immunity.

5.3.4. *Thyroid Function*

Thyroid mass, histology, and hormones are other potential biomarkers for toxic effects of HAHs in Great Lakes wildlife (Moccia *et al.*, 1986; Fox *et al.*, 1997). In the 1980s, adult herring gulls in the Great Lakes had larger thyroid glands than those from a less-contaminated colony on the Atlantic coast. In Lake Ontario, where gull tissues have been sampled over time, goiter has decreased at the same time that organochlorine contamination has decreased. At highly contaminated Great Lakes colonies, there was a high prevalence of microfollicular epithelial hyperplasia, which was not observed at the reference colony. Concentrations of circulating thyroid hormones also might be affected by environmental contaminants. In both Caspian tern and herring gull chicks during 1992 and 1993, Grasman *et al.* (1996) found significant differences in plasma thyroxine concentrations among sites and years, but no associations with organochlorine contaminants were found. However, due to cost constraints, organochlorine contamination was measured in pooled egg samples from each site rather than in individual birds. Measurement of individual contamination might reveal associations between organochlorines and thyroxine.

5.3.5 *Immunosuppression*

A recent series of studies has shown that organochlorine contaminants are associated with immunosuppression in young herring gulls and Caspian terns at highly contaminated Great Lakes sites. Extensive laboratory studies have shown that one of the most characteristic effects of exposure to PCBs and other pollutants found in the Great Lakes is immunosuppression (Thomas and Faith, 1985; Luster *et al.*, 1987). In adult animals, acute exposure to coplanar PCBs and TCDD can suppress antibody-mediated immunity. The developing immune system is particularly sensitive to contaminants, and such effects may be long-lasting (Thomas and Faith, 1985; Vos *et al.*, 1989). In younger animals, chronic or perinatal exposure to coplanar PCBs and TCDD causes atrophy of the thymus gland and suppression of various T lymphocyte functions. Increased susceptibility to infectious diseases potentially is an important mechanism by which contaminants could produce mortality and population effects in fish-eating birds of the Great Lakes.

In a multi-year study (1992-1994), antibody- and T-cell-mediated immune function was measured in herring gull and Caspian tern chicks at Great Lakes sites across a gradient of organochlorine contamination (Grasman, 1995; Grasman *et al.*, 1996). T-cell-mediated immunity was assessed with the phytohemagglutin (PHA) skin test. In both species, there was a strong exposure-response relationship between suppression of PHA response and organochlorine contamination. Suppression was most severe (30-45%) in Lake Ontario and Saginaw Bay for both species and western Lake Erie for herring gulls. In a parallel investigation, thymic atrophy was associated with increasing liver EROD activity in 4 week old herring gull chicks from nine Great Lakes and one Atlantic coast colony. Although thymic mass was not associated with any single organochlorine, the association between thymic atrophy and high EROD activity strongly suggests that the complex contaminant mixtures found in these birds affect T-cell-mediated immunity. Although antibody-mediated immunity (antibody production following immunization with sheep red blood cells) varied between sites, no associations with contaminants were evident. Associations between contaminants and altered white blood cell counts also were observed in herring gull adults and Caspian tern chicks. These studies with Great Lakes birds suggest that T-cell-mediated immunity is more sensitive than antibody-mediated immunity to HAH exposure in developing and young animals.

Similar effects on immunosuppression have been seen in captive harbor seals *(Phoca vitulina)* fed HAH contaminated fish (De Swart *et al.*, 1994; Ross *et al.*, 1995), and in wild bottlenose dolphins *(Tursiops truncatus)* exposed to organochlorine compounds (Lahvis *et al.*, 1995). In laboratory animals, HAH-induced immunosuppression increases susceptibility to viral, bacterial, and parasitic diseases (Friend and Trainer, 1970; Vos and Luster, 1989). Few studies have examined the relationship between environmental contamination and the prevalence of particular diseases in Great Lakes birds. Ludwig *et al.* (1996) reported the prevalence of severe eye infections *(Pasteurella sp.)* that co-occurred in cormorant chicks with malformed eyes or eyelids during 1986-1991. Using data reported in this paper, we have found a strong correlation between the prevalence of eye disease and egg PCB concentrations ($r$=0.81, $P$=0.029, n=7 sites).

## 5.4. CONTINUED POPULATION-LEVEL PROBLEMS

Recent studies have shown continued associations between environmental contaminants and GLEMEDS and other reproductive problems in Great Lakes bald eagles. Between 1983 and 1988, Kozie and Anderson (1991) found that bald eagles nesting on the shores of Lake Superior had a mean nest success of 57% and a mean of 0.8 young/occupied nest, compared to 77% nest success and 1.3 young/occupied nest for bald eagles nesting in inland Wisconsin. Reproduction of Lake Superior bald eagles had increased from 10% nest success and 0.1 young/occupied nest during 1961-1970, but reproduction during 1983-1988 was still significantly lower for Lake Superior bald eagles compared to inland nesters. Contaminant concentrations, especially PCBs, were higher in dead nestling and immature bald eagles from Lake Superior compared to inland bald eagles of similar ages. Blood plasma PCB concentrations were 7.5 times greater in bald eagle nestlings on the Great Lakes shorelines as compared to nestlings inland (Bowerman, 1993). Reproduction within 8 km of the Great Lakes was only 0.71 young/occupied nest as compared to 1.05 young/occupied nest further inland (Bowerman, 1993). Pairs of nesting bald eagles on Lake Michigan and Lake Huron shorelines exhibit reproductive failure within 5 years of establishment, and a high turnover rate in the adult population is suspected (Government of Canada, 1991). Bald eagle chicks with bill deformities also have been observed in the Great Lakes Basin: two during the 1960s, four during the 1980s, three during 1993, and one during 1995 (Bowerman, 1993; Bowerman *et al.*, 1994; D. Best, U.S. Fish and Wildlife Service, East Lansing, MI. pers. comm.). A recent ecological risk assessment has concluded that PCBs and particularly TCDD-equivalents represent a significant reproductive risk to bald eagles living along the shorelines of the Great Lakes and rivers open to Great Lakes anadromous fish (Bowerman *et al.*, 1995).

Although populations of herring gulls and Caspian terns have recovered on a basin-wide scale, the four species of Great Lakes terns (common, Caspian, Forster's, and black) are exhibiting population-level effects in some areas. Long-term banding studies show that Caspian terns hatched on contaminated U.S. colonies (Green Bay and Saginaw Bay) return to breed at a rate less than half of that for terns hatched on cleaner Canadian colonies (Ludwig, 1979; Mora *et al.*, 1993). Indeed, the breeding population on these U.S. colonies is supported by recruitment of young adult terns hatched on Canadian colonies. Similar population-level effects likely are occurring in other species, but there are less long-term banding data on which to test this hypothesis. J. Ludwig (pers. comm.) has observed low band recovery rates for black-crowned night herons in Saginaw Bay. Of 1700 black-crowned night herons banded in Saginaw Bay during 1987-89 (the three years after the 1986 flood that flushed PCB-laden sediments into the bay), only 2 band returns (0.1%) have been received for adults. A normal band return rate for this species is 2%, suggesting high mortality of juvenile black-crowned night herons from Saginaw Bay.

## 6. Conclusion

Extensive research spanning three decades has demonstrated numerous effects of environmental contaminants on fish-eating birds of the Great Lakes. During the 1960s and early 1970s, DDT and PCBs were the most important contaminants affecting Great Lakes wildlife on a regional scale. DDE-induced eggshell thinning was shown conclusively to be a causative factor in population declines of bald eagles and double-crested cormorants. The effects of PCBs and dioxins were less clear, although GLEMEDS, which is similar to chick edema disease, was probably caused by PCBs and (or) dioxins. Since the ban on many organochlorine compounds, populations of some affected birds have recovered on the regional scale, but contaminants still are associated with reproductive and physiological effects, especially in heavily polluted areas. While the negative impacts of DDT have decreased, PCBs are still a cause for concern, especially since highly toxic PCB congeners with dioxin-like activity are accumulating in fish-eating birds. Current population-level effects are understudied, but long-term investigations of several species have raised concerns about reduced recruitment into the breeding population at highly contaminated sites. Reduced recruitment is more difficult to detect but no less significant than the more obvious reproductive failures observed in the 1960s and 1970s. Recent research investigating physiological biomarkers has helped strengthen cause-effect relationships between contaminants and organismal and population-level effects. Biomonitoring using biomarkers and population-level measures in fish-eating birds will continue to be important for assessing the effects of contaminants on the health of the Great Lakes ecosystem.

## Acknowledgments

We would like to thank Dr. Klaus D. Elgert, Dr. Roy L. Kirkpatrick, Dr. Prakash S. Nagarkatti, Dr. Michael R. Vaughan, and Mr. Michael Gilbertson for reviewing various drafts of this manuscript.

## References

Anderson, D.W., and Hamerstrom, F.: 1967, 'The recent status of Wisconsin cormorants', *Passenger Pigeon* **29**, 3-15.

Boersma, D.C., Ellenton, J.A., and Yagminas, A.: 1986, 'Investigation of the hepatic mixed-function oxidase system in herring gull embryos in relation to environmental contaminants', *Environ. Toxicol. Chem.* **5**, 309-318.

Bowerman, W.W.: 1993, *Regulation of bald eagle* (Haliaeetus leucocephalus) *productivity in the Great Lakes Basin: An ecological and toxicological approach*, Ph.D. thesis, Michigan St. Univ., E. Lansing, Michigan.

Bowerman, W.W., Kubiak, T.J., Holt, J.B., Evans, D.L., Eckstein, R.G., Sindelar, C.R., Best, D.A., and Kozie, K.D.: 1994, 'Observed abnormalities in mandibles of nestling bald eagles *Haliaeetus leucocephalus*', *Bull. Environ. Contam. Toxicol.* **53**, 450-457.

Bowerman, W.W., Giesy, J. P., Best, D.A,. and Kramer, V.J.: 1995, 'A review of factors affecting productivity of bald eagles in the Great Lakes region: Implications for recovery', *Environ. Health Perspect.* **103**, 51-59.

Broley, C.L.: 1951, 'Bald eagle nesting failures continue', *Audubon Mag.* **53**, 208.

Broley, C.L.: 1952, 'Broley reports on eagles', *Audubon Mag.* **54**, 71-72.

Colborn, T.E.: 1991, 'Epidemiology of Great Lakes bald eagles', *J. Toxicol. Environ. Health* **33**, 395-453.

De Swart, R.L., Ross, P.S., Vedder, L.J., Timmerman, H.H., Heisterkamp, S., Van Loveren, H., Vos, J.G., Reijnders, P.J.H., and Osterhaus, A.D.M.E.: 1994, 'Impairment of immune function in harbor seals *(Phoca vitulina)* feeding on fish from polluted waters', *Ambio* **23**, 155-159.

Ecological Research Services, Inc.: 1991, 'Toxic contaminants in certain Great Lakes fish and wildlife and their relevance to hydrorelicensing decisions on the AuSable, Manistee, and Muskegon Rivers, Michigan', Summary report submitted to Consumers Power Co., Jackson, Michigan.

Edwards, M.: 1970, '43. Uninhabited island in Lake Ontario', *Audubon Field Notes* **24**, 767-768.

Ellenton, J.A., Brownlee, L.J. and Hollebone, B.R.: 1985, 'Aryl hydrocarbon hydroxylase levels in herring gull embryos from different locations on the Great Lakes', *Environ. Toxicol. Chem.* **4**, 615-622.

Fox, G.A.: 1993, 'What have biomarkers told us about the effects of contaminants on the health of fish-eating birds in the Great Lakes? The theory and a literature review', *J. Great Lakes Res.* **19**, 722-736.

Fox, G.A., Gilman, A.P., Peakall, D.B., and Anderka, F.W.: 1978, 'Behavioral abnormalities in nesting Lake Ontario herring gulls', *J. Wildl. Manage.* **42**, 477-483.

Fox, G.A., Kennedy, S.W., Norstrom, R.J., and Wigfield, D.C.: 1988, 'Porphyria in herring gulls: A biochemical response to chemical contamination of Great Lakes food chains', *Environ. Toxicol. Chem.* **7**, 831-839.

Fox, G.A., Collins, B., Hayakawa, E., Weseloh, D.V., Ludwig, J.P., Kubiak, T.J. and Erdman, T.C.: 1991, 'Reproductive outcomes in colonial fish-eating birds: A biomarker for developmental toxicants in Great Lakes food chains. II. Spatial variation in the occurrence and prevalence of bill defects in young double-crested cormorants in the Great Lakes, 1979-1987', *J. Great Lakes Res.* **17**, 158-167.

Fox, G.A., Trudeau, S., Won, H., and Grasman, K.A.: 1998, 'Monitoring the elimination of persistent toxic substances from the Great Lakes: Chemical and physiological evidence from adult herring gulls,' *Environ. Monit. Assessment. J.* This Issue.

Friend, M., and Trainer, D.O.: 1970, 'Polychlorinated biphenyl: Interaction with duck hepatitis virus', *Science* **170**, 1314-1316.

Giesy, J.P., Ludwig, J.P., and Tillitt, D.E.: 1994, 'Deformities in birds of the Great Lakes region', *Environ. Sci. Technol.* **28**, 128-135.

Gilbertson, M.: 1974a, 'Pollutants in breeding herring gulls in the lower Great Lakes', *Can. Field Nat.* **88**, 273-280.

Gilbertson, M.: 1974b, 'Seasonal changes in organochlorine compounds and mercury in common terns of Hamilton Harbour, Ontario', *Bull. Environ. Contam. Toxicol.* **12**, 726-732.

Gilbertson, M.: 1975, 'A Great Lakes Tragedy', *Nature Canada* **4**, 22-25.

Gilbertson, M.: 1983, 'Etiology of chick edema disease in herring gulls in the lower Great Lakes', *Chemosphere* **12**, 357-370.

Gilbertson, M.: 1988, 'Epidemics in birds and mammals caused by chemicals in the Great Lakes', in: M.S. Evans. (ed.) *Toxic Contaminants and Ecosystem Health: A Great Lakes Focus.* John Wiley & Sons, Inc., New York. pp. 133-152.

Gilbertson, M.: 1989, 'Effects on fish and wildlife', *Halogenated Biphenyls, Terphenyls, Naphthalenes, Dibenzodioxins, and Related Products.* R.D. Kimbrough and A.A. Jenson. (eds.) Elesevier Science Publishers, New York. pp. 103-127.

Gilbertson, M., and Reynolds, L.M.: 1972, 'Hexachlorobenzene (HCB) in the eggs of common terns in Hamilton Harbour, Ontario', *Bull. Environ. Contam. Toxicol.* **7**, 371-373.

Gilbertson, M., and Hale, R.: 1974a, 'Characteristics of the breeding failure of a colony of herring gulls on Lake Ontario', *Can. Field Nat.* **88**, 356-358.

Gilbertson, M., and Hale, R.: 1974b, 'Early embryonic mortality in a herring gull colony in Lake Ontario', *Can. Field Nat.* **88**, 354-356.

Gilbertson, M., Morris, R.D., and Hunter, R.A.: 1976, 'Abnormal chicks and PCB residue levels in eggs of colonial waterbirds on the lower Great Lakes (1971-1975)', *Auk* **93**, 434-442.

Gilbertson, M., and Fox, G.A.: 1977, 'Pollutant-associated embryonic mortality of Great Lakes herring gulls', *Environ. Pollut.* **12**, 211-216.

Gilbertson, M., Kubiak, T., Ludwig, J., and Fox, G: 1991, 'Great Lakes embryo mortality, edema, and deformities syndrome (GLEMEDS) in colonial fish-eating birds: Similarity to chick-edema disease', *J. Toxicol. Environ. Health* **33**, 455-520.

Government of Canada: 1991, *Toxic Chemicals in the Great Lakes and Associated Effects*, Supply and Services Canada.

Grasman, K.A.: 1995, *Immunological and hematological biomarkers for contaminants in fish-eating birds of the Great Lakes*, Ph.D. thesis, Virginia Polytech. Instit. St. Univ., Blacksburg, Virginia.

Grasman, K.A., Fox,G.A., Scanlon, P.F., and Ludwig, J.P.: 1996, 'Organochlorine-associated immunosuppression in prefledling Caspian terns and herring gulls from the Great Lakes: An ecoepidemiological study', *Environ. Health Perspect.* **104(Suppl. 4)**, 829-842.

Grier, J.W.: 1982, 'Ban of DDT and subsequent recovery of reproduction in bald eagles', *Science* **218**, 1232-1235.

Harris, H.J., Erdman, T.C., Ankley, G.T., and Lodge, K.B. 1993. 'Measures of reproductive success and polychlorinated biphenyl residues in eggs and chicks of Forster's terns on Green Bay, Lake Michigan, Wisconsin--1988', *Arch. Environ. Contam. Toxicol.* **25**:304-314.

Hebert, C. E., Norstrom, R.J., Simon, M., Braune, B.M., Weseloh, D.V., and Macdonald, C.R.: 1994, 'Temporal trends and sources of PCDDs and PCDFs in the Great Lakes: Herring gull egg monitoring, 1981-1991', *Environ. Sci. Technol.* **28**, 1268-1277.

Hickey, J.J., Keith, J.A., and Coon, F.B.: 1966, 'An exploration of pesticides in a Lake Michigan ecosystem', *J. Appl. Ecol.* **3(Suppl.)**, 141-154.

Hickey, J.J., and Anderson, D.W.: 1968, 'Chlorinated hydrocarbons and eggshell changes in raptorial and fish-eating birds', *Science* **162**, 271-273.

Hoffman, D.J., Rattner, B.A., Sileo, L., Docherty, D., and Kubiak, T.J.: 1987, 'Embryotoxicity, teratogenicity, and aryl hydrocarbon hydroxylase activity in Forster's terns on Green Bay, Lake Michigan', *Environ. Res.* **42**, 176-184.

Hoffman, D.A., Smith, G.J., and Rattner, B.A.: 1993, 'Biomarkers of contaminant exposure in common tern and black-crowned night herons in the Great Lakes', *Environ. Toxicol. Chem.* **12**, 1095-1103.

Hoffman, D.A., Rice, C.P., and Kubiak, T.J.: 1996, 'PCBs and dioxins in birds', in: *Environmental Contaminants in Wildlife: Interpreting Tissue Concentrations*, W.N. Beyer, G.H. Heinz, and A.W. Redmon-Norwood. (eds.) Lewis Publishers, Inc., New York, pp. 165-207.

International Joint Commission: 1991, *Cleaning up our Great Lakes: A report on toxic substances in the Great Lakes basin ecosystem*, International Joint Commission, Windsor, Ontario.

Jones, P.D., Ankley, G.T., Best, D.A., Crawford, R., DeGalan, N., Giesy, J.P., Kubiak, T.J., Ludwig, J.P., Newstead, J.L., Tillitt, D.E., Verbrugge, D.A.: 1993a, 'Biomagnification of bioassay derived 2,3,7,8-tetrachlorodibenzo-*p*-dioxin equivalents', *Chemosphere* **26**, 1203-1212.

Jones, P.D., Giesy, J.P., Newstead, J.L., Verbrugge, D.A., Beaver, D.L., Ankley, G.T., Tillitt, D.E., Lodge, K.B., and Niemi, G.J.: 1993b, '2,3,7,8-tetrachlorodibenzo-*p*-dioxin equivalents in tissues of birds at Green Bay, Wisconsin', *Arch. Environ. Contam. Toxicol.* **24**, 345-354.

Keith, J.A.: 1966, 'Reproduction in a population of herring gulls *(Larus argentatus)* contaminated by DDT', *J. Appl. Ecol.* **3 (Suppl.)**, 57-60.

Kennedy, S.W., Lorenzen, A., James, C.A., and Collins, B.T.: 1993, 'Ethoxyresorufin O-deethylase and porphyrin analysis in chicken embryo hepatocyte cultures with a fluorescence multiwell plate reader', *Anal. Biochem.* **211**, 101-112.

Kennedy, S.W., Lorenzen, A., and Jones, S.P.: 1994., 'Sensitivity of various primary avian embryo hepatocyte cell cultures to cytochrome P450 1A induction by TCDD, TCDF, and PCBs', *Organohalogen Compounds* **21**, 475-480.

Kennedy, S.W., Lorenzen, A., Jones, S.P., Hahn, M.E., and Stegeman, J.J.: 1996, 'Cytochrome P4501A induction in avian hepatocyte cultures: A promising approach for predicting the sensitivity of avian species and to toxic effects of halogenated aromatic hydrocarbons', *Toxicol. Appl. Pharmacol.* **141**, 214-231.

Kozie, K.D., and Anderson, R.K.: 1991, 'Productivity, diet, and environmental contaminants in bald eagles nesting near the Wisconsin shoreline of Lake Superior', *Arch. Environ. Contam. Toxicol.* **20**, 41-48.

Kubiak, T.J., Harris, H.J., Smith, L.M., Schwartz, T.R., Stalling, D.L., Trick, J.A., Sileo, L., Docherty, D.E., and Erdman, T.C.: 1989, 'Microcontaminants and reproductive impairment of the Forster's tern on Green Bay, Lake Michigan-- 1983', *Arch. Environ. Contam. Toxicol.* **18**, 706-727.

Lahvis, G., Wells, R.W., Kuehl, D.W., Stewart, J.L., Rhinehart, H.L., and Via, C.S.: 1995, 'Decreased lymphocyte responses in free-ranging bottlenose dolphins *(Tursiops truncatus)* are associated with increased concentrations of PCBs and DDT in peripheral blood', *Environ. Health Perspect.* **103**, 67-72.

Ludwig, J.P.: 1979, 'Present status of the Caspian tern population of the Great Lakes', *Michigan Academian* **12**, 69-77.

Ludwig, J.P., and Ludwig, C.E.: 1969, 'The effect of starvation on insecticide contaminated herring gulls removed from a Michigan colony', *Proc. 12$^{th}$ Conf. Great Lakes Res.*, 53-64.

Ludwig, J.P., Auman, H.J., Kurita, H., Ludwig, M.E., Campbell, L.M., Giesy, J.P., Tillitt, D.E., Jones, P., Yamashita, N., Tanabe, S., and Tatsukawa, R.: 1993, 'Caspian tern reproduction in the Saginaw Bay ecosystem following a 100-year flood event', *J. Great Lakes Res.* **19**, 96-108.

Ludwig, J.P., Auman, H.J., Weseloh, D.V., Fox, G.A., Giesy, J.P., and Ludwig, M.E.: 1995, 'Evaluation of the effects of toxic chemicals in Great Lakes cormorants: Has causality been established?', *Colonial Waterbirds* **18 (special publ. 1)**, 53-60.

Ludwig, J.P., Kurita-Matsuba, H., Auman, H.J., Ludwig, M.E., Summer, C.L., Giesy, J.P., Tillitt, D.E., and Jones, P.D.: 1996, 'Deformities, PCBs, and TCDD-equivalents in double-crested cormorants *(Phalacrocorax auritus)* and Caspian terns *(Hydroprogne caspia)* of the Upper Great Lakes 1986-1991: Testing a cause-effect hypothesis', *J. Great Lakes Res.* **22**, 172-197.

Luster, M.I., Blank, J.A., and Dean, J.H.: 1987, 'Molecular and cellular basis of chemically induced immunotoxicity', *Ann. Rev. Pharmacol. Toxicol.* **27**, 23-49.

Metcalfe, C.D., and Haffner, G.D.: 1995, 'The ecotoxicology of coplanar polychlorinated biphenyls', *Environ. Rev.* **3**, 171-190.

Mineau, P., Fox, G.A., Norstrom, R.J., Weseloh, D.V., Hallet, D.J., and Ellenton, J.A.: 1984, 'Using the herring gull to monitor levels and effects of organochlorine contamination in the Canadian Great Lakes', in: *Toxic Contaminants in the Great Lakes*, J.O. Nriagu and J.S. Simmons. (eds.) John Wiley & Sons, Inc., New York, pp. 425-452.

Moccia, R.D., Fox, G.A., and Britton, A.: 1986, 'A quantitative assessment of thyroid histopathology of herring gulls *(Larus argentatus)* from the Great Lakes and a hypothesis on the causal role of environmental contaminants', *J. Wildl. Dis.* **22**, 60-70.

Mora, M.A., Auman, H.J., Ludwig, J.P., Giesy, J.P., Verbrugge, D.A., and Ludwig, M.E.: 1993, 'Polychlorinated biphenyls and chlorinated insecticides in plasma of Caspian terns: Relationships with age, productivity, and colony site tenacity in the Great Lakes', *Arch. Environ. Contam. Toxicol.* **24**, 320-331.

Peakall, D.B.: 1988, 'Known effects of pollutants on fish-eating birds in the Great Lakes of North America', in: *Toxic Contamination in Large Lakes, Vol. 1.:*

*Chronic Effects of Contaminants in Large Lakes,* N.W. Schmidtke. (ed.) Lewis Publishers, Inc., Chelsea, Michigan, pp. 39-54.

Peakall, D.B., Fox, G.A., Gilman, A.D., Hallet, D.J., and Norstrom, R.J.: 1980, 'Reproductive success of herring gulls as an indicator of Great Lakes water quality', in: *Hydrocarbons and Halogenated Hydrocarbons in the Aquatic Environment*, B.K. Afhgan and D. MacKay (eds.) Plenum Press, New York, pp. 337-344.

Postupalsky, S.: 1971, 'Toxic chemicals and declining bald eagles and cormorants in Ontario', *Can. Wildl. Serv. Manuscript Rep.* **#20**, 45 pp.

Postupalsky, S.: 1978, 'The bald eagle returns', *Natural History* **87**, 62-63.

Price, I.M., and Weseloh, D.V.: 1986, 'Increased numbers and productivity of double-crested cormorants, *Phalacrocorax auritus*, on Lake Ontario', *Canadian Field Naturalist* **100**, 474-482.

Quilliam, H.R.: 1965, *History of the birds of Kingston, Ontario.* Privately printed, Kingston, Ontario, 210 pp.

Ratcliffe, D.A.: 1967, 'Decrease in eggshell weight in certain birds of prey', *Nature* **215**, 208-210.

Rattner, B.A., Melancon, M.J., Custer, T.W., Hothem, R.L., King, K.A., LeCaptain, L.J., Spann, J.W., Woodin, B.R., and Stegeman, J.J.: 1993, 'Biomonitoring environmental contamination with pipping black-crowned night heron embryos: Induction of cytochrome P450', *Environ. Toxicol. Chem.* **12**, 1719-1732.

Rattner, B.A., Hatfield, J.S., Melancon, M.J., Custer, T.W., and Tillitt, D.E.: 1994, 'Relation among cytochrome P450, Ah-active PCB congeners and dioxin equivalents in pipping black-crowned night-heron embryos', *Environ. Toxicol. Chem.* **13**, 1805-1812.

Ross, P.S., de Swart, R.L., Reijnders, P.J.H., Van Loveren, H., Vos, J.G., and Osterhaus, A.D.M.E.: 1995, 'Contaminant-related suppression of delayed-type hypersensitivity and antibody responses in harbor seals fed herring from the Baltic Sea', *Environ. Health Perspect.* **10,** 162-167.

Safe, S.: 1990, 'Polychlorinated biphenyls (PCBs), dibenzo-*p*-dioxins (PCDDs), dibenzofurans (PCDFs), and related compounds: Environmental and mechanistic considerations which support the development of toxic equivalency factors (TEFs)', *Crit. Rev. Toxicol.* **21**, 51-88.

Sileo, L., Karstad, L., Frank, R., Holdrinet, M.V.H., Addison, E., and Braun, H.E.: 1977, 'Organochlorine poisoning of ring-billed gulls in southern Ontario', *J. Wildl. Dis.* **13**, 313-322.

Spear, P.H., Moon, T.W., and Peakall, D.B.: 1985, 'Liver retinoid concentrations in natural populations of herring gulls *(Larus argentatus)* contaminated by 2,3,7,8-tetrachlorodibenzo-*p*-dioxin and in ring doves *(Streptopelia risoria)* injected with a dioxin analogue', *Can. J. Zool.* **64**, 204-208.

Spear, P.H., Bourbonnais, D.H., Norstrom, R.J., and Moon, T.W.: 1990, 'Yolk retinoids (vitamin A) in eggs of the herring gull and correlations with polychlorinated dibenzo-*p*-dioxins and dibenzofurans', *Environ. Toxicol. Chem.* **9**, 1053-1061.

Stow, C.A.: 1995, 'Great Lakes herring gull egg PCB concentrations indicate approximate steady-state conditions', *Environ. Sci. Technol.* **29**, 2893-2897.

Thomas, P.T., and Faith, R.I.: 1985, 'Adult and perinatal immunotoxicity induced by halogenated aromatic hydrocarbons', in: *Immunotoxicology and Immunopharmacology*, J.H. Dean, M.I. Luster, A.E. Munson, and H. Amos. (eds.) Raven Press, New York, pp. 305-313.

Tillitt, D.E., Ankley, G.T., and Giesy, J.P.: 1989, 'Planar chlorinated hydrocarbons (PCHs) in colonial fish-eating waterbird eggs from the Great Lakes', *Marine Environ. Res.* **28**, 505-508.

Tillitt, D.E., Giesy, J.P., and Ankley, G.T.: 1991, 'Characterization of the H4IIE rat hepatoma cell bioassay as a tool for assessing toxic potency of planar halogenated hydrocarbons (PHHs) in environmental samples', *Environ. Sci. Tech.* **25**, 187-192.

Tillitt, D.E., Ankley, G.T., Giesy, J.P., Ludwig, J.P., Kurita-Matsuba, H., Weseloh, D.V., Ross, P.S., Bishop, C.A., Sileo, L., Stromborg, K.L., Larson, J., and Kubiak, T.J:. 1992, 'Polychlorinated biphenyl residues and egg mortality in double-crested cormorants from the Great Lakes', *Environ. Toxicol. Chem.* **11**, 1281-1288.

Velleux, M, and Endicott, D.: 1994, 'Development of a mass balance model for estimating PCB export from the lower Fox River to Green Bay', *J. Great Lakes Res.* **20**, 416-434.

Vos, J.G., and Luster, M.I.: 1989, 'Immune alterations', in: *Halogenated Biphenyls, Terphenyls, Naphthalenes, Dibenzodioxins, and Related Products*, R.D. Kimbrough and A.A. Jensen. (eds.) Elsevier Science Publishers, New York, pp.295-322.

Vos, J., Van Loveren, H., Wester, P., and Vethaak, D.: 1989, 'Toxic effects of environmental chemicals on the immune system', *TIPS* **10**, 289-292.

Weseloh, D.V., Mineau, P., and Hallet, D.J.: 1979, 'Organochlorine contaminants and trends in reproduction in Great Lakes herring gulls, 1974-1978', *Trans. N. Am. Wildl. Nat. Resour. Conf.* **44**, 543-557.

Weseloh, D.V., Teeple, S.M., and Gilbertson, M.: 1983, 'Double-crested cormorants of the Great Lakes: Egg-laying parameters, reproductive failure, and contaminant residues in eggs, Lake Huron 1972-1973', *Can. J. Zool.* **61**, 427-436.

Weseloh, D.V., Custer, T.W., and Braune, B.M.: 1989, 'Organochlorine contaminants in eggs of common terns from the Canadian Great Lakes, 1981', *Environ. Pollut.* **59**, 141-160.

Weseloh, D.V., Ewins, P.J., Struger, J., Mineau, P., Bishop, C.A., Postupalsky, S., and Ludwig, J.P.: 1995, 'Double-crested cormorants of the Great Lakes: Changes in population size, breeding distribution and reproductive output between 1913 and 1991', *Colonial Waterbirds* **18 (special publ. 1)**, 48-59.

Yamashita, N., Tanabe, S., Ludwig, J.P., Kurita, H., Ludwig, M.E., and Tasukawa, R.: 1993, 'Embryonic abnormalities and organochlorine contamination in double-crested cormorants *(Phalacrocorax auritus)* and Caspian terns *(Hydroprogne caspia)* from the upper Great Lakes in 1988', *Environ. Pollut.* **79**, 163-173.

Zile, M.H.: 1992, 'Vitamin A homeostasis endangered by environmental pollutants', *Proc. Soc. Exper. Biol. Med.* **201**, 141-153.

# MONITORING THE ELIMINATION OF PERSISTENT TOXIC SUBSTANCES FROM THE GREAT LAKES; CHEMICAL AND PHYSIOLOGICAL EVIDENCE FROM ADULT HERRING GULLS

G.A.FOX[1], S.TRUDEAU[1], H.WON[1] and K.A.GRASMAN[2]
[1]*Canadian Wildlife Service, National Wildlife Research Centre, Hull, Quebec, K1A 0H3, Canada;* [2]*Department of Biological Sciences, Wright State University, Dayton, OH, 45435, USA*

**Abstract.** To assess progress towards virtual elimination of PCBs, DDE, dieldrin and Mirex and their associated physiological effects, we compared their concentrations in pooled livers of adult herring gulls *(Larus argentatus)* repeatedly sampled at 8 Great Lakes colonies and a reference colony on the Atlantic coast between 1974 and 1993. We measured the relative thyroid mass and concentrations of highly carboxylated porphyrins and retinyl palmitate in the liver of each individual. PCBs, dieldrin and mirex declined in 7 of 8 colonies while DDE decreased in six. The greatest decreases occurred pre-1985. PCBs and DDE did not decrease in gulls from Middle Island in western L. Erie. Middle Island and Saginaw Bay had the highest concentrations of PCBs of 11 Great Lakes colonies in the 1990s. Thyroids of gulls from Great Lakes colonies were slightly enlarged but the degree of enlargement has decreased over time. In 1991, gulls from Great Lakes colonies had slight to moderately elevated concentrations of highly carboxylated porphyrins. In the early 1990s, hepatic stores of retinyl palmitate were very seriously depleted in gulls from the Detroit River, western basin of Lake Erie, and Lake Ontario, reflecting decreased availability and altered storage. We conclude that PCBs and/or other persistent toxic substances in the food of herring gulls have not been virtually eliminated.

## 1. Introduction

The general condition of the boundary waters of the Great Lakes deteriorated in the first half of this century and worsened following the post World War II chemical revolution. After extensive studies in the 1960s, the governments of Canada and the United States signed the Great Lakes Water Quality Agreement (GLWQA) in 1972. In the 1970s, there was a growing scientific appreciation of the seriousness of the injury to wildlife resources associated with the releases of persistent toxic substances to the Great Lakes. In 1978, the two governments renegotiated the GLWQA and incorporated a policy which stated that the discharge of any and all persistent toxic substances be virtually eliminated. Today's challenge is to determine whether the regulatory and remedial actions to virtually eliminate discharges of these substances have been sufficient to stop the injury.

Monitoring temporal changes in concentrations of organochlorine contaminants and their associated physiological effects in resident herring gulls *(Larus argentatus)* is one approach to assessing progress towards attainment of the GLWQA's goals of remediation of water quality and virtual elimination of persistent toxic substances and their associated effects. Mature herring gulls are year-round residents in the Great Lakes basin, nest colonially in all of the Great Lakes, and the species has been extensively studied throughout the holarctic region (see Mineau *et al.* 1984). In 1974, the Canadian Wildlife Service initiated a surveillence project to monitor chemical contaminants and their effects in this species (Gilbertson 1974) and has amassed extensive data. Although a wide variety of items has been identified in the diet of Great Lakes herring gulls, fish are the food items most frequently encountered and the greatest contributor to contaminant burdens. The most important species are the alewife *(Alosa pseudoharengus)*, rainbow smelt *(Osmerus mordax)*, freshwater drum *(Aplodinotus grunniens)*, sunfishes *(Lepomis sp.)* and yellow perch *(Perca flavescens)* (Fox *et al.* 1990, Ewins *et al.* 1994) . These species differ in their temporal and geographic availability, their position in the food web, and in their mass, lipid and contaminant content; all factors that contribute to the deposition of lipophilic contaminants in gull tissues.

Our repeated sampling of herring gulls in eight Great Lakes colonies, including four Areas of Concern (AOCs) and a reference site on the Atlantic coast, provides an opportunity to evaluate spatial differences and temporal changes in the bioavailability of persistent toxic substances and a suite of sensitive physiological responses. Here we report the concentrations of 1,1-dichloro-2,2-bis(*p*-chlorophenyl)ethylene (DDE), polychlorinated biphenyls (PCBs), dieldrin, and mirex measured in the pooled livers of adult herring gulls, and three organochlorine-sensitive physiological measures (biomarkers) of impairment (Fox 1993); (i) the mass of the thyroids relative to body mass as a measure of thyroid enlargement (goitre), (ii) concentrations of highly carboxylated porphyrins in the liver (Fox *et al.* 1988) as a measure of disruption of the heme biosynthetic pathway, and (iii) concentrations of hepatic retinyl palmitate as a measure of the capacity of the liver to store vitamin A (Zile 1992).

## 2. Study Area

Herring gulls were collected in colonies located in the Bay of Fundy (Kent Island: 1980,1982,1991); in Black Bay in western Lake Superior (Granite Island: 1980, 1982,1991); at the mouth of the Fox River in Green Bay, Lake Michigan (Cat Island: 1983, Lone Tree Island:1983,1985 and Kidney Island:1985,1991); in Saginaw Bay, Lake Huron (Channel Shelter Island:1982,1985,1991); the Detroit River (Fighting Island:1985,1991); in the western basin of Lake Erie (Middle Sister Island:1980,1993 and Middle Island:1983, 1985, 1991); in the western basin of Lake Ontario (Hamilton Harbour:1985,1991); and in the eastern basin of Lake Ontario (Scotch Bonnet Island:1974,1991) (Figure I).

**Figure I** Locations of Colonies of Great Lakes Herring Gulls Monitored for Chemically-Induced Physiological Changes

## 3. Methods

At each site, approximately 10 adult gulls (equal numbers of each gender when possible) were trapped on their nests, under Federal permit, mid-way through incubation. They were weighed, killed by decapitation and their livers removed, weighed, and rapidly frozen for shipment to the laboratory where they were stored frozen for various lengths of time.

Thyroids were removed, fixed, trimmed and their mass measured to the nearest mg. The relative thyroid mass was calculated as mg mass of both thyroids/kg body mass (Moccia *et al.* 1986).

Highly carboxylated porphyrins were determined in approximately 1 gm aliquots of liver by the technique of Kennedy *et al.* (1986), or Kennedy and James (1993). Briefly, porphyrins were extracted with perchloric acid-methanol (pre-1991 collections) or acetonitrile-hydrochloric acid (1991 and 1993 collections), and concentrated on disposable octadecylsilyl columns. Porphyrins were separated by HPLC employing gradient elution with a sodium phosphate-methanol mobile phase on a 3 cm long $C_{18}$ column and detected with a spectrofluorometric detector. The detection limit of 20 fmol for both methods is sufficiently sensitive to detect the earliest indication of halogenated aromatic hydrocarbon-induced porphyria.

Another approximately 0.5 g sample of liver was ground and dehydrated in anhydrous sodium sulphate. Retinyl acetate was added as an internal standard and retinyl palmitate and other vitamin A compounds extracted with dichloromethane:methanol (1:9). The extract was centrifuged and an aliquot of the supernatant was filtered and analysed by reverse-phase HPLC using an ODS column with solvent and flow programing (100% methanol at 1 ml/min for first 6 min.; 70% methanol-30% dichloromethane, at 1 ml/min. 6.0 to 6.1 min., and at 2 ml/min. 6.1 to 13.0 min.; 100% methanol at 2 ml/min. 13.0 to 13.5 min.; 13.5 min. wash with 100% methanol at 1 ml/min) and U.V.-visible detection (326 nm). Changes in the column characteristics and solvent composition have decreased the minimal detectable amount of retinyl palmitate from 100 $\mu$g/g liver (pre-1985), to 12 $\mu$g/g (1985) and 1.2 $\mu$g/g in the 1990s.

All samples of pooled liver homogenate were assayed for organochlorine contaminants in Environment Canada's Analytical Service Laboratory at the National Wildlife Research Centre, Hull, Quebec by the same analyst using high resolution gas chromatography with electron capture detection as reported by Peakall *et al.* (1986) between 1984 and 1994. It should be noted that there is considerable variation in diet composition of Great Lakes herring gulls among colonies and between years (Fox *et al.* 1990, Ewins *et al.* 1994). These dietary differences, which reflect differing availability of food items and the opportunistic nature of the gull's foraging activity, are likely to result in considerable individual variation in contaminant exposure which we cannot assess with pooled analyses. Biological variability in contaminant exposure will be unique to each collection and is likely to be higher at the most contaminated sites. The standard deviations about the mean for individual analyses of 10 livers collected in 1985 (Fox *et al.* 1988) ranged from a low of ± 55% (PCBs) to a high of ±132% (mirex) for Saginaw Bay birds, and from ± 36% (dieldrin) to ±46% (PCBs) for Lake Ontario birds.

In our experience, the distributions of biomarker measurements are frequently skewed, therefore we used the median as an unbiased measure of central tendency and Spearman rank coefficient as a measure of correlation. A Kruskall-Wallis one-way ANOVA on ranks (Zar 1984, Glantz 1992) was used to determine whether among-year differences in medians for biomarker responses were statistically significant, and where differences were found, Dunn's method was used to identify the differing years (Glantz 1992). Since pooled analyses provide no measure of variability, it was not possible to establish the statistical significance of temporal differences in contaminant concentrations in gull livers.

## 4. Results and Discussion by Location

The results for all locations are summarized in Table I, and discussed by location. Results are also discussed in relation to all 11 Great Lakes colonies sampled in 1991/93. The severity of the physiological abnormalities is also expressed in qualitative terms, relative to the most severe cases we have encountered in our studies of this species to date.

### 4.1 KENT ISLAND IN THE BAY OF FUNDY

Kent Island is remote from any point sources and the contaminant loading of the local food chain is predominantly from the atmosphere. This site is thought to serve as an adequate reference colony. Concentrations of PCBs, DDE, dieldrin and mirex were 64%, 66% 43% and 76% lower, respectively, in pooled livers of gulls collected in 1991 at this site than in the 1980 collection. The 1980 concentrations of these contaminants were lower than most of the 11 Great Lakes colonies sampled in the 1990s. PCB and DDE concentrations in the pooled livers declined by 58% between 1980 and 1982, in contrast to 30% and 19%, respectively, between 1982 and 1991.

Highly carboxylated porphyrin concentrations in gulls from this location have remained at a concentration we regard as normal (Fox *et al.* 1988) between 1980 and 1991. The relative thyroid mass decreased ($P<0.05$) by 26% between 1980 and 1982, but did not change between 1982 and 1991, possibly suggesting that

**Table I.** Temporal Variation in Concentration of Persistent Toxic Substances in Pooled Livers (mg/kg, wet wt) and Medians of the Relative Thyroid Mass (mg/kg) and Concentrations of Highly Carboxylated Porphyrins (pmol/g) and Vitamin A ($\mu$g/g) in Individual Livers of Adult Herring Gulls. NA = Not Available.

| Site | Year | PCBs | DDE | Dieldrin | Mirex | Porphyrin | Thyroid | Vit. A |
|---|---|---|---|---|---|---|---|---|
| Kent Island | 1980 | 10.60 | 1.72 | 0.070 | 0.050 | < 5 | 128 | 687 |
| | 1982 | 4.51 | 0.73 | 0.040 | 0.035 | <10 | 95 | 1575 |
| | 1991 | 3.81 | 0.59 | 0.017 | 0.012 | <11 | 95 | 4576 |
| Granite Island | 1980 | 31.3 | 7.30 | 0.110 | 0.120 | 14.5 | 296 | 347 |
| | 1982 | 14.2 | 3.05 | 0.250 | 0.170 | 14.5 | NA | 542 |
| | 1991 | 11.1 | 1.98 | 0.133 | 0.042 | 38 | 174 | 961 |
| Green Bay | 1983 | 37.7 | 6.90 | 0.720 | 0.097 | 70 | 242 | 247 |
| | 1985 | 35.0 | 5.60 | 0.320 | 0.060 | 130 | NA | 458 |
| | 1991 | 9.3 | 0.90 | 0.146 | 0.018 | 110 | 193 | 503 |
| Saginaw Bay | 1982 | 44.0 | 5.12 | 0.310 | 0.140 | 110 | 169 | 596 |
| | 1985 | 33.0 | 4.10 | 0.220 | 0.062 | 128 | NA | 1014 |
| | 1991 | 31.0 | 6.50 | 0.123 | 0.117 | 111 | 145 | 585 |
| Fighting Island | 1985 | 89.0 | 3.80 | 0.210 | 0.130 | 62 | NA | 125 |
| | 1991 | 26.3 | 1.50 | 0.069 | 0.034 | 68 | 163 | 104 |
| Middle Sister | 1980 | 68.8 | 3.52 | 0.220 | 0.130 | 43 | 328 | <100 |
| Island. | 1993 | 29.5 | 1.46 | 0.070 | 0.030 | 45 | 194 | 95 |
| Middle Island | 1983 | 45.7 | 2.15 | 0.240 | 0.060 | 30 | 252 | <100 |
| | 1985 | 40.0 | 1.70 | 0.130 | 0.051 | 64 | NA | 37 |
| | 1991 | 44.3 | 2.20 | 0.101 | 0.059 | 75 | 179 | 117 |
| Hamilton | 1985 | 33.0 | 4.20 | 0.150 | 1.100 | 80 | NA | 115 |
| Harbour | 1991 | 21.0 | 3.20 | 0.075 | 0.067 | 62 | 164 | 183 |
| Scotch Bonnet | 1974 | 101 | 14.5 | 0.540 | 4.09 | 190 | 257 | <100 |
| Island | 1991 | 28.3 | 7.40 | 0.076 | 1.27 | 46 | 125 | 201 |

hypertrophy does not occur at these low contaminant levels. The relative thyroid mass of gulls from Kent Island was the lowest of any collection from the early 1990s. Vitamin A and its precursors are much more abundant in marine than freshwater environments (Simpson *et al.* 1981). Between 1980 and 1991 the concentration of retinyl palmitate in the livers of Kent Island gulls has increased ($P<0.001$) by over 566%, and were the highest of any collection in the 1990s, and at least 4x any Great Lakes colony. This increase may reflect the decreasing interference of contaminants with vitamin A storage, and/or a change in primary productivity or diet.

### 4.2. GRANITE ISLAND, NORTH SHORE OF LAKE SUPERIOR

The major loading of contaminants to Lake Superior is via atmospheric deposition (Eisenrich *et al.* 1981). Concentrations of PCBs, DDE, and mirex were 64%, 73%, and 65% lower, respectively, and of dieldrin 33% higher in pooled livers of gulls collected in 1991 at this site than in the 1980 collection. Between 1980 and 1982, PCBs and DDE concentrations in gulls from this colony decreased by 55% and 58%, respectively, in contrast to 22% and 35% between 1982 and 1991. These temporal changes are almost identical to those observed in gulls from Kent Island, the other site where atmospheric sources are thought to predominate.

Highly carboxylated porphyrin concentrations in gull livers from this site were normal in 1980 and 1982 but increased ($P<0.01$) by 162% between 1982 and 1991. This slight elevation in porphyrins, which is not attributable to any known porphyrinogen we have measured to date, could be interpreted as an indication of increasing concentrations of an unrecognised porphyrinogenic substance in Lake Superior, but such a conclusion is speculative in the absence of more extensive analytical data. The relative thyroid mass of gulls from Granite Island decreased by 41% between 1980 and 1991 (from moderate to slight enlargement). Concentrations of retinyl palmitate in the liver have increased ($P<0.01$) by 177% between 1980 and 1991 (from moderate to slight depletion) at a calculated linear (average) rate of 28% per year 1980-1982 and 8.6% per year, 1982-1991. The highest concentration of retinyl palmitate in 119 individuals from 11 Great Lakes sites collected in the 1990s, was measured in an individual from this colony, and the median concentration for this site was the second highest observed.

### 4.3. DREDGE SPOIL ISLANDS IN LOWER GREEN BAY, LAKE MICHIGAN

The discharge of PCBs from pulp and paper, and paper recycling mills has contaminated the sediments, water column, and biota of the Fox River, the largest tributary to Green Bay, resulting in continuing export of PCBs into Green Bay

(Hermanson *et al.* 1991) and in the designation of Green Bay as an Area of Concern. Sediment resuspension is the single greatest factor influencing PCB export to lower Green Bay (Velleux and Endicott, 1994). When the sediments are intensely recycled as a result of high flow resuspension events, PCBs are released from the estimated in-place inventory of 30,000 kg in the lower Fox River. Under low and moderate flow conditions, resuspension is minimal, and PCB concentrations in the river decline to relatively low levels (Velleux and Endicott 1994).

Concentrations of PCBs, DDE, dieldrin and mirex were 75%, 87%, 80% and 82% lower, respectively, in pooled livers of gulls collected in 1991 at this site than in the 1983 collection. The concentration of PCBs and DDE in pooled livers of gulls from this site decreased by 7% and 19%, respectively, between 1983 and 1985, and by 83% and 84% between 1985 and 1991. The calculated linear (average) rate of apparent decrease in contaminant burden between 1985 and 1991 in lower Green Bay was greater than that measured in any other colony. In 1991, the DDE and mirex concentrations in pooled livers of herring gulls from lower Green Bay were the lowest of 11 Great Lakes sites sampled, while their dieldrin concentration was the second highest. River flow at Rapid Croche, as measured by the U.S. Geological Survey, differed significantly among our sampling intervals. The mean monthly flow for the 70 months between our 1985 and 1991 samples was 3725 cfs, and only 43% of the months exceeded the median flow of 4200 cfs. This is markedly lower than for the interval between our 1983 and 1985 samples, in which 82% of the 23 months exceeded median flow (mean = 6193 cfs). Thirty-five percent of the 40 months preceeding our 1983 sample exceeded the median flow (mean=4730 cfs). These temporal differences in river flow, and therefore PCB transport, are consistent with changes in PCB concentrations in herring gull livers.

Our findings are very similar to those observed for Forster's terns *(Sterna forsteri)* nesting in lower Green Bay. In 1983, Kubiak *et al.* (1989) associated the reproductive impairment of this species with dioxin-like activity of co-planar PCB congeners. However, when studied on Renard (= Kidney) Island in 1988 (Harris *et al.* 1993), reproductive success had improved and the PCB content of eggs decreased by 67%. Harris *et al.* (1993) attributed this improvement to low river flows. The calculated linear (average) decrease of 13.4% per year observed in PCB content of eggs of migratory Forster's terns is remarkably similar to the 12.3% calculated for livers of resident herring gulls, confirming the importance of local PCB contamination.

Despite the marked decrease in hepatic total PCB concentration between 1985 and 1991, the moderate hepatic concentrations of highly carboxylated porphyrins of gulls from this site have only declined ($P$>0.05) 15% over this period and remain among the highest of the 11 sites sampled in the 1990s. We believe that the porphyrinogenic potency of bioavailable PCBs at this site is very high (Fox *et al.* in prep). Relative thyroid mass has decreased ($P<0.01$) by 20% between 1983 and 1991 (from moderate to slight enlargement). Concentrations of retinyl palmitate in the liver of these gulls have increased ($P>0.05$) by 104% between 1983 and 1991 (from severe to moderate depletion) at a calculated linear (average) rate of 42.7% per year 1983-1985, and 1.6% per year between 1985 and 1991.

### 4.4. CHANNEL SHELTER ISLAND IN SAGINAW BAY

Saginaw Bay serves as a sink for particle-associated hydrophobic pollutants originating from a 22,500 km$^2$ watershed (Jaffe *et al.* 1985) contaminated with PCBs and other persistent lipophilic toxic chemicals from the automotive, chemical and agricultural industries. Recent studies have concluded that the vast majority of loadings of contaminants to the inner bay occur during a small number of climatic events which mobilize soils from run-off over terrestrial surfaces and resuspend in-stream sediments from upstream (Moll *et al.* 1995, Verbrugge *et al.* 1995). The Saginaw Bay ecosystem is one of the most polluted of the 43 Areas of Concern designated for remediation by the United States and Canadian governments.

Concentrations of PCBs, dieldrin and mirex were 30%, 60%, 16% lower, respectively, and DDE 27% higher in pooled livers of gulls collected in 1991 at this site than in the 1982 collection. Although PCB and DDE concentrations in pooled gull livers decreased by 25% and 20% , respectively, between 1982 and 1985, PCBs decreased by 6% while DDE increased by 58% between 1985 and 1991. The hepatic concentration of PCBs in gulls sampled in Saginaw Bay was the highest of the 11 Great Lakes colonies examined in 1991. The increase in DDE concentration between 1985 and 1991 may reflect the transport and resuspension of sediments and contaminants released by a 100 year record flood event in September of 1986 (Ludwig *et al.* 1993). This flood event removed as much as 200 cm of sediments from locations near and immediately downstream from various highly PCB-contaminated facilities along the river margins. In the five years that followed, the reproduction of Caspian terns *(Sterna caspia)* collapsed and then slowly recovered. Reproductive abnormalities included embryonic mortality, teratogensis and wasting syndrome (Ludwig *et al.* 1993).

Highly carboxylated porphyrin concentrations in gulls from this colony in 1982 and 1991 were identical and moderately elevated, the highest of the 11 Great Lakes colonies sampled in the 1990s. Relative thyroid mass only decreased ($P>0.05$) by 14.2% in these gulls between 1982 and 1991, and the thyroids remain slightly enlarged. Concentrations of retinyl palmitate in the livers of these gulls increased by 70%, 1982-1985, but decreased by 42% between 1985 and 1991, resulting in no net change. Although retinyl palmitate concentrations are still moderately depleted, it is noteworthy that concentrations measured in gulls from this colony were second only to Granite Island among Great Lakes colonies. During the period of this study, the waters surrounding this colony were very eutrophic and such conditions are likely to increase the bioavailability of retinoids to the food chain.

### 4.5. FIGHTING ISLAND IN DETROIT RIVER

As the most industrialized of the Great Lakes connecting channels, the Detroit River is an Area of Concern designated by the United States and Canadian governments. The Detroit River is the largest tributary to Lake Erie; 77% of its chemical loading remains within the shallow western basin (Koslowski *et al.* 1994).

Concentrations of PCBs, DDE, dieldrin and mirex were 70%, 60% 67% and 74% lower, respectively, in pooled livers of gulls collected in 1991 at this site than in the 1985 collection. In 1991, the thyroids of gulls from this colony were slightly enlarged. Despite the marked decrease in PCB concentrations, the slightly elevated liver concentrations of highly carboxylated porphyrins have increased by 10% ($P>0.05$) over this period, possibly suggesting that a contaminant other than PCBs is contributing, or that the porphyrinogenic potency of the bioavailable PCBs is increasing. The concentrations of retinyl palmitate stored in the liver decreased ($P>0.05$) by 17% between 1985 and 1991 and are severely depleted. The dietary availability of vitamin A and its precursors to herring gulls in the Detroit River region may have been reduced as a result of alterations to phytoplankton densities brought about by the introduction of the zebra mussel *(Dreissena polymorpha)* to Lake St. Clair in 1986 (Griffiths *et al.* 1991), but this explanation is speculative.

### 4.6. MIDDLE SISTER ISLAND, WESTERN BASIN OF LAKE ERIE

Middle Sister Island is located centrally in the western basin of Lake Erie, within easy foraging distance of the Detroit River. As with gulls nesting on Fighting Island (Ewins *et al.* 1994), gulls nesting on this colony feed on a wide variety of foods.

Concentrations of PCBs, DDE, dieldrin and mirex were 57%, 59% 68% and 77% lower, respectively, in pooled livers of gulls collected in 1993 at this site than in the 1980 collection. The livers of gulls collected in the 1990s from Middle Sister Island and Fighting Island contained almost identical concentrations of PCBs, DDE, dieldrin and mirex, suggesting that gulls from these colonies feed in the same food web. The slightly elevated highly carboxylated porphyrin concentrations have not changed ($P>0.05$) between 1980 and 1993. The relative thyroid mass has decreased ($P<0.001$) by 41%, from moderate to slight enlargement. However, the relative thyroid mass of gulls from this colony was the second largest of the 11 Great Lakes colonies sampled in the 1990s. Stored retinyl palmitate concentrations have increased but are still very severely depleted and one of eight individuals in 1993 had levels below the 1.2 $\mu$g/g detection limit.

### 4.7. Middle Island in Western Basin of Lake Erie

Middle Island is located in the eastern extreme of the western basin, close to the larger, deeper and less productive central basin of Lake Erie and therefore less influenced by the Detroit River. Gulls nesting on this colony fish very locally and scavenge offal from commercial fishing boats. Their diet contains more fish than any other Great Lakes colony examined (Ewins *et al.* 1994)

Unlike gulls sampled on Middle Sister Island, the concentrations of PCBs, DDE and mirex in gulls collected on Middle Island were similar in 1991 and 1983. In contrast, dieldrin concentrations were 58% lower in 1991 than in 1983. In 1991, liver PCB concentrations in gulls from this colony were the highest of the 11 Great Lakes colonies sampled. The slightly elevated concentrations of highly carboxylated porphyrins in livers of gulls from this site increased ($P$<0.001) 150% between 1983 and 1991. This increase is not reflected in an increase in the concentration of total PCBs in the pooled livers and may reflect an increase in the porphyrinogenicity of the bioavailable PCBs or the influence of another unidentified porphyrinogenic contaminant. The relative thyroid mass decreased ($P$<0.001) by 29% between 1983 and 1991 (from moderate to slight enlargement). Although hepatic retinyl palmitate concentrations have increased ($P$<0.01) by at least 216%, between 1983 and 1991, they are still very severely depleted with four of 20 individuals below the 1.2 $\mu$g/g detection limit.

### 4.8. Hamilton Harbour in Western Lake Ontario

This shallow natural embayment is part of a watershed composed of complex urban, industrial and agricultural lands. Two large, coke-burning steel mills use a very large quantity of water for cooling purposes and return it to the harbour. Ling *et al.* (1993) applied a QWASI fugacity/aquivalence model and concluded that water quality in this Area of Concern is primarily controlled by current land-based contaminant inputs, particularly industrial emissions, rather than in-place sediments.

Concentrations of PCBs, DDE, dieldrin and mirex were 36%, 24%, 50% and 39% lower, respectively, in pooled livers of gulls collected in 1991 at this site than in the 1985 collection. Although the porphyrin concentration in the livers of these gulls decreased ($P$>0.05) by 22% between 1985 and 1991, they remain slightly elevated. Although still severely depleted, the concentration of retinyl palmitate stored in their livers increased ($P$>0.05) by 59% between 1985 and 1991. In 1991, thyroids of these gulls were only slightly enlarged. These improvements may reflect the considerable efforts by governments and industry in treating wastes and in preventing pollution in this Area Of Concern over this interval.

### 4.9. SCOTCH BONNET ISLAND IN EASTERN LAKE ONTARIO

In terms of diversity and concentrations of persistent toxic substances, Lake Ontario is recognised to be the most contaminated of the Great Lakes. It receives inflow from the rest of the Great Lakes basin plus the largest volume of persistent toxic substances in urban runoff in Canada. The Niagara River contributes 90% of the water to the lake and is the major source of contaminated sediments (Strachan and Edwards 1984). Since the late 1800s, the chemical industry on the United States side of the Niagara River has been attracted to this location by cheap electricity from the power generation at Niagara Falls, and has discharged toxic substances (i) directly to the river in massive quanitites of wastewater, (ii) in a large but unknown quantity via municipal sewage treatment plants, and (iii) via groundwater contaminated with leachate from hundreds of landfill sites.

Scotch Bonnet Island is the only study site where gulls were sampled in the mid-1970s. The liver contaminant concentrations, severity of porphyria and goitre, and abnormality of vitamin A storage measured in these birds were greater than those from any collection made 1980-1993 from any site. Concentrations of PCBs, DDE, dieldrin and mirex were 72%, 49%, 86% and 69% lower, respectively, in pooled livers of gulls collected in 1991 at this site than in the 1974 collection. Of the 11 Great Lakes colonies sampled in the 1990s, the pooled livers of gulls from Scotch Bonnet Island had the highest concentrations of DDE and mirex. Over the same 17 year period, the liver concentrations of highly carboxylated porphyrins have decreased ($P$<0.001) by 76%, from moderate to slighly elevated. Concentrations of retinyl palmitate in the liver have increased by at least 100% ($P$<0.001), but remain severely depleted. The relative thyroid mass has decreased ($P$<0.001) by 51% (from moderate to slight enlargement). The thyroids of gulls from this colony were the least enlarged of the 11 Great Lakes colonies sampled in the 1990s.

## 5. General Discussion and Conclusions

The herring gull has proven an extremely useful indicator species with which to (i) identify chemicals and their sources (Norstrom *et al.* 1980, Hebert *et al.* 1994), (ii) measure spatial and temporal trends in contaminant concentrations (Mineau *et al.*

1984, Weseloh *et al.* 1990, Hebert *et al.* 1994, Stow 1995), (iii) identify gross pathological manifestations of toxicant-induced injury (Gilbertson and Fox 1977, Moccia *et al.* 1986, Gilbertson *et al.* 1991), and (iv) measure biochemical and physiological impairments, the precursors of disease (Fox *et al.* 1988, reviewed in Fox 1993, Grasman *et al.* 1996, this issue).

During the period of this study, concentrations of PCBs, dieldrin and mirex in pooled livers of adult herring gulls decreased at 7 of 8 sites, whereas DDE concentrations only decreased at 6 sites. Middle Island in Lake Erie was the most frequent exception. It was not possible to determine the statistical significances of the temporal changes in contaminant concentrations we measured, but they are consistent with the findings of others. The biological significance of these changes is reflected in the changes in the degree of impairment observed. Stow (1995) found that the concentrations of PCBs in herring gulls eggs collected at various Great Lakes colonies between 1978 and 1992 have decreased exponentially but, with the exception of colonies in Lake Erie, are no longer perceptibly decreasing. According to Hebert *et al.* (1994), concentrations of various polychlorinated dibenzo-*p*-dioxin congeners in pooled herring gull eggs collected from Great Lakes colonies decreased between 1981 and 1984 but remained relatively stable from 1984 to 1991. Suns *et al.* (1993) found significant reductions over time of PCBs and total DDT in young-of-the-year spottail shiners *(Notropis hudsonius)* at 44% and 36%, respectively, of the 36 Great Lakes nearshore sites sampled between 1975 and 1990, and mirex concentrations in samples collected in the late 1980s were generally lower than samples from the 1970s. Exponential curves provided the best fit for 25 of the 29 spottail shiner data sets where significant correlations with time were found. DDE concentrations in lake trout *(Salvelinus namaycush)* from Lake Ontario decreased rapidly from 1977 to 1980 but then remained relatively constant between 1980 and 1988, while PCB concentrations gradually declined with a half-life of roughly 10 years (Borgmann and Whittle 1991).

Suns *et al.* (1993) found considerable variability among collections of spottail shiners within each waterbody suggesting site-specific differences in PCB bioavailability and concluded that "to meet the requirements of the Great Lakes Water Quality Agreement, further containment of watershed and other point-source inputs of PCBs is required." Our data for lower Green Bay, Saginaw Bay and Middle Island strongly support this conclusion. As with other recent studies in Saginaw Bay (Ludwig *et al.* 1993) and Green Bay (Harris *et al.* 1993), our findings suggest that disturbance of contaminated in-place sediments by episodic meterological events present an on-going threat to breeding fish-eating birds.

From the data on herring gull eggs collected from Middle Island, Mineau *et al.* (1984) drew attention to the long half-life of PCBs in the western basin of Lake Erie,

1974-1979, despite known rapid sedimentation and the lake's short flushing time. Our data suggest that concentrations of PCBs, DDE, dieldrin and mirex have not decreased in the livers of gulls from this location between 1983 and 1991, and the 1991 concentration of PCBs in gulls from this site was markedly higher than the samples from Middle Sister Island, Fighting Island (Detroit River), and exceeded that in livers from Saginaw Bay. More importantly, highly toxic non-*ortho* congeners comprised 5.4% of the planar PCBs (congeners 37, 77, 105, 118, 126 and 169) in the livers of the Middle Island gulls, but only 0.3% of the planar PCBs in the sample from the Detroit River (Fox *et al.*, in prep). Middle Island is equidistant from the mouth of the Detroit and Maumee Rivers. Tributaries other than the Detroit River contribute 76% of the total sediments deposited in the western basin (Oliver and Bourbonniere 1985), and in recent years, the Maumee River has contributed over one million tonnes of suspended sediments annually (Roberts and Veley 1995). Water collected off the mouth of the Maumee River in April 1994, contained 15.8 ng PCBs/L; the highest concentration measured for 9 open water sites in Lake Erie (L'Italien and Struger 1995). However, the highest PCB concentrations measured in the tissues of the burrowing mayfly *(Hexagenia)* collected from 13 locations in the Detroit River and the western basin in 1994 were measured in the collection from Monroe, at the mouth of the Raisin River and not in the collections from Maumee State Park or the Detroit River (Corkum *et al.* 1995). It is noteworthy that (i) bald eagle *(Haliaeetus leucocephalus)* nestlings with crossed bills were found in a nest on the Woodtick Peninsula north of Toledo in 1993 and in a nest on the Raisin River in Monroe in 1993 and 1995 (Bowerman *et al.* 1994, D.A. Best, United States Fish and Wildlife Service, East Lansing, MI, pers. comm.), and (ii) the T-cell-mediated immune response of prefledged herring gull chicks in the Monroe colony, at the mouth of the Raisin River, was severely depressed in 1992 (Grasman *et al.* 1996). If the relative importance of the Raisin and/or Maumee Rivers continues to increase as a source of PCBs in the western basin, there may be further enrichment of the toxic congeners which may be accompanied by additional manifestations of toxicity in fish-eating birds.

In the 17 years between 1974 and 1991, there have been marked decreases in the concentrations of PCBs (72%), DDE (49%) and dieldrin (86%) in pooled livers of herring gulls sampled on Scotch Bonnet Island in eastern Lake Ontario. Similarly, between 1975 and 1990 there were significant decreases in the concentrations of PCBs (81%), DDE (86%) and dieldrin (31%) in eggs of red-breasted mergansers *(Mergus serrator)* nesting on islands in northwestern Lake Michigan near the mouth of Green Bay (White and Cromartie 1977, Heinz *et al.* 1994). The marked differences in the degree of decline of DDE and dieldrin between these locations over a comparable time period suggests substantial continuing inputs of dieldrin to Lake

Michigan and of DDE to Lake Ontario. Agricultural and non-health uses of DDT were suspended in the United States in 1972, and soon thereafter in Canada. Of the 11 Great Lakes colonies sampled in the the 1990s, livers of gulls from Scotch Bonnet Island in Lake Ontario had the highest concentrations of DDE. Suns *et al.* (1993) found the highest concentrations of DDE in spottail shiners from the Niagara River, the major source of contaminated sediments to Lake Ontario. Hebert *et al.* (1994) reported DDE concentrations of 57 and 486 ppm in two orchard soils collected near Grimsby, Ontario in the northern Niagara Peninsula in 1989. When farm-raised mallards *(Anas platyrhynchos)* were released onto a sewage lagoon on the south shore of Lake Ontario near Grimsby in 1991, their accumulations of DDE were much greater, and of PCBs and mirex much lower, than those of mallards released into Hamilton Harbour over the same period (Gebauer and Weseloh 1993). These observations suggest that DDE-laden run-off from these fruit growing areas continues to reach Lake Ontario via a number of tributaries.

In our pooled samples of liver, PCB and DDE concentrations were significantly correlated ($r_s = 0.55$, $P<0.007$). Both the PCB ($r_s= 0.51$, $P=0.012$) and DDE ($r_s= 0.48$, $P=0.02$) concentrations were significantly correlated with the median highly carboxylated porphyrin concentrations. However, since DDE is not porphyrinogenic, this relationship must reflect the correlation between PCBs and DDE. The median thyroid mass of 17 collections was positively correlated with hepatic PCB ($r_s = 0.74$, $P<0.001$) but markedly less with DDE ($r_s = 0.42$, $P=0.09$) concentrations. The median hepatic retinyl palmitate concentration was negatively correlated with PCB ($r_s = -0.65$, $P<0.001$) but not DDE ($r_s = -0.15$, $P=0.49$) concentrations. On this basis, we conclude that PCBs, rather than DDE, were responsible for the bulk of the physiological effects we measured.

In the early 1990s, the thyroids of gulls from all the Great Lakes sites sampled were slightly enlarged, but the severity of this enlargement (goitre) has decreased at all sites during the period of this study, consistent with the decline in organochlorine residues. The incidence and severity of microfollicular hyperplasia previously seen in thyroids of Great Lakes herring gulls (Moccia *et al.* 1986) has not yet been assessed in the 1991 collections. PCBs, DDE, dieldrin and mirex are known to affect the avian thyroid (reviewed in Moccia *et al.* 1986).

In the early 1990s, livers of gulls from all the Great Lakes sites sampled contained slightly or moderately elevated concentrations of highly carboxylated porphyrins. Overall, during the period of this study, the concentrations of porphyrins increased significantly in collections from Middle and Granite Islands. The concentrations of porphyrins did not change in gulls from Saginaw Bay, or Fighting and Middle Sister Islands, but decreased significantly in the collections from Scotch Bonnet Island. Recent work in our laboratory has shown a high correlation between the highly carboxylated porphyrin concentration in a gull's liver and PCB concentrations, particularly the mono-*ortho* substituted PCB congeners 105 and 118. Other organochlorine contaminants were not correlated with the liver porphyrins and together these findings indicate a high degree of specificity in the compounds responsible for the observed porphyrinogenicity. Similarly, there was a high correlation between the highly carboxylated porphyrin concentration in the gull's liver and the ability of an extract from that liver to induce porphyrin accumulation in chick embryo hepatocytes *in vitro* (Kennedy *et al.* 1998). The spatial and temporal variations in porphyrin accumulation we observed suggest that PCBs may not be the only contaminant contributing to this effect in Great Lakes herring gulls or that the porphyrinogenic potency of the bioavailable PCB congener mix differs among sites and over time. Candidate chemicals of unknown porphyrinogenicity include toxaphene, the heavily used lampricide 3-trifluoromethyl-4-nitrophenol (TFM) and particularly the substituted diphenyl ethers and dibenzo-*p*-dioxins recently identified as contaminants in the formulated product (Hewitt *et al.* 1996) and/or its degradation products (Munkittrick *et al.* 1994).

Fish-eating birds obtain their essential requirement of vitamin A from forage fish and invertebrates, who obtain it from certain plankton (Norman and Greenwood 1963), and store it in their livers predominantly as retinyl palmitate. The concentration of retinyl palmitate in the liver of gulls reflects (i) dietary intake which is a function of primary productivity plus the availability and lipid content of forage fish in the vicinity of the colony in which they reside, and (ii) the ability of their liver to trap and store the bioavailable vitamin. The herring gull is an opportunistic piscivore, and the amount and species composition of fish consumed varies with location and season (Fox *et al.* 1990, Ewins *et al.* 1994). This opportunistic nature of herring gull diet will be reflected in variability in vitamin intake and contaminant exposure. It is well-established that exposure to a variety of xenobiotic chemicals including PCBs, DDE, dieldrin, TCDD, and PAHs can deregulate vitamin A metabolism and deplete vitamin A stores (reviewed in Zile 1992, Spear *et al.* 1986).

The concentration of retinyl palmitate increased in the livers of gulls from 5 of 8 Great Lakes sites over the period of this study, but was significant only in gulls from Granite and Scotch Bonnet Islands. Concentrations of retinyl palmitate remained very similar in collections from Saginaw Bay and Middle Sister Island. The availability of phosphorus limits primary productivity in most large lakes but not oceans (Russell-Hunter 1970, Simpson *et al.* 1981). With the exception of the oligotrophic/mesotrophic waters surrounding Granite Island, the waters surrounding the Great Lakes colonies sampled in this study were eutrophic or mesotrophic throughout much of the 1980s. Although the bioavailability of vitamin A for gulls nesting on Granite Island may have been lower than for gulls nesting at the other

Great Lakes colonies, hepatic retinyl palmitate concentrations in some individuals from Granite Island have overlapped those in gulls from the marine environment. This suggests that the bioavailability of vitamin A in their diet is adequate. Despite the eutrophic/mesotrophic state of the Detroit River and the western basin of Lake Erie, retinyl palmitate concentrations in the livers of gulls nesting on Middle Island, Middle Sister Island, and Fighting Island have remained very low, relative to Granite Island, throughout this period. The apparent inability of these gulls to store vitamin A may reflect their high contaminant burdens or reduced phytoplankton densities resulting from remedial efforts to eliminate eutrophication which have greatly reduced phosphorus loadings in the western basin of Lake Erie, lower Green Bay and Lake Ontario over the 1980s (Nicholls and Hopkins 1993, Harris and Holden 1990, Lake Ontario Committee 1993). Phytoplankton densities have been further reduced by the accidental introduction of Dreissenid mussels. The decline in phytoplankton densities in the 1980s has contributed to the decline in the biomass and lipid content of alewife in Lake Ontario (Lake Ontario Committee 1993) and in smelt in Lake Erie (Whittle *et al.* 1995), The significantly greater occurrence of nondetectable concentrations of retinyl palmitate in the livers of gulls from Lake Erie colonies in 1991, relative to all other Great Lakes colonies combined, is consistent with these recent limitations on dietary availability of vitamin A. The ability of the highly contaminated gulls nesting in Saginaw Bay and Green Bay to store more vitamin A than those nesting in Lake Erie or Lake Ontario colonies suggests differences in primary productivity and/or quantity, nutritional quality and species of fish consumed.

In summary, our data suggest that the bioavailability of PCBs, DDE, dieldrin and mirex has decreased at most, but not all of the sites studied, and that, particularly for PCBs and DDE, most of the improvement occurred before 1985. At some sites, there have been corresponding improvements in one or more physiological measures. At all sites there has been a decrease in the severity of goitre, but thyroids are still slightly enlarged relative to gulls from the Atlantic coast. In contrast, it would appear that in the early 1990s, concentrations of PCBs and possibly other persistent toxic substances in the food of herring gulls were still sufficient to cause biochemical impairments at these Great Lakes sites, and therefore discharges have not yet been virtually eliminated. Giesy *et al.* (1994) also concluded that "all of the populations of colonial fish-eating waterbirds on the Great Lakes are displaying symptoms of exposure to chlorinated chemicals at the biochemical level". Finally, herring gull chicks at the more contaminated sites suffer from altered immune function correlated with exposures to TCDD-equivalents (Grasman *et al.* 1996) indicating that, in addition to such biomarker changes as mild goitre, slight to moderate elevations in highly carboxylated porphyrins and moderate to very severe depletion of stored vitamin A in adult birds, there are significant pathobiological lesions in juvenile birds in the Great Lakes basin.

## Acknowledgments

We are grateful to the many individuals who have assisted with the collection and preservation of gull tissues for this study over the years. In particular, we wish to thank D.V. Weseloh, T. Erdman, J.P. Ludwig, T. Kubiak and the Bowdoin Scientific Station for logistic support, and S. Kennedy, K. Williams and G. Sans-Cartier for biochemical measurements. Flow data for the Fox River at Rapid Croche was provided by H. Harris and H. Garn. Constructive comments on an earlier version of this manuscript were provided by P. Ewins, A. Niimi, R. Norstrom and two anonymous reviewers. Support was provided for this work under Environment Canada's Great Lakes Action Plan.

## References

Borgmann, U. and Whittle, D.M.: 1991, 'Contaminant concentration trends in Lake Ontario lake trout *(Salvelinus namaycush)*: 1977-1988', *J. Great Lakes Res.* **17**,368-381.

Bowerman, W.W., IV, Kubiak, T.J., Holt, J.B.Jr., Evans, D.L., Eckstein, R.G., Sindelar, C.R., Best, D.A. and Kozie, K.D.: 1994, 'Observed abnormalities in mandibles of nesting bald eagles *Haliaeetus leucocephalus*', *Bull. Environ. Contam. Toxicol.* **53**,450-457.

Corkum, L.D., Ciborowski, J.J.H. and Lazar, R.: 1995, 'Spatial distributions and contaminant burdens in *Hexagenia* (Ephemeroptera) populations in western Lake Erie and the Detroit River', Abstracts of the 38th conference of the International Association for Great Lakes Research, East Lansing, MI, May 28-June 1, 1995.

Eisenreich, S.J., Looney, B.B. and Thornton, J.D.: 1981, 'Airborne organic contaminants in the Great Lakes ecosystem ', *Environ. Sci. Technol.* **15**,30-38.

Ewins, P.J., Weseloh, D.V., Groom J.H., Dobos, R.Z. and Mineau P.: 1994, 'The diet of herring gulls *(Larus argentatus)* during winter and early spring on the lower Great Lakes', In; *Aquatic Birds in the Trophic Web of Lakes*, Kerekes, J.J. (ed.) Hydrobiologia **279/280**,39-55.

Fox, G.A.: 1993, 'What have biomarkers told us about the effects of contaminants on the health of fish-eating birds in the Great Lakes? The theory and a literature review', *J. Great Lakes Res.* **19**,722-736.

Fox, G.A., Allan, L.J., Weseloh, D.V. and Mineau, P.: 1990, 'The diet of herring gulls during the nesting period in the Canadian waters of the Great Lakes', *Can. J. Zool.* **68**,1075-1085.

Fox, G.A., Kennedy S.W., Norstrom R.J. and Wigfield, D.C.: 1988, 'Porphyria in herring gulls: a biochemical response to chemical contamination of Great Lakes food chains', *Environ. Toxicol. Chem.* **7**,831-839.

Gebauer, M.B. and Weseloh, D.V.: 1993, 'Accumulation of organic contaminants in sentinel mallards utilizing confined disposal facilities at Hamilton Harbour, Lake Ontario, Canada', *Arch. Environ. Contam. Toxicol.* **25**,234-243.

Giesy, J.P., Ludwig, J.P. and Tillitt, D.E.: 1994, 'Deformities in birds of the Great Lakes region: Assigning causality', *Environ. Sci. Technol.* **28**,128-135.

Gilbertson, M.: 1974, 'Pollutants in breeding herring gulls in the lower Great Lakes', *Can. Field-Nat.* **88**,273-280.

Gilbertson, M. and Fox, G.A.: 1977, 'Pollutant-associated embryonic mortality of Great Lakes herring gulls', *Environ. Pollut.* **12**,211-216.

Gilbertson, M., Kubiak, T., Ludwig, J. and Fox, G.: 1991, 'Great Lakes embryo mortality, edema and deformities syndrome (GLEMEDS) in colonial fish-eating birds: similarity to chick edema disease', *J. Toxicol. Environ. Health* **33**, 455-520.

Glantz, S.A.: 1992, *Primer of Biostatistics*. 3rd Ed., McGraw-Hill, Inc., New York.

Grasman, K.A., Fox, G.A., Scanlon, P.F. and Ludwig, J.P.: 1996, 'Organochlorine-associated immunosuppression in prefledgling Caspian terns and herring gulls from the Great Lakes: An ecoepidemiological study', *Environ. Health Perspect.* **104(Suppl. 4)**,829-842.

Griffiths, R.W., Schloesser, D.W., Leach, J.H. and Kovalak, W.P.: 1991, 'Distribution and dispersal of the zebra mussel *(Dreissena polymorpha)* in the Great Lakes region', *Can. J. Fish. Aquat. Sci.* **48**,1381-1388.

Harris, H.J., Erdman, T.C., Ankley, G.T. and Lodge, K.B.: 1993, 'Measures of reproductive success and polychlorinated biphenyl residues in eggs and chicks of Forster's terns on Green Bay, Lake Michigan, Wisconsin - 1988', *Arch. Environ. Contam. Toxicol.* **25**,304-314.

Harris, H.J. and Holden, C.: 1990, The State of the Bay, A Capsule Review. The Institute of Land and Water Studies, The University of Wisconsin - Green Bay, Green Bay.

Hebert, C.E., Norstrom, R.J., Simon, M., Braune, B.M., Weseloh, D.V. and Macdonald, C.R.: 1994, 'Temporal trends and sources of PCDDs and PCDFs in the Great Lakes: Herring gull egg monitoring, 1981-1991', *Environ. Sci. Technol.* **28**,1268-1277.

Hebert, C.E., Weseloh, D.V., Kot, L. and Glooschenko, V.: 1994, 'Organochlorine contaminants in a terrestrial foodweb on the Niagara Peninsula, Ontario, Canada 1987-89', *Arch. Environ. Contam. Toxicol.* **26**,356-366.

Heinz, G.H., Miller, D.S., Ebert, B.J. and Stromborg, K.L.: 1994, 'Declines in organochlorines in eggs of red-breasted mergansers from Lake Michigan, 1977-1978 versus 1990', *Environ. Monit. Assess.* **33**,175-182.

Hermanson, M.H., Christensen E.R., Buser, D.J., and Chen, L-M.: 1991, 'Polychlorinated biphenyls in dated sediment cores from Green Bay and Lake Michigan', *J. Great Lakes Res.* **17**,94-108.

Hewitt, M., Servos, M., Van Der Kraak, G., Solomon, K., and Munkittrick, K.R.: 1996, 'Characterization of bioactive 3-trifluoromethyl-4-nitrophenol (TFM) lampricide formulation impurities'. *Organohalogen Compounds* **29**,434-438.

Jaffe, R., Stemmler, E.A., Eitzer, B.D. and Hites, R.A.: 1985, 'Anthropogenic, polyhalogenated, organic compounds in sedentary fish from Lake Huron and Lake Superior tributaries and embayments', *J. Great Lakes Res.* **11**,156-162.

Kennedy, S.W., Fox, G.A., Trudeau, S., Bastien, L.J. and Jones S.P.: 1998, 'Highly carboxylated porphyrin concentrations: a biochemical marker of PCB exposure in herring gulls'. *Mar. Environ. Res.* (in press).

Kennedy, S.W., Wigfield, D.C. and Fox, G.A.: 1986, 'Tissue porphyrin determination by high-speed high-performance liquid chromatography', *Anal. Biochem.* **157**,1-7.

Kennedy, S.W. and James C.A.: 1993, 'Improved method to extract and concentrate porphyrins from liver tissue for analysis by high-performance liquid chromatography', *J. Chromatog., Biomed. Appl.* **619**,127-132.

Koslowski, S.E., Metcalfe, C.D., Lazar, R., and Haffner, G.D.: 1994, 'The distribution of 42 PCBs, including three coplanar congeners, in the food web of the western basin of Lake Erie', *J. Great Lakes Res.* **20**,260-270.

Kubiak, T.J., Harris, H.J., Smith, L.M., Schwartz, T.R., Stalling, D.L., Trick, J.A., Sileo, L., Docherty, D.E. and Erdman, T.C.: 1989, 'Microcontaminants and reproductive impairment of the Forster's tern on Green Bay, Lake Michigan - 1983', *Arch. Environ. Contam. Toxicol.* **18**,706-727.

Lake Ontario Committee: 1993, *Signs of Change in the Lake Ontario Ecosystem.* Ontario Minsitry of Natural Resources and the New York Department of Environmental Conservation.

Ling, H., Diamond, M. and Mackay, D.: 1993, 'Application of the QWASI fugacity/aquivalence model to assessing sources and fate of contaminants in Hamilton Harbour', *J. Great Lakes Res.* **19**,582-602.

L'Italien S. and Struger, J.: 1995, 'Pesticide concentrations in water and suspended sediment in Lake Erie'. Abstracts of the 38th Concerence of the International Association for Great Lakes Research, East Lansing, MI, May 20 -June1, 1995.

Ludwig, J.P., Auman, H.J., Kurita, H., Ludwig, M.E., Campbell, L.M., Giesy, J.P., Tillitt, D.E., Jones, P., Yamashita, N., Tanabe, S. and Tatsukawa, R.: 1993, 'Caspian tern reproduction in the Saginaw Bay ecosystem following a 100-year flood event', *J. Great Lakes Res.* **19**,96-108.

Mineau, P., Fox, G.A., Norstrom, R.J., Weseloh, D.V., Hallett, D.J. and Ellenton, J.A.: 1984, 'Using the herring gull to monitor levels and effects of organochlorine contamination in the Canadian Great Lakes', in; *Toxic Contaminants in the Great Lakes*, J.O. Nriagu and M.S. Simmons (eds.), J. Wiley and Sons, New York. pp. 425-452.

Moccia, R.D., Fox, G.A. and Britton, A.: 1986, 'A quantitative assessment of thyroid histopathology of herring gulls *(Larus argentatus)* from the Great Lakes and a hypothesis on the causal role of environmental contaminants', *J. Wildlife Dis.* **22**,60-70.

Moll, R.A., Jude, D., Rossmann, R., Kantak, G.V., Barres, J., DeBoe, S., Giesy, J. and Tuchman, M.: 1995, 'Movement and loadings of inorganic contaminants through the lower Saginaw River', *J. Great Lakes Res.* **21**,17-24.

Munkittrick, K.R., Servos, M.R., Parrott, J.L., Martin V., Carey, J.H., Flett, P.A. and Van Der Kraak, G.J.: 1994, 'Identification of lampricide formulations as a potent inducer of MFO activity in fish', *J. Great Lakes Res.* **20**,355-365.

Nicholls, K.H. and Hopkins, G.J.: 1993, 'Recent changes in Lake Erie (north shore) phytoplankton: cumulative impacts of phosphorus loading reductions and the zebra mussel introduction', *J. Great Lakes Res.* **19**,637-647.

Norman, M. and Greenwood, P.H.: 1963, *A History of Fishes*. Ernest Brown Ltd, London, UK.

Norstrom, R.J., Hallet, D.J., Onuska, F.I. and Comba, M.E.: 1980, 'Mirex and its degradation products in Great Lakes herring gulls', *Environ. Sci. Technol.* **14**,860-866.

Oliver, B.G. and Bourbonniere, R.A.: 1985, 'Chlorinated contaminants in surficial sediments of Lakes Huron, St. Clair, and Erie: Implications regarding sources along the St. Clair and Detroit Rivers', *J. Great Lakes Res.* **11**,366-372.

Peakall, D.B., Norstrom, R.J., Rahimtula, A.D. and Butler, R.D.: 1986, 'Characterization of mixed-function oxidase systems of the nestling herring gull and its implications for bioeffects monitoring', *Environ. Toxicol. Chem.* **5**,379-385.

Roberts, J.W. and Veley, R.J.: 1995, 'Loads, yields, and transport relations of nutrients, chloride, and suspended sediments in three Lake Erie tributaries in

Ohio', Abstracts of the 38th conference of the International Association of Great Lakes Research, East Lansing, MI, May 28-June 1, 1995.

Russell-Hunter, W.D.: 1970, *Aquatic Productivity: An Introduction to Some Basic Aspects of Biological Oceanography and Limnology.* The Macmillan Co., New York.

Simpson, K.L., Katayama, T. and Chichester, C.O.: 1981, 'Carotenoids in fish feeds', in; J.C. Baurenfeind (ed), *Carotenoids as Colorants and Vitamin A Precursors*, Academic Press, New York, pp. 463-538.

Spear, P.A., Moon, T.W. and Peakall, D.B.: 1986, 'Liver retinoid concentrations in natural populations of herring gulls *(Larus argentatus)* contaminated by 2,3,7,8-tetrachlorodibenzo-*p*-dioxin and in ring doves *(Streptopelia risoria)* injected with a dioxin analogue', *Can. J. Zool.* **64**,204-408.

Stow, C.A.: 1995, 'Great Lakes herring gull egg PCB concentrations indicate approximate steady-state conditions', *Environ. Sci. Technol.* **29**,2893-2897.

Strachan, W.M.J. and Edwards, C.J.: 1984, 'Organic pollutants in Lake Ontario', In: J.O. Nriagu and M.S. Simmons (eds.), *Toxic Contaminants in the Great Lakes*, J. Wiley and Sons, New York, pp. 425-452.

Suns, K.R., Hitchin, G.G. and Toner, D.: 1993, 'Spatial and temporal trends of organochlorine contaminants in spottail shiners from selected sites in the Great Lakes (1975-1990)', *J. Great Lakes Res.* **19**,703-714.

Velleux, M.and Endicott, D.: 1994, 'Development of a mass balance model for estimating PCB export from the lower Fox River to Green Bay', *J. Great Lakes Res.* **20**,416-434.

Verbrugge, D.A., Giesy, J.P., Mora, M.A., Williams, L.L., Rossmann, R., Moll, R.A. and Tuchman, M.L.: 1995, 'Concentrations of disolved and particulate polychlorinated biphenyls in water from the Saginaw River, Michigan', *J. Great Lakes Res.* **21**,219-235.

Weseloh, D.V., Mineau, P. and Struger, J.: 1990, 'Geographical distribution of contaminants and productivity measures of herring gulls in the Great Lakes: Lake Erie and connecting channels 1978/9', *Sci. Total Environ.* **91**,141-159.

White, D.H. and Cromartie, E.: 1977, 'Residues of environmental pollutants and shell thinning in merganser eggs', *Wilson Bull.* **89**,532-542.

Whittle, D.M., Kiriluk, R.M., Carswell, A.A. and Keir, M.J.: 1995, 'Recent changes in the dynamics and rate of contaminant accumulation in the Lake Erie aquatic community'. Abstracts of the 38th conference of the International Association for Great Lakes Research, East Lansing, MI, May 28-June 1, 1995.

Zar, J.H.: 1984, *Biostatistical Analysis.* Prentice-Hall Inc., Englewood Cliffs, NJ.

Zile, M.H.: 1992, 'Vitamin A homeostasis endangered by environmental pollutants'. *Proc. Soc. Expt. Biol. Med.* **201**,141-153.

# SPATIAL AND TEMPORAL TRENDS IN ORGANOCHLORINE CONTAMINATION AND BILL DEFORMITIES IN DOUBLE-CRESTED CORMORANTS (*PHALACROCORAX AURITUS*) FROM THE CANADIAN GREAT LAKES

D.P. RYCKMAN[1], D.V. WESELOH[2 ‡], P. HAMR [3], G.A. FOX[4], B. COLLINS[4], P.J. EWINS[5], and R.J. NORSTROM[4]

*[1]729 Eaglemount Cres, Mississauga, Ontario, L5C 1H2, [2] Canadian Wildlife Service, Ontario Region, Environment Canada, 4905 Dufferin St., Downsview, Ontario, M3H 5T4, [3]Appleby College, 540 Lakeshore Rd. West, Oakville, Ontario, L6K 3P1, [4]Canadian Wildlife Service, Environment Canada, National Wildlife Research Centre, Ottawa, Ontario, K1A 0H3, [5]World Wildlife Fund, 90 Eglington Ave. E., Suite 504, Toronto, Ontario, M4P 2Z7.*

**Abstract.** The levels of organochlorine contaminants (OCs) in the eggs of double-crested cormorants (*Phalacrocorax auritus*) from the Canadian Great Lakes, Lake Nipigon and Lake-of-the-Woods were monitored between 1970 and 1995. PCBs and *p,p'*-DDE were present at the highest concentrations. Significant declines in OC concentrations on the Great Lakes were observed over this period for Lake Ontario, Lake Superior, Georgian Bay and North Channel but not Lake Erie where levels remained relatively stable. In the early 1970s, the greatest OC levels were generally observed in cormorant eggs from nesting sites in Georgian Bay and North Channel of Lake Huron. Between 1984 and 1995 mirex and PCB levels were consistently highest in samples from Lakes Ontario and Erie, respectively. Similar levels of PCDDs and PCDFs were observed from all regions of the Canadian Great Lakes in 1989. In general, OC levels in cormorant eggs between 1984-95 were ranked as follows: Lake Erie>Lake Ontario≥Lake Superior>Lake Huron. In 1995, eggshell thickness in Canadian Great Lakes cormorants, ranged from 0.423 to 0.440 mm and was on average only 2.3% thinner than pre-DDT era values. Between 1988 and 1996, 31 cormorant chicks with bill defects were observed at 16 different colonies (21% of all colonies surveyed) in Lakes Ontario and Superior, Georgian Bay and North Channel, and the main body of Lake Huron. No bill deformities were observed at reference sites in northwestern Ontario (Lake Nipigon and Lake-of-the-Woods). For the period 1988-96, the prevalence of bill defects in cormorant chicks (0.0 to 2.8/10,000 chicks) did not differ significantly ($P > 0.05$) among most regions in the Canadian Great Lakes. Georgian Bay was the only region to show a significant decrease in the prevalence of bill defects between the periods 1979-87 and 1988-95.

**Key words:** Double-crested cormorant, *Phalacrocorax auritus*, Great Lakes, organochlorine contaminants, bill deformities, eggshell thickness.

‡ Author to whom correspondence should be addressed. Email: Chip.Weseloh@ec.gc.ca.

## 1. Introduction

Concentrations of organochlorine contaminants (OCs) in eggs of double-crested cormorants (*Phalacrocorax auritus*), hereafter referred to as cormorants, from the Great Lakes region were first monitored in the late 1960s. At that time, cormorant eggs were found to be heavily contaminated with *p,p'*-DDE, PCBs and mercury, (Postupalsky, 1978; Bishop *et al.*, 1992). The cormorant's sensitivity to DDE-induced eggshell thinning (Gress *et al.*,1973; Vermeer and Peakall, 1977), which resulted in crushed and defective eggshells (Postupalsky, 1978; Weseloh *et al.*, 1983; Price and Weseloh, 1986) resulted in low breeding success and serious population declines on the Great Lakes. By the late 1970s, shell thickness had increased in response to declining DDE levels and breeding populations began to increase on the Great Lakes (Scharf and Shugart, 1981; Ludwig, 1984; Price and Weseloh, 1986). By 1991, the numbers of nesting cormorants had increased more than 400 fold to 38,000 pairs from their numbers in the 1970s and had an average annual increase of 29% (Weseloh *et al.*, 1995). Tyson *et al.* (1997) recently evaluated the continent wide status of the double-crested cormorant and determined that during the 1990s, up to 1997, the Great Lakes population had increased at 22% per year and now stood at approximately 93,000 pairs.

Fox *et al.* (1991a) provided the first extensive account of the incidence and prevalence of congenital deformities in cormorant populations within and outside the Great Lakes basin. Although there are published accounts of congenital deformities for at least seven fish-eating bird species nesting on the Great Lakes (Fox *et al.*, 1991b), deformities have been most frequently observed in cormorants (Fox *et al.*, 1991a). The cormorant's susceptibility to bill deformities and the effects of eggshell thinning, coupled with its abundance throughout the Great Lakes, makes it an excellent species for long-term monitoring of developmental toxicants in that aquatic food web.

In this report we provide geographical and temporal trends in OC concentrations in cormorant eggs collected from sites in the Great Lakes Basin between 1970 and 1995. In doing so, we have reviewed both older, published data and new data from recent samples. We also update the status of eggshell thinning and bill deformities, two biological effects associated with contamination, as well as present temporal trends in the incidence and prevalence of bill deformities in young of this species in relation to the data presented by Fox *et al.* (1991a) for the period 1979 to 1987.

## 2. Methods

### 2.1 EGG COLLECTION

Eggs were collected, one per four egg (complete) clutch, during one or more years between 1970 and 1995 from 18 nesting sites in the Canadian Great Lakes (Figure 1, Table 1). For the purpose of statistical analysis, nesting sites were grouped according to lakes or geographic regions of lakes as follows: Lake Ontario, Lake Erie, Georgian Bay (Lake Huron), North Channel (Lake Huron), Lake Superior, Lake Nipigon and Lake-of-the-Woods (Figure 1). Upon collection, eggs were placed in a container of water to confirm their freshness. Eggs falling to the bottom of the container and lying on the long axis were determined not to have been incubated and therefore to have been recently laid. In the early 1970s, inviable, dented or cracked eggs were sometimes included with sample collections. Eggs were placed in a cushioned egg box and stored at 5°C for three to six weeks. They were then weighed, measured, opened into hexane-rinsed glass jars, covered with aluminum foil-lined lids, and frozen at -20°C until analysis.

### 2.2 CHEMICAL ANALYSIS

Between 1970 and 1983, eggs were analyzed by the Ontario Research Foundation (or ORTECH Corporation), Mississauga, Ontario, except for 1971 samples from Lake Erie, collected by the Ontario Ministry of Natural Resources, which were analyzed by the Ontario Department of Agriculture. After 1983, eggs were analyzed at the Canadian Wildlife Service (CWS) laboratory at the National Wildlife Research Centre (NWRC), Hull, Quebec. Eggs were analyzed annually except for 1984 samples which were not analyzed at NWRC until 1991. Eggs collected up to 1981 were analyzed individually, but starting in 1983, chemical analyses were performed on pooled samples; based on a 10-13 egg pool (Turle and Collins, 1992).

Overall, 31 OC compounds were measured in the samples, however not all OCs were screened in every year and/or every sample. Details of the extraction, cleanup and high resolution gas chromatography-electron capture detector analysis of OCs can be found in Norstrom and Won (1985). In early samples (1970s), PCBs were quantified based on Aroclor 1260, which was used as the standard during that period. Beginning in 1984, 42 individual PCB congeners were determined in eggs, and PCBs were also reported as the sum concentration of congeners measured individually in samples, as described in Norstrom *et al.* (1988). Congeners used for "Sum of PCB congeners" were: PCBs 28, 31, 42, 44, 49, 52, 60, 64, 66, 70, 87, 97, 99, 101, 105, 110, 118, 128, 129, 137, 138, 141, 146, 151, 153, 158, 170, 171, 172, 174, 180, 182,

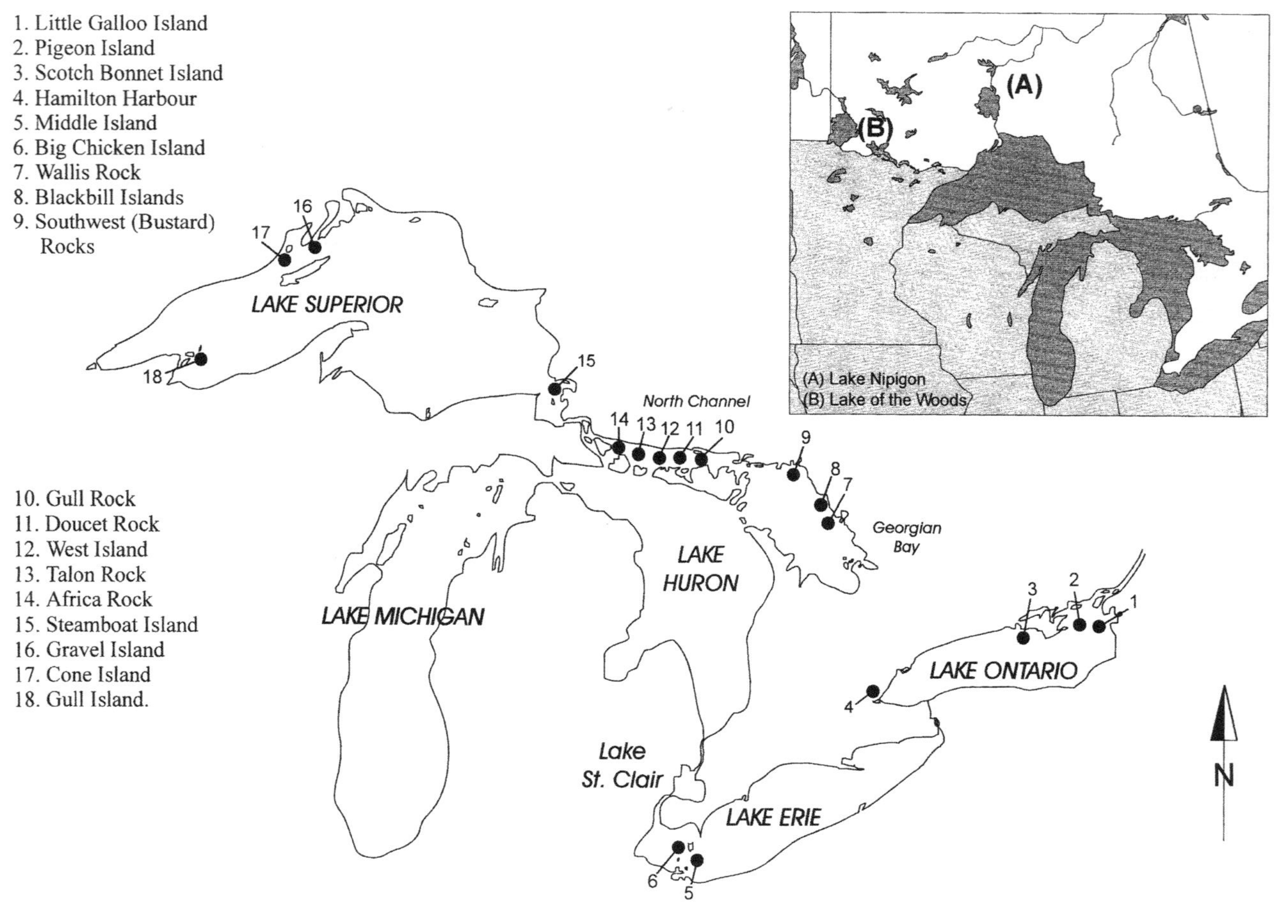


**Figure I.** Location of Great Lakes Colonies from which Eggs were Collected.

183, 185, 194, 195, 200, 201, 203 and 206; in addition to PCBs 74 and 149 in 1984 and 1995; and PCB 47 in 1989; IUPAC numbering system, (Ballschmiter and Zell, 1980). To maintain temporal comparability in our data set, all reference to PCBs in our report are to concentrations expressed as Aroclor 1260, unless otherwise stated. Polychlorinated dioxin (PCDD) and polychlorinated dibenzofuran (PCDF) congeners were determined by gas chromatography-mass spectrometry according to the procedures given in Norstrom *et al.* (1990) and Norstrom and Simon (1991).

Throughout this report, PCBs, DDE, dieldrin, mirex, heptachlor epoxide (HE) and hexachlorobenzene (HCB) are referred to as 'major' contaminants since they were found at the highest concentrations. All contaminants, except PCDDs and PCDFs, are expressed as $\mu$g/g (ppm), wet weight. PCDFs and PCDDs are expressed as pg/g (ppt), wet weight. Detection limits used in the analytical determination of OC contaminants varied with the laboratory and methodology used but the following can be used as a guide: chlorinated benzenes: 0.001 $\mu$g/g; OC pesticides 0.005 $\mu$g/g; polychlorinated biphenyls: 0.01 $\mu$g/g; TCDD/TCDF: 1-2 pg/g; PnCDD/PnCDF: 2-3 pg/g, HxCDD/HxCDF: 3-4 pg/g; HpCDD: 6-8 pg/g; OCDD: 8-12 pg/g.

Contaminant data from the early 1970s, appearing in Postupalsky (1978), were re-analyzed statistically and included here to complete the data set to give entire geographical and temporal patterns. Additionally, all but the most recent data have appeared previously in Weseloh *et al.* (1983); Government of Canada (1991); Bishop *et al.* (1992) and/or Pettit *et al.* (1994), but have not been fully analyzed or interpreted.

### 2.3 Bill Deformities

Cormorant nesting sites in Lake Ontario and the Canadian portions of Lake Superior and Lake Huron as well as two reference sites in northwestern Ontario (Lake Nipigon and Lake-of-the-Woods), were visited in late June to early July between 1988 and 1996. With the exception of Lake Superior, the only region surveyed in 1996, all regions were surveyed up to 1994. During each visit, the total numbers of chicks and apparently occupied nests at each colony were counted and the numbers of chicks with bill defects were noted. Bill defects, as opposed to other types of abnormalities (club feet, supernumerary digits, and eye and skeletal abnormalities), were used as they were the most prominent and highly visible deformity in cormorant chicks and are assumed to be clearly developmental rather than the result of post-hatching trauma(Fox *et al.*, 1991a). Counting and examining of chicks was done usually by one to two observers, using hand held counters. Each observer moved slowly and systematically through the colony and was responsible for counting and examining chicks in sections that were delineated visually prior to the start of each survey.

Nesting colonies on Lake Erie were not surveyed for deformities because nearly all the cormorants breeding there nest in trees, making it difficult to examine young accurately. The nesting sites surveyed were divided into geographical regions as follows:

1. Lake Ontario: colonies located on Little Galloo I. (U.S.), Pigeon I., Salmon I., Snake I., Hamilton Harbour, Gull I. and High Bluff I..
2. Lake Huron (main body of lake): colonies located on Chantry I., Wheeler Rf., Manitoba Rf., Mayflower I. and the Fishing Is..
3. Georgian Bay: colonies located on North and South Watcher Is., North and South Tribune Is., South Limestone I., Wallis Rk., an island southwest of Hawk I., Colin Rocks, Birnie Group, Blackbill Is., Castle Rks, Southwest (Bustard) Rks., islands southwest of Mary I., an island southeast of Hooper I., Erie Shingle, Halfmoon I., Gull I., Southeast Rk., an island east of Flat Rks, West Rk., Heywood Rk., Elgin Rk., Gull Rks and James I.
4. North Channel: colonies located on Elm I., Gull Rks, Egg I., Cousins I. and adjacent rocks, Doucet Rk., West I. and adjacent rocks, Batture I., Herbert I., Africa Rk., Kalulah Rk., Kangaroo Rk., Ivor Rks, Mouse I. and Bird I..
5. Lake Superior: colonies located on Steamboat Is. (one each in western and eastern Lake Superior), Welcome Is., Deadman I., island off Pic River, Druid Rk., Condon I., Evelyn Rk., Mood (Hood) I., Granite I. and adjacent rocks, Gravel I., Cone I., islands west of Victoria Cove, Clark I., island east of Clark I., Rolland Rks and Hamilton I.
6. Northwestern Ontario: Colonies located in Lake Nipigon (small island east of Logan I.) and Lake-of-the-Woods (Dream I., Knight I., Crowduck I., The Three Sisters, and rocks northwest of Elm I.).

For more precise location of these colonies see Fox *et al.* (1991a), Bishop *et al.* (1992) and Pettit *et al.* (1994).

Terminology used throughout this paper with regard to bill defects is as presented in Fox *et al.* (1991a) and is summarized below. The term 'visit' refers to a colony-year and represents the reproductive effort of a single colony in a single breeding season. 'Occurrence' is the number of visits to colonies in which abnormal individuals were observed. 'Incidence' is the likelihood of encountering an abnormal chick on any visit and was calculated as the 'occurrence' of deformities divided by the total number of 'visits'. The 'prevalence' of defects in hatched chicks was calculated for each region as the sum of all chicks with bill defects found in all visits within a region divided by the sum of all chicks examined in all visits within that region, expressed on the basis of 10,000 young examined. Data reported on bill defects in cormorant chicks for years 1979-1987 were taken from Fox *et al.* (1991a).

## 2.4 EGGSHELL THICKNESS

Shell thickness was measured from eggs which were collected in 1989 and 1995 for chemical analysis and stored at the Canadian Wildlife Service Tissue Bank, NWRC. Shell thickness was measured using a Starrett No.1010 ball micrometer accurate to 0.001 mm. As discussed by Postupalsky (1978), the usefulness of this method is affected by the varying degrees of thickness between different parts of the shell, the whitish chalky covering with which cormorant eggs are normally covered, and the separation of the egg membrane from the shell proper. To compensate for these problems, shell thickness was measured at three points around the equator of the egg and included only areas with firmly attached membranes and no or minimal chalky covering. The mean of these three measurements was then used as the average thickness for that egg.

## 2.5 DATA ANALYSIS

In order to facilitate comparisons with published accounts of contaminants in colonial waterbirds and other biota on the Great Lakes, residue levels are presented as arithmetic means. All data sets were first tested for homoscedasticity (Bartlett's Test for Homogeneity of Variance) and for deviations from normal distributions (Kolmogorov-Simirnov One-sample Test, using Lilliefores 2-tailed Probability Test). In some instances data were logarithmically transformed, $\log_{10}(x+1)$, to meet the assumptions for parametric testing. When these assumptions could not be met, the appropriate non-parametric procedure(s) was employed. Temporal trends in egg contaminant levels were assessed using 'Spearman Rank Correlation' analysis. For temporal trend analysis, the mean, rather than the individual values of the 1970-1981 (individual) samples at each region was used to compare to the 1983-1995 pooled samples. Due to the limited temporal data for some compounds in each region, a minimum of four years of data was used as a criterion for inclusion for temporal trend analysis. Statistical comparisons of incidence and prevalence of bill deformities among all regions for the years 1988-1996 were performed using a Randomization Test (Sokal and Rolf, 1981), and Ryan's Multiple Comparison Procedure (Ryan, 1959). Other statistical methods and conventions follow Zar (1984). Throughout this paper the term 'significant' is used only in the statistical sense ($P < 0.05$).

## 3. Results

### 3.1 ORGANOCHLORINE CONTAMINANTS

Of all OCs detected, PCBs (PCB 1260 and sum of PCB congeners) and DDE levels were consistently found in the highest concentrations in cormorant eggs from the Canadian Great Lakes, Lake Nipigon and Lake-of-the-Woods between 1970 and 1995, with mean (or pooled) colony values ranging from 1.94-35.52 and 1.95-18.56 $\mu$g/g wet weight, respectively (Table I). The range values for dieldrin (0.04-0.55), mirex (0.01-0.87), HE (0.02-0.26), HCB (0.01-0.18), photomirex (0.005-0.28), oxy-chlordane (0.02-0.13) and *trans*- (0.001-0.03) and *cis*-nonachlor (0.01-0.09), were considerably lower (Table I). For all regions, levels of *cis*-chlordane (range 0.002 - 0.4), *trans*-chlordane (range <0.001 - 0.005), 1,2,3,4-tetrachlorobenzene (1,2,3,4-TCB) (range 0.001 - 0.005), 1,2,3,5-/1,2,4,5-TCB (range 0.001 - 0.006) pentachlorobenzene (range 0.001 - 0.03) octachlorostyrene (range 0.006 - 0.03), *alpha*-hexachlorocyclohexane (*alpha*-HCH) (range 0.003 - 0.02), *beta*-HCH (range 0.002 - 0.02), and *gamma*-HCH (range 0.001 - 0.003) were present at the lowest concentrations and were not included here.

In 1989, seven PCDD congeners were detected at low levels (below 30 pg/g) in samples from all sites (Table II), with 2,3,7,8-TCDD, 1,2,3,7,8-PnCDD and 1,2,3,6,7,8-HxCDD detected in the highest concentrations. In 1989, the major PCDF detected was 2,3,4,7,8-PCDF ranging from 8 to 21 pg/g, with the other four PCDFs measured consistently near the detection limit (1 to 5 pg/g). Higher concentrations of PCDDs and PCDFs were observed at Pigeon Island, Lake Ontario in 1991 compared to 1989 (Table II); with levels increasing by a range of 92 to 237%.

Spatial Trends

During the 1970-72 period, levels of HCB, DDE, dieldrin, HE and PCBs differed significantly (Kruskal-Wallis test, $\chi^2 \geq 13.16$, $P \leq <0.01$) among the four regions compared (Table I). Contaminant levels were typically highest for Georgian Bay and North Channel colonies, followed by Lake Ontario and Lake Nipigon. Contaminant levels were lowest in Lake Erie. In 1979, eggs sampled from Lake Erie had significantly greater levels of PCBs $F = 15.29$, $P <0.001$) than eggs from North Channel and Lake Nipigon and greater levels of mirex ($t = 2.20$, $P = 0.04$) than in Lake Nipigon; levels of DDE, dieldrin, HCB and HE were similar ($F \leq 2.4$, $P \geq 0.11$). In 1981, Lake Ontario had significantly higher mirex levels ($t = 4.05$, $P < 0.01$), than Lake Erie (Table I).

**Table I.** Data Base Structure, Concentrations ($\mu$g/g, Wet Weight) of Organochlorine Contaminants, and Percent Lipid Values from Double-Crested Cormorant Eggs from the Great Lakes Basin, 1970-1995.

| | | YEAR | | | | | | | | | | |
|---|---|---|---|---|---|---|---|---|---|---|---|---|
| | SAMPLE REGION | 1970-72 | 1975 | 1979 | 1981 | 1983[P] | 1984[P] | 1989[P] | 1990[P] | 1995[P] | *r* | *P* |
| **DATA BASE STRUCTURE** | | | | | | | | | | | | |
| **Number of Eggs** | Lake Ontario | *(7, 1)* | | | *(20, 2)* | | *(10, 1)* | *(20, 2)* | *(20, 2)* | *(30, 3)* | | |
| | Lake Erie | *(18, 1)* | | *(10, 1)* | *(10, 1)* | *(10, 1)* | *(10, 1)* | *(12, 1)* | | *(10, 1)* | | |
| **Collected, and** | Georgian Bay | *(23, 3)* | | | | | | *(10, 1)* | | *(10, 1)* | | |
| | North Channel | *(21, 3)* | *(2, 1)* | *(9, 1)* | | | *(10, 1)* | *(12, 1)* | | *(10, 1)* | | |
| **Number of Colonies** | Lake Superior | | | | | *(10, 1)* | *(10, 1)* | *(20, 2)* | | *(10, 1)* | | |
| | Lake Nipigon | *(47, 1)* | | *(10, 1)* | | | | | | | | |
| **Visited** | Lake-of-the-Woods | *(1, 1)* | | | | *(10, 1)* | *(10, 1)* | | | | | |
| **ORGANOCHLORINE CONTAMINANT** | | | | | | | | | | | | |
| **Hexachlorobenzene** | Lake Ontario | 0.15 ± 0.06† [a] | | | 0.10 ± 0.22 [a] | | 0.03 | 0.015 | 0.02 | 0.01 | -0.82, | 0.002** |
| | Lake Erie | 0.01 ± 0.00 [d] | | 0.05 ± 0.03 [a] | 0.04 ± 0.02 [a] | 0.03 | 0.04 | 0.02 | | 0.02 | -0.14, | 0.76 |
| | Georgian Bay | 0.18 ± 0.30 [ab] | | | | | | 0.01 | | 0.01 | -0.44, | 0.38 |
| | | *(16, 3)* | | | | | | | | | | |
| | North Channel | 0.02 ± 0.02 [bc] | 0.01 | 0.04 ± 0.01 [a] | | | 0.02 | 0.03 | | 0.01 | 0.11, | 0.78 |
| | Lake Superior | | | | | 0.03 | 0.03 | 0.015 | | 0.01 | -0.98, | 0.005** |
| | Lake Nipigon | 0.05 ± 0.12 [c] | | 0.07 ± 0.03 [a] | | | | | | | NT | |
| | Lake-of-the-Woods | | | | | 0.03 | 0.02 | | | | NT | |
| ***p,p'*-DDE** | Lake Ontario | 9.37 ± 2.19 | | | 4.77 ± 2.28 [a] | | 6.41 | 2.56 | 3.21 | 2.40 | -0.74, | 0.01** |
| | Lake Erie | 6.36 ± 4.62 [c] | | 4.52 ± 2.33 [a] | 3.59 ± 3.09 [a] | 2.69 | 4.62 | 3.36 | | 2.40 | -0.71, | 0.07 |
| | Georgian Bay | 18.56 ± 16.97 [a] | | | | | | 2.03 | | 2.80 | -0.80, | 0.03** |
| | North Channel | 16.11 ± 8.45 [a] | 7.1 | 2.45 ± 2.77 [a] | | | 3.19 | 1.95 | | 2.26 | -0.96, | <0.001** |
| | Lake Superior | | | | | 2.46 | 4.35 | 3.8 | | 2.83 | 0.05, | 0.94 |
| | Lake Nipigon | 10.41 ± 5.53 [b] | | 4.03 ± 2.13 [a] | | | | | | | NT | |
| | Lake-of-the-Woods | 6.36 | | | | 2.55 | 3.38 | | | | NT | |
| **Dieldrin** | Lake Ontario | 0.27 ± 0.16 [ab] | | | | | 0.29 | 0.1 | 0.09 | 0.05 | -0.74, | 0.02** |
| | Lake Erie | 0.23 ± 0.14 [b] | | 0.26 ± 0.16 [a] | | 0.25 | 0.39 | 0.06 | | 0.08 | -0.43, | 0.40 |
| | Georgian Bay | 0.47 ± 0.43 [a] | | | | | | 0.04 | | 0.06 | -0.95, | 0.001** |
| | North Channel | 0.47 ± 0.34 [a] | 0.52 | 0.28 ± 0.16 [a] | | | 0.35 | 0.05 | | 0.07 | -0.78, | 0.01** |
| | Lake Superior | | | | | 0.4 | 0.55 | 0.14 | | 0.09 | -0.87, | 0.05 |
| | Lake Nipigon | 0.45 ± 0.23 [a] | | 0.20 ± 0.09 [a] | | | | | | | NT | |
| | Lake-of-the-Woods | 0.53 | | | | 0.19 | 0.13 | | | | NT | |

**Table I (cont'd.).** Data Base Structure, Concentrations ($\mu$g/g, Wet Weight) of Organochlorine Contaminants, and Percent Lipid Values from Double-Crested Cormorant Eggs from the Great Lakes Basin, 1970-1995.

| | | YEAR | | | | | | | | | | |
|---|---|---|---|---|---|---|---|---|---|---|---|---|
| | SAMPLE REGION | 1970-72 | 1975 | 1979 | 1981 | 1983[p] | 1984[p] | 1989[p] | 1990[p] | 1995[p] | $r$ | $P$ |
| ORGANOCHLORINE CONTAMINANT | | | | | | | | | | | | |
| Heptachlor | Lake Ontario | 0.04 ± 0.01 [b] | | | | | 0.07 | 0.03 | 0.06 | 0.02 | -0.47, | 0.20 |
| | Lake Erie | 0.02 ± 0.01 [c] | | 0.09 ± 0.09 [a] | | 0.05 | 0.09 | 0.02 | | 0.03 | -0.03, | 0.96 |
| | Georgian Bay | 0.26 ± 0.49 [a] | | | | | | 0.02 | | 0.03 | -0.95, | 0.001** |
| Epoxide | North Channel | 0.05 ± 0.04 [b] | 0.05 | 0.06 ± 0.04 [a] | | | 0.08 | 0.02 | | 0.03 | -0.30, | 0.43 |
| | Lake Superior | | | | | 0.09 | 0.15 | 0.045 | | 0.03 | -0.87, | 0.05 |
| | Lake Nipigon | 0.18 ± 0.07 [ab] | | 0.12 ± 0.21 [a] | | | | | | | NT | |
| | Lake-of-the-Woods | 0.03 | | | | 0.04 | 0.05 | | | | NT | |
| Mirex | Lake Ontario | | | | 0.87 ± 0.63 [a] | | 0.72 | 0.35 | 0.5 | 0.29 | -0.69, | 0.03** |
| | Lake Erie | | | 0.11 ± 0.09 [a] | 0.05 ± 0.03 [b] | 0.05 | 0.07 | 0.07 | | 0.03 | -0.43, | 0.40 |
| | Georgian Bay | | | | | | | 0.04 | | 0.09 | NT | |
| | North Channel | | | 0.15 ±0.12 | | | 0.06 | 0.01 | | 0.03 | NT | |
| | Lake Superior | | | | | 0.03 | 0.03 | 0.04 | | 0.05 | NT | |
| | Lake Nipigon | | | 0.04 ± 0.03 [b] | | | | | | | NT | |
| | Lake-of-the-Woods | | | | | 0.05 | 0.02 | | | | NT | |
| PCB 1260 | Lake Ontario | 18.30 ± 8.11 | | | 25.46 ± 12.88 | | 10.44 | 6.69 | 8.39 | 5.32 | -0.85, | 0.001** |
| | Lake Erie | 18.81 ± 12.25 | | 35.52 ± 18.81 | 33.80 ± 9.67 | 29.3 | 14.56 | 13.38 | | 15.46 | -0.68, | 0.09 |
| | Georgian Bay | 24.05 ± 12.78 | | | | | | 5.22 | | 4.51 | -0.47, | 0.28 |
| | North Channel | 27.25 ± 10.91 | 9.9 | 17.70 ± 7.51 | | | 3.93 | 2.78 | | 4.74 | -0.88, | 0.002** |
| | Lake Superior | | | | | 10 | 4.01 | 5.1 | | 2.52 | -0.82, | 0.09 |
| | Lake Nipigon | 16.07 ± 8.77 | | 6.05 ± 3.43 [b] | | | | | | | NT | |
| | Lake-of-the-Woods | | | | | 4.6 | 1.94 | | | | NT | |
| Sum of PCB | Lake Ontario | | | | | | 11.25 | 6.33 | 7.96 | 4.83 | -0.64, | 0.09 |
| | Lake Erie | | | | | | 12.91 | 11.01 | | 11.35 | NT | |
| | Georgian Bay | | | | | | | 4.8 | | 4.01 | NT | |
| Congeners | North Channel | | | | | | 4.57 | 2.85 | | 4.25 | NT | |
| | Lake Superior | | | | | | 5.15 | 4.95 | | 2.59 | NT | |
| | Lake Nipigon | | | | | | | | | | NT | |
| | Lake-of-the-Woods | | | | | | | | | | NT | |

**Table I (cont'd.).** Data Base Structure, Concentrations ($\mu g/g$, Wet Weight) of Organochlorine Contaminants, and Percent Lipid Values from Double-Crested Cormorant Eggs from the Great Lakes Basin, 1970-1995.

| | | YEAR | | | | | | | | | | |
|---|---|---|---|---|---|---|---|---|---|---|---|---|
| | SAMPLE REGION | 1970-72 | 1975 | 1979 | 1981 | 1983P | 1984P | 1989P | 1990P | 1995P | *r* | *P* |
| **ORGANOCHLORINE CONTAMINANT** | | | | | | | | | | | | |
| **Oxy-Chlordane** | Lake Ontario | | | | | | 0.06 | 0.03 | 0.04 | 0.03 | -0.41, | 0.32 |
| | Lake Erie | | | 0.07 ± 0.03 | | 0.09 | 0.05 | 0.04 | | 0.03 | -0.35, | 0.24 |
| | Georgian Bay | | | | | | | 0.03 | | 0.04 | NT | |
| | North Channel | | | 0.05 ± 0.02 | | | 0.05 | 0.02 | | 0.03 | -0.44, | 0.15 |
| | Lake Superior | | | | | 0.12 | 0.13 | 0.10 | | 0.04 | -0.62, | 0.27 |
| | Lake Nipigon | | | 0.04 ± 0.03 | | | | | | | NT | |
| | Lake-of-the-Woods | | | | | 0.04 | 0.02 | | | | NT | |
| **Photomirex** | Lake Ontario | | | | | | 0.28 | 0.13 | 0.20 | 0.11 | -0.54, | 0.16 |
| | Lake Erie | | | 0.03 ± 0.03 | | | 0.02 | 0.03 | | 0.02 | NT | |
| | Georgian Bay | | | | | | | 0.01 | | 0.03 | NT | |
| | North Channel | | | 0.05 ±0.06 | | | 0.02 | 0.005 | | 0.01 | NT | |
| | Lake Superior | | | | | | 0.02 | 0.03 | | 0.02 | NT | |
| | Lake Nipigon | | | 0.01 ± 0.02 | | | | | | | NT | |
| | Lake-of-the-Woods | | | | | | 0.001 | | | | NT | |
| ***Trans*-Nonachlor** | Lake Ontario | | | | | | 0.03 | 0.02 | 0.01 | 0.01 | -0.41, | 0.32 |
| | Lake Erie | | | | | 0.01 | 0.03 | 0.02 | | 0.02 | 0.20, | 0.80 |
| | Georgian Bay | | | | | | | 0.01 | | 0.01 | NT | |
| | North Channel | | | | | | 0.02 | 0.01 | | 0.01 | NT | |
| | Lake Superior | | | | | 0.01 | 0.03 | 0.02 | | 0.01 | 0.20, | 0.80 |
| | Lake Nipigon | | | | | | | | | | NT | |
| | Lake-of-the-Woods | | | | | 0.01 | 0.001 | | | | NT | |
| ***Cis*-Nonachlor** | Lake Ontario | | | | | | 0.06 | 0.02 | 0.02 | 0.02 | -0.24, | 0.58 |
| | Lake Erie | | | | | | 0.05 | 0.03 | | 0.02 | NT | |
| | Georgian Bay | | | | | | | 0.03 | | 0.02 | NT | |
| | North Channel | | | | | | 0.07 | 0.02 | | 0.01 | NT | |
| | Lake Superior | | | | | | 0.09 | 0.05 | | 0.02 | NT | |
| | Lake Nipigon | | | | | | | | | | NT | |
| | Lake-of-the-Woods | | | | | | 0.01 | | | | NT | |

**Table I (cont'd.).** Data Base Structure, Concentrations ($\mu g/g$, Wet Weight) of Organochlorine Contaminants, and Percent Lipid Values from Double-Crested Cormorant Eggs from the Great Lakes Basin, 1970-1995.

| | | YEAR | | | | | | | | | | |
|---|---|---|---|---|---|---|---|---|---|---|---|---|
| | **SAMPLE REGION** | 1970-72 | 1975 | 1979 | 1981 | 1983ᵖ | 1984ᵖ | 1989ᵖ | 1990ᵖ | 1995ᵖ | *r* | *P* |
| **ORGANOCHLORINE CONTAMINANT** | | | | | | | | | | | | |
| **DDE:PCB Ratio** | Lake Ontario | 0.51 | | | 0.18 | | 0.61 | 0.38 | 0.38 | 0.45 | | |
| | Lake Erie | 0.34 | | 0.13 | 0.11 | 0.09 | 0.32 | 0.27 | | 0.16 | | |
| | Georgian Bay | 0.77 | | | | | | 0.39 | | 0.62 | | |
| | North Channel | 0.59 | 0.72 | 0.14 | | | 0.81 | 0.70 | | 0.48 | | |
| | Lake Superior | | | | | 0.25 | 1.1 | 0.75 | | 1.12 | | |
| | Lake Nipigon | 0.70 | | 0.67 | | | | | | | | |
| | Lake-of-the-Woods | | | | | 0.55 | 1.74 | | | | | |
| **% Lipid** | Lake Ontario | 3.46 ± 1.29 | | | 4.24 ± 0.45 | | 5.2 | 5.17 | 4.45 | 4.4 | | |
| | Lake Erie | 8.26 ± 1.75 | | 4.15 ± 0.26 | 4.06 ± 0.65 | 4.7 | 4.8 | 4.53 | | 5.1 | | |
| | Georgian Bay | 5.77 ± 2.73 | | | | | | 4.55 | | 4.3 | | |
| | North Channel | 6.25 ± 1.80 | 3.8 | 4.50 ± 0.50 | | | 4.8 | 4.33 | | 5.3 | | |
| | Lake Superior | | | | | 4.7 | 4.8 | 4.8 | | 5 | | |
| | Lake Nipigon | 4.18 ± 0.96 | | 4.26 ± 0.52 | | | | | | | | |
| | Lake-of-the-Woods | 3.7 | | | | 5.5 | 4.3 | | | | | |

Some of the data have appeared previously in Postupalsky (1978), Weseloh *et al.* (1983), Bishop *et al.* (1992) and Pettit (1994). Number of eggs collected and number of colonies visited for each year(s) in brackets, unless otherwise indicated; ** indicates the probability (*P*)of a significant declining trend.

p analysis performed on a pooled sample, based on 10-12 egg pool.

† arithmetic mean ± s.d. for individual samples.

NT no temporal analysis carried out on these data.

Subscript letters (a, b, c or d) indicate the results of statistical comparisons among geographic regions for each contaminant for the periods 1970-72, 1979 and 1981. Means with similar letters are not significantly different from one another.

Colonies and years visited are as follows:

Lake Ontario: Scotch Bonnet I., 1972; Little Galloo I., 1981, 1990 and 1995; Pigeon I., 1981, 1984, 1989-90 and 1995; Hamilton Harbour, 1989 and 1995.

Lake Erie: Big Chicken I., 1972, 1979, 1981 and 1983-84; East Sister I., 1989 and 1995.

Georgian Bay: Blackbill Is., 1970-71 and 1989; Southwest (Bustard) Rks., 1972; Wallis Rk., 1971-72 and 1995.

North Channel: Doucet Rk., 1971-72 and 1984; Talon Rk., 1971; Gull Rk., 1972; West I., 1989; Africa Rk., 1975, 1979 and 1995.

Lake Superior: Gravel I., 1983 and 1989; Gull I., 1984; Cone I., 1989; Steamboat I. E., 1995.

Lake Nipigon: East Logan I., 1970-71 and 1979.

Lake-of-the-Woods: The Three Sisters Is., 1970 and 1983-84.

**Table II.** Concentrations (pg/g, Wet Weight) of Polychlorinated Dibenzo-*p*-Dioxin (PCDD) and Polychlorinated Dibenzofuran (PCDF) Congeners in Double-Crested Cormorant Eggs Collected from Colonies on the Great Lakes, 1989 and 1991.

| | SAMPLE LOCATION | | | | | | | |
|---|---|---|---|---|---|---|---|---|
| | Lake Superior | | Lake Huron | | Lake Erie | Lake Ontario | | |
| | GI | CI | WI | BI | ESI | HH | PI | PI |
| **YEAR** | 1989 | 1989 | 1989 | 1989 | 1989 | 1989 | 1989 | 1991 |
| **NUMBER OF SAMPLES** | 10[p] | 10[p] | 12[p] | 10[p] | 12[p] | 10[p] | 12[p] | 5 |
| % Lipid | 4.4 | 4.7 | 4.3 | 4.5 | 4.5 | 4.6 | 5.7 | 5.3 ± 0.8[a] |
| **COMPOUND** | | | | | | | | |
| **PCDD Congener** | | | | | | | | |
| 2378-TCDD | 9 | 12 | 14 | 18 | 20 | 18 | 18 | 36 ± 10.4 |
| 12378-PnCDD | 14 | 21 | 21 | 27 | 22 | 21 | 17 | 32 ± 5.4 |
| 123678-HxCDD | 18 | 25 | 17 | 21 | 25 | 19 | 11 | 37 ± 15.6 |
| 1234678-HpCDD | 8 | 21 | 13 | 8 | 8 | 11 | 8 | 27 ± 16.9 |
| OCDD | 10 | 22 | 16 | 4* | 11 | 12 | 13 | 25 ± 13.2 |
| 123478-HxCDD | 1* | 3* | 2* | 3* | 2* | 2* | 2* | 6 ± 2.8 |
| 123789-HxCDD | 5 | 8 | 4 | 4 | 5 | 6 | 3* | 12 ± 6.1 |
| **PCDF Congener** | | | | | | | | |
| 2378-TCDF | 1* | 2* | 2 | ND | ND | 1* | 1* | ND |
| 23478-PnCDF | 11 | 8 | 12 | 20 | 14 | 21 | 14 | 29 ± 8.8 |
| 123478-/123467-HxCDF | 5 | 4 | 2* | 3* | 2* | 7 | 5 | 10 ± 3.4 |
| 123678-HxCDF | 3* | 3* | 2* | 2* | 1* | 4 | 2* | 4 ± 1.2 |
| 234678-HxCDF | 2* | 2* | 2* | 2* | 1* | 4 | 3* | 3* ± 0.9 |

**Key:** GI=Gravel I., CI=Cone I., Lake Superior; WI=West I., North Channel, BI=Blackbill I., Georgian Bay, Lake Huron; ESI=East Sister I. Lake Erie; HH=Hamilton Harbour, PI=Pigeon I, Lake Ontario.

p pooled sample, one analysis

a arithmetic mean ± s.d. for individual samples.

* trace levels

ND not detected.

TCDD = tetrachlorodibenzo-*p*-dioxin

PnCDD = pentachlorodibenzo-*p*-dioxin

HxCDD = hexachlorodibenzo-*p*-dioxin

HpCDD = heptachlorodibenzo-*p*-dioxin

OCDD = octachlorodibenzo-*p*-dioxin

TCDF = tetrachlorodibenzofuran

PnCDF = pentachlorodibenzofuran

HxCDF = hexachlorodibenzofuran

Geographical trends could not be assessed statistically from 1983 onwards because of the use of pooled samples. In 1983 and 1984, Lake-of-the-Woods displayed lower PCB, dieldrin, and HE levels compared to colonies from the Great Lakes. Mirex levels from Lake-of-the-Woods were similar to Great Lakes levels in 1983 but were lower than Great Lakes levels in 1984. Overall, Lake Ontario colonies displayed much higher mirex levels, with recorded concentrations five to 29 times higher than those from other regions. Between 1983 and 1995, levels of PCBs were consistently highest from Lake Erie colony sites, with recorded concentrations 1.4 to 8 times higher than other regions. Since the early 1980s, regional differences in OC levels in cormorant eggs, with the exception of mirex and PCBs, have become less evident, especially in the most recent samples. When generally ranked, concentrations of 'major' contaminants in cormorant eggs between 1984 and 1995 for the Great Lakes were as follows: Lake Erie > Lake Superior ≥ Lake Ontario > North Channel = Georgian Bay.

Temporal Trends

Eggs from Lake Ontario displayed significant declines for levels of HCB ($r = -0.82$, $P = 0.002$), DDE ($r = -0.74$, $P = 0.01$), dieldrin ($r = -0.74$, $P = 0.02$), mirex ($r = -0.69$, $P = 0.03$), and PCBs ($r = -0.85$, $P = 0.001$) (Table I). Lake Erie displayed no significant declining trends between 1970-72 and 1995. Eggs from Georgian Bay and North Channel displayed significant declines in DDE ($r \geq -0.80$, $P \leq 0.03$), and dieldrin ($r \geq -0.78$, $P \leq 0.01$), concentrations (Table I). Georgian Bay also displayed significant declines in HE levels ($r = -0.95$, $P = 0.001$) and North Channel displayed significant declines in PCB levels ($r = -0.88$, $P = 0.002$). For Lake Superior, a significant decline was observed for levels of HCB ($r = -0.98$, $P = 0.005$).

## 3.2 Bill Deformities

In the Great Lakes, a total of 146,485 cormorant chicks was examined during 308 visits to 76 colonies between June 1988 and July 1996 (Table III). Thirty-one chicks with bill defects were observed at 16 colonies (21% of colonies). Lake Ontario had the highest number of deformities (11) as well as the largest percentage of colonies with deformed chicks (43%). At reference sites in northwestern Ontario, for the same period, 4,850 cormorant chicks were examined during 10 visits to 6 colonies (Table III), with no chicks with bill defects observed.

No significant difference in the prevalence of bill deformities (number of deformities/10,000 young examined) was observed among the six regions examined between 1988-96 (Table III). The prevalence of bill defects ranged from 2.8 in Lake Ontario to 1.2 in Lake Superior. A significant difference was observed in the

**Table III.** Geographical Variation in Observations of Bill Defects in Cormorant Chicks from the Great Lakes and Northwestern Ontario, 1988-96.

| GEOGRAPHIC REGION | Period | No. of Colonies | Visits (a) | Occurrences (b) | Incidence (b/a) | | Colonies with Abnormal Chicks | | No. of Chicks with Defects (c) | No. of Chicks Examined (d) | Prevalence (c x 10,000/d) | |
|---|---|---|---|---|---|---|---|---|---|---|---|---|
| | | | | | | | N | % | | | | |
| **Lake Ontario** | 1988-94 | 7 | 22 | 6 | 0.27 | A | 3 | 43 | 11 | 38,898 | 2.8 | A |
| **Lake Huron** | 1988-94 | 6 | 24 | 3 | 0.13 | AB | 2 | 33 | 3 | 11,611 | 2.6 | A |
| North Channel | 1988-94 | 15 | 73 | 6 | 0.08 | AB | 5 | 33 | 9 | 35,022 | 2.6 | A |
| Georgian Bay | 1988-94 | 30 | 118 | 6 | 0.05 | B | 5 | 17 | 7 | 47,381 | 1.5 | A |
| **Lake Superior** | 1988-96 | 18 | 61 | 1 | 0.02 | B | 1 | 6 | 1 | 8,723 | 1.2 | A |
| **NW Ontario** | 1988-94 | 6 | 10 | 0 | 0.0 | B | 0 | 0 | 0 | 4,850 | 0.0 | A |

**Key:** Letters (A and B) indicate the results of statistical comparisons of the incidence and prevalence of bill deformities among regions for the period 1988-96. Incidence and prevalence data with similar letters are not significantly different ($P>0.05$) from one another.

incidence of deformities (no. occurrences/ no. visits) among the six regions. The incidence of bill defects in cormorant colonies in Lake Ontario was significantly higher than in Georgian Bay, Lake Superior and the reference sites in northwestern Ontario (Table III).

To assess temporal changes in the incidence and prevalence of bill defects in cormorant chicks, data from individual regions for years 1988-96 (Table III) were compared statistically with data from years 1979-87, reported in Fox *et al.* (1991a). The only significant difference observed was for Georgian Bay prevalence data (Randomization Test, $P = 0.03$), which declined significantly between these two periods (1.5 in 1988-95, compared to 6.1 in 1979-87).

### 3.3 EGGSHELL THICKNESS

In 1995, no significant spatial differences in shell thickness were observed among seven cormorant colonies on the Great Lakes (One-way Analysis of Variance, $F = 0.76$, $P = 0.63$). Overall in 1995, cormorant eggshells from the Great Lakes were 2.2% thinner when compared to a 1947 (pre-DDT) mean eggshell thickness of 0.440 mm, reported for eggs in museums collected from the Great Lakes in Ontario (Anderson and Hickey, 1972; Table IV). The colony with the thickest eggshells in 1995, from Georgian Bay, had a mean shell thickness equal to that reported in 1947.

**Table IV.** Mean Shell Thickness in Eggs of Great Lakes Double-Crested Cormorants, 1989 and 1995.

| SAMPLE REGION | Year | N | Eggshell Thickness (mm) Mean and Standard Error | % Thinning [a] |
|---|---|---|---|---|
| Lake Superior | 1989 | 25 | 0.400 ± 0.008 | 9.1 |
| | 1995 | 12 | 0.423 ± 0.013 | 3.9 |
| North Channel | 1989 | 21 | 0.415 ± 0.007 | 5.7 |
| | 1995 | 11 | 0.429 ± 0.011 | 2.5 |
| Georgian Bay | 1989 | 10 | 0.424 ± 0.006 | 3.6 |
| | 1995 | 13 | 0.440 ± 0.010 | 0 |
| Lake Erie | 1989 | 12 | 0.398 ± 0.009 | 9.5 * |
| | 1995 | 13 | 0.430 ± 0.011 | 2.3 |
| Lake Ontario | 1989 | 42 | 0.421 ± 0.005 | 4.3 |
| | 1995 | 39 | 0.429 ± 0.034 | 2.5 |

[a] Change from 1947 mean thickness for Great Lakes cormorant eggshells of 0.440 ± 0.013 mm (Anderson and Hickey, 1972).
* Significant change in shell thickness between 1989 and 1995.
Colonies visited are as follows:
Lake Superior: Gravel I., 1989; Cone I., 1989; Steamboat I. E., 1995.
Lake Huron: North Channel: West Island, 1989; Africa Rock, 1995,: Georgian Bay: Blackbill Is., 1989; Wallis Rk., 1995.
Lake Erie: Easter Sister I., 1989, 1995.
Lake Ontario: Hamilton Harbour, 1989, 1995; Pigeon I., 1989, 1995; Little Galloo I., 1995.

Cormorant eggshell thickness values from 1995 were compared with 1989 values from similar sites. Lake Erie (East Sister I.) was the only region to show a significant change, with shell thickness increasing between 1989 and 1995 (Student's t-test, $t = 2.24$, $P = 0.04$; Table IV).

No significant relationships between shell thickness and DDE levels were found in Great Lakes cormorant eggs for 1989 and/or 1995 (Linear Regression Analysis, $r^2 \leq 0.07$, $P \geq 0.60$).

## 4. Discussion

### 4.1 Contaminant Residues In Eggs

Cormorant eggs from the Great Lakes between 1970 and 1995 were contaminated with a variety of OCs, with PCBs and DDE found in the highest concentrations. Gilbertson and Reynolds (1974), Vermeer and Peakall (1977), Postupalsky (1978) and Weseloh *et al.* (1983) have shown that PCB, DDE and dieldrin levels in Great Lakes cormorant eggs, particularly from Georgian Bay and North Channel, Lake Huron, were the highest known for this species in North America between the late 1960s and early 1980s (Kury, 1969; Vermeer and Reynolds, 1970; Faber and Hickey, 1973; Gress *et al.*, 1973; Zitko, 1976; Ohlendorf *et al.*, 1978; Henny *et al.*, 1982; Elliott *et al.*, 1989; Pearce *et al.*, 1989). Cormorant eggs collected from Lake Michigan in 1977-78 (Heinz *et al.*, 1985) and 1988 (Yamashita *et al.*, 1993), had levels of DDE, PCB, HCB and dieldrin similar to this study for comparable years. Cormorant eggs from a declining population in southern California in 1969 (Gress *et al.*, 1973) had DDE levels comparable to levels on the Upper Great Lakes (Georgian Bay and North Channel) for the same period, but had considerably lower PCB levels. Cormorants breeding on the Canadian prairies in 1984-85 had similar overall mean DDE egg residues, but lower PCB, HCB and dieldrin levels, compared to Great Lakes cormorants for similar years (Somers *et al.*, 1993). Considerably lower DDE and PCB levels (0.5 and 0.01 $\mu$g/g, wet weight, respectively), have been reported in marine double-crested cormorant eggs from the Pacific northwest coast of North America (Elliot *et al.*, 1989; Henny *et al.*, 1989), compared to populations form the Great Lakes (this study) and eastern Canada (Pearce *et al.*, 1989) for similar years.

Declines observed in OC levels in cormorant eggs generally match declines observed in other fish eating bird species on the Great Lakes (Government of Canada, 1991; Bishop *et al.*, 1992; Ewins *et al.*, 1994; Petitt *et al.*, 1994; Pekarik and Weseloh, in press). Since the early 1970s, decreases in OC residues in cormorant eggs have been most evident for PCBs, DDE and dieldrin. Concentrations of these three

OCs decreased 71-83%, 74-86%, and 82-87%, respectively, between the period 1970-72 and 1995 on Lakes Ontario and Huron, compared with 18%, 62% and 65%, respectively, for Lake Erie, where levels have remained more stable.

In this study, average DDE:PCB ratios for 1984-95 were, when compared with average values for 1970-72, similar for Lake Ontario (0.48 compared to 0.51), lower for Lake Erie (0.25 compared to 0.34), and Georgian Bay (0.50 compared to 0.77) and higher for North Channel (0.66 compared to 0.59). The highest average DDE:PCB ratio between 1984-95 was observed for Lake Superior (1.12), reflecting the greater reductions in PCBs than DDE on this lake when compared to the lower Great Lakes, especially Lake Erie, for this period. Temporal analysis of DDE:PCB ratios in Great Lakes herring gulls (*Larus argentatus*) eggs have reported similar trends. Greater reductions in PCB residues than DDE residues have occurred in herring gull eggs sampled from the upper Great Lakes and Lake Ontario, compared to Lake Erie. (Hughes *et al.*, in press).

In 1989, patterns of PCDD and PCDF contamination in cormorant eggs were similar among all Great Lakes colonies and were similar to levels reported for Lake Michigan cormorant eggs in 1988 (Yamashita *et al.*,1993). Differing patterns have been observed in other fish-eating bird species. Herring gull eggs sampled from Saginaw Bay (Lake Huron) and Lake Ontario have consistently had higher concentrations of 2,3,7,8-TCDD, compared to other areas on the Great Lakes (Hebert *et al.*, 1994). These high levels have been attributed to past production of 2,4,5-trichlorophenol at chemical manufacturing plants beside the Niagara River and in the Saginaw River watershed (Hebert *et al.*, 1994). Similarly, Caspian tern (*Sterna caspia*) eggs from Lakes Ontario and Huron had higher 2,3,7,8-TCDD concentrations compared to samples from Lake Michigan colonies in 1991 (Ewins *et al.*, 1994).

Struger and Weseloh (1985) have suggested that migratory bird species might serve as valid indicator species, along with non-migratory species on their breeding grounds on the Great Lakes. Unlike herring gulls, which are residents as adults on the Great Lakes (Mineau *et al.*, 1984), cormorants are migratory, arriving on their breeding colonies only about one month prior to egg laying. Somers *et al.* (1993) reports that OC residues in eggs of cormorants breeding on the Canadian prairies largely reflect levels accumulated from dietary sources at wintering grounds and staging areas. This does not appear to be the case for Great Lakes cormorants. PCBs and mirex levels were consistently higher in cormorant eggs from eastern Lake Erie and Lake Ontario colonies, respectively, between 1984 and 1995. This colony specific distribution of PCB and mirex residues are indicative of contaminants accumulated from dietary sources while adult birds are on their breeding colonies. High PCB levels in eastern Lake Erie have also been reported in sediments (Frank *et al.*, 1977; Oliver and Bourbonniere, 1985), young-of-the-year spottail shiners

(*Notropis hudsonius*) (Suns *et al.*, 1981, 1991), and herring gulls (Weseloh *et al.*, 1990) and have been attributed to point souces on the Detroit River. Sites on Lake Ontario (Oswego, New York) and the Niagara River have been identified as major past sources of mirex on the Great Lakes (Kaiser, 1978). Higher mirex residues have also been found in herring gull and Caspian tern eggs sampled from Lake Ontario when compared to other regions of the Great Lakes (Ewins *et al.*, 1994; Bishop *et al.*, 1992; Pettit *et al.*, 1994).

Mean (or pooled) concentrations of HCB, DDE, HE, dieldrin, mirex and PCB in cormorant and herring gull eggs collected from the same colony sites on the Great Lakes between 1979 and 1995 were compared (this study, Table V). Results indicated that for all six compounds examined, concentrations of individual OCs were significantly correlated ($r \geq 0.55$, $P \leq 0.006$, Pearson Correlation Analysis) between these two species. We suggest, in the case of cormorants on the Great Lakes, the relative contaminant loads which they accumulate while resident there far surpasses anything they accumulate while they are away from the Great Lakes. As a result, year to year fluctuations in their contaminant loads display patterns similar to those in the herring gull, a year-round resident on the Great Lakes. This confirms that for some situations migratory birds can be used as valid indicators of changes in contaminant levels on their breeding grounds.

**Table V.** Correlation Between Mean (or Pooled) Concentrations of Six 'Major' Contaminants in Eggs of Double-Crested Cormorants and Herring Gulls Collected from Geographically Similar Sites on the Great Lakes for the Same Years.

| | HCB | DDE | Dieldrin | HE | Mirex | PCB 1260 |
|---|---|---|---|---|---|---|
| r | 0.551 | 0.621 | 0.805 | 0.629 | 0.812 | 0.776 |
| (n) | (23) | (23) | (20) | (20) | (21) | (22) |
| P | 0.006 | 0.002 | < 0.001 | 0.003 | < 0.001 | < 0.001 |

**Key:** Colonies and years visited are as follows:

**Double-Crested Cormorant:**

Lake Ontario: Little Galloo I., 1981, 1990 and 1995; Pigeon I., 1981, 1984, 1989, 1990 and 1995; Hamilton Harbour, 1989 and 1995.

Lake Erie: Big Chicken I., 1979, 1981, 1983 and 1984; East Sister I., 1989 and 1995.

Lake Huron: North Channel: Africa Rk., 1975, 1979 and 1995; Doucet Rk., 1984; West I., 1989.

Lake Superior: Gravel I., 1983 and 1989; Gull I., 1984; Steamboat I. E., 1995.

**Herring Gull:**

Lake Ontario: Little Galloo I., 1981; Pigeon I., 1981; Hamilton Harbour, 1989 and 1995; Snake I., 1984, 1989-90 and 1995.

Lake Erie: Middle I., 1979, 1981, 1983-84; 1989 and 1995.

Lake Huron: North Channel: Double I., 1975, 1979, 1984, 1989 and 1995.

Lake Superior: Papoose I., 1983; Granite I., 1984 and 1989; Agawa Rock, 1995.

## 4.2 CONTAMINANT EFFECTS

In the early 1970s, Great Lakes cormorant eggs were between 18 and 30% thinner (Postupalsky, 1978), compared to the 1947 shell thickness mean for Great Lakes cormorant eggs reported by Anderson and Hickey (1972). Since then, the thickness of eggshells in cormorant colonies on the Great Lakes has improved greatly in response to declining DDE egg residues. Current DDE, PCBs and other OC egg residues appeared to have had little to no impact on cormorant productivity at the population level on the Great Lakes as reproductive rates in this species remain high (Weseloh *et al.*, 1995). A concentration of 10 $\mu$g/g DDE has been associated with 20% eggshell thinning in cormorants, the value considered critical for this species (Pearce *et al.*, 1979). In this report, cormorant shell thickness between zero and 3.9% thinner than normal was associated with mean DDE levels of 2.4 to 2.8 $\mu$g/g. Ewins *et al.* (1992) found no evidence of hatching impairment at a mean eggshell thinning of 6% in herring gulls on Lake Huron. As well, no evidence of shell thinning was observed in Caspian terns at mean concentrations of 4 $\mu$g/g DDE, wet weight (Ewins *et al.*, 1994). Thus, it appears that eggshell thinning is no longer a major reproductive factor for these colonial fish-eating waterbirds on the Great Lakes.

Although environmental declines in levels of OCs have generally occurred, adverse reproductive and developmental effects continue to be observed in wildlife populations on the Great Lakes (Kubiak *et al.*, 1989; Yamashita *et al.*, 1993; Bishop *et al.*, 1994; Larson *et al.* 1996; Ludwig *et al.* 1996). Between 1979 and 1987, the prevalence of bill deformities in cormorant chicks was considerably higher in Green Bay, Lake Michigan (52.1/10,000) compared to other regions on the Great Lakes (2.7 to 12.3/10,000) (Table VI). During this period, bill deformities were observed more frequently and in more colonies in the Great Lakes basin than in colonies on the prairies and northwestern Ontario (Fox *et al.*, 1991a). Larson *et al.* (1996) also reported a substantially higher rate of bill deformities in cormorant chicks examined from upper Green Bay, Lake Michigan (44 bill defects in 5759 chicks examined for a prevalence of 76.4 per 10,000 chicks) compared to chicks examined from colonies from Lake Winnipegosis, Manitoba (14 bill defects in 24,736 chicks examined for a prevalence of 5.6 per 10,000 chicks). Comparisons of historical and current data (Table VI) suggest a numerical stability, with regards to the incidences of bill deformities in Great Lakes cormorants between 1979 and 1994. Georgian Bay was the only region where significant decreases in the prevalence of bill deformities in young cormorants were observed. Although Lake Michigan colonies were not surveyed in this present study, a comparison of previously published data suggests that the prevalence of bill deformities has not changed for this region either. Bill deformities in cormorant chicks for Green Bay between 1986 and 1991, reported by

Larsen *et al.* (1996) and Ludwig *et al.* (1996), indicated a prevalence of 51.0 per 10,000 chicks for this region (Table VI), which is identical to that reported in Fox *et al.* (1991a) for the period 1979 to 1987.

**Table VI.** Geographical and Temporal Variation in the Prevalence of Bill Defects in Double-Crested Cormorant Chicks in Great Lakes and Reference Sites.

| | | Prevalence per 10,000 chicks (No. Examined in No. Colonies) | | |
|---|---|---|---|---|
| | Region | 1979 - 1987 (a) | 1986 - 1991 (b+c) | 1988 - 1994 (d) |
| Great Lakes | Green Bay | 52.1 (11,520 in 11) | 51.0 (13,734) | - |
| | Lake Ontario | 3.5 (2,882 in 2) | - | 2.8 (38,898 in 7) |
| | Beaver-Mackinac | 12.3 (3,258 in 4) | 10.3 (13,645) | - |
| | Georgian Bay | 6.1 (3,250 in 5) | 1.8 (5,516) | 1.5 (47,381 in 30) * |
| | Lake Superior | 5.5 (1,825 in 5) | - | 1.2 (8,723 in 18) |
| | North Channel | 2.7 (7,342 in 12) | 2.5 (15,873) | 2.6 (35, 022 in 15) |
| Reference Sites | NW Ontario | 2.4 (4,184 in 17) | - | 0.0 (4,850 in 6) |
| | Prairies | 0.6 (16, 778 in 18) | - | - |
| | Lake Winnipegosis | - | 5.7 (24,736) | - |

**Sources:** (a) Fox *et al.* (1991); (b) Ludwig *et al.* (1996); (c) Larsen *et al.* (1996); (d) this study.
* Significant decline ($P$=0.03) between 1979-87 and 1988-1994.

Recent studies have implicated planar halogenated hydrocarbons (PHHs), particularly PCBs, and not OC pesticides, as the causative agent for most reproductive symptoms observed in fish-eating bird species on the Great Lakes (Kubiak *et al.*, 1989; Tillitt *et al.*, 1992; Fox 1993; Giesy *et al.* 1994; Ludwig *et al.* 1995, 1996). Yamashita *et al.* (1993) have shown that double-crested cormorant eggs from colonies with the highest sum of PCDD, PCDF and PCB equivalents also had the highest live-deformity rates (such as crossed-bills). Tillett *et al.* (1991) observed a dose-response relationship between total PCB concentrations and bioassay-derived TCDD equivalents (TCDD-EQs) and mortality rates in double-crested cormorant eggs from the same Great Lakes colonies. Ludwig *et al.* (1996) observed a high correlation between TCDD-EQs and deformity rates in cormorant and Caspian tern chicks, live and dead eggs, and dead egg rates. In this present study, no significant relationship was observed for total PCB concentrations and bill deformities at colony sites on the Great Lakes. However, both Tillett *et al.*, (1991) and Ludwig *et al.* (1996) found TCDD-EQs to be more highly correlated to egg mortality and deformity rates than total concentrations of PCBs. In the present study we did not analyze for all the PCDD and PCDF congeners and, hence, were unable to calculate TCDD-EQs and therefore could not examine the relationship between TCDD-EQs and rates of bill deformities among Great Lakes sites. However, the eggs are available in the CWS tissue bank and that study remains to be done.

## Acknowledgments

We would like to thank Craig Hebert, and Laird Shutt and two anonymous reviewers for suggestions on earlier versions of this paper. We would also like to express our appreciation to the many volunteers, contractors and staff who have assisted us over the years with cormorant egg collections and chick counts. They include Larry Benner, Hans Blokpoel, Birgit Braune, Neil Burgress, Rob Dobos, Jim Gardner, Martin Gebaner, Martin McNicholl, Doug McRae, Bob Miller and staff at the New York State Department of Environmental Conservation, Jan Neuman, Yves Oullette, Karen Pettit and Peter Ross to name a few. Technical Operations Division of the National Water Research Institute at the Canada Centre for Inland Waters provided logistical support. Funding for much of the Canadian work was provided by Environment Canada's Great Lakes Action Plan. Access permission to various colonies was kindly granted by : Philips Petroleum, J. Day, C. Moss, Ontario Ministry of Natural Resources, Metro Toronto Regional Conservation Authority, and the Hamilton Harbour Commissioners. We also appreciate the efforts of the staff at the National Wildlife Research Centre in preparing and analyzing eggs over the years.

## References

Anderson, D. W., and Hickey, J. J.: 1972, 'Eggshell changes in certain North American birds', *Proc. Int. Ornithol. Congr.* **15**, 514-540.

Anderson, D. W., Hickey, J. J., Risebrough, R. W., Hughes, D. F., and Christensen, R. E.: 1969, 'Significance of chlorinated hydrocarbon residues to breeding pelicans and cormorants', *Canadian Field-Naturalist* **83**, 91-112.

Ballschmiter, K., and Zell, M.: 1980, 'Analysis of polychlorinated biphenyls (PCBs) by glass capillary gas chromatography. Composition of technical Aroclor- and Clophen- PCB mixtures', *Fresenius Z. Anal. Chem.* **302**, 20-31.

Bishop, C. A., Brooks, R. J., Ng, P., Kennedy, S., Stegeman, J. J., Norstrom, R. J., and Petitt, K. E.: 1994, 'Effects of chlorinated hydrocarbons on snapping turtles in Areas of Concern in the Great Lakes.' Program and Abstracts of the annual meeting of the Society of Environmental Toxicology and Chemistry **15**, 317.

Bishop, C. A., Weseloh, D. V., Burgess, N. M., Struger, J., Norstrom, R. J., and Logan, K. A.: 1992, 'An atlas of contaminants in eggs of fish-eating colonial birds of the Great Lakes (1970-1988), Vol. 1 Accounts by species and locations', Technical Report Series No. 152, Canadian Wildlife Service, Ontario Region.

Elliott, J. E., Noble, D. G., Norstrom, R. J., and Whitehead, P. E.: 1989, 'Organochlorine contaminants in seabird eggs from the Pacific Coast of Canada 1971-1986', *Environ. Monit. Assess.* **12**, 67-82.

Ewins, P. J., Weseloh, D. V., and Mineau, P.: 1992, 'Geographical distribution of contaminants and productivity measures of herring gulls in the Great Lakes: Lake Huron 1980', *J. Great Lakes Res.* **18**:316-330.

Ewins, P. J., Weseloh, D. V. (Chip), Norstrom, R. J., Legierse, K., Auman, H. J., and Ludwig, J. P.: 1994, 'Caspian Terns on the Great Lakes: organochlorine contamination, diet, and population changes, 1972-91', Can. Wildl. Ser. Occas. Pap. No. 85.

Faber, R. A., and Hickey, J. J.: 1973, 'Eggshell thinning, chlorinated hydrocarbons, and mercury in inland aquatic bird eggs, 1969 and 1970', *Pesticides Monitoring J.* **7**, 27-36.

Fox, G. A.: 1993, 'What have biomarkers told us about the effects of contaminants on the health of fish-eating birds in the Great Lakes? The theory and a literature review', *J. Great Lakes Res.* **19**, 722-736.

Fox, G. A., Collins, B., Hayakawa, E., Weseloh, D. V., Ludwig, J. P., Kubiak, T. J., and Erdman T. C.: 1991a, 'Reproductive outcomes in fish-eating birds: a biomarker for developmental toxicants in Great Lakes food chains: II. Spatial variation in the occurrence and prevalence of bill defects in young double-crested cormorants in the Great Lakes, 1979-1987', *J. Great Lakes Res.* **17**, 158-167.

Fox, G. A., Weseloh, D. V., Kubiak, T. J., and Erdman, T. C.: 1991b, 'Reproductive outcomes in fish-eating birds: a biomarker for developmental toxicants in Great Lakes food chains: I. Historical and ecotoxicological perspectives', *J. Great Lakes Res.* **17**, 153-157.

Frank, R., Holdrinet, M., and Braun, H. E.: 1977, 'Organochlorine insecticides and PCBs in sediments of Lake St. Clair (1970 and 1974) and Lake Erie (1971)', *Sci. Total Environ.* **8**, 205-227.

Giesy, J. P., Ludwig J. P., and Tillitt, D. E.: 1994, 'Deformities in birds of the Great lakes region: assigning causality', *Environ. Sci. Technol.* **28**, 128A-136A.

Gilbertson, M., and Reynolds, L. M.: 1974, 'DDE and PCB in Canadian birds, 1969 to 1972', Can. Wildl. Ser. Occas. Pap. No. 19.

Government of Canada.: 1991, 'Toxic chemicals in the Great Lakes and associated effects. Volume 1: Contaminant levels and trends', Environment Canada, Fisheries and Oceans, Health and Welfare, Canada. Ottawa, Ontario. 487 pp.

Gress, F., Risebrough, R. W., Anderson, D. W., Kiff, L. F., and Jehl, J. R., Jr.: 1973, 'Reproductive failures of Double-crested cormorants in southern California and Baja California', *Wilson Bull.* **85**, 197-208.

Hebert, C. E., Norstrom, R. J., Simon, M., Braune, B. M., Weseloh, D. V., and Macdonald, C. R.: 1994, 'Temporal Trends and sources of PCDDs and PCDFs in the Great Lakes: Herring gull egg monitoring. 1981-1991', *Environ. Sci. Technol.* **28**, 1268-1277.

Heinz, G. H., Erdman, T. C., Haseltine, S. D., and Stafford, C.: 1985, 'Contaminant levels in colonial waterbirds from Green Bay and Lake Michigan, 1975-1980', *Environ. Monit. Assess.* **5**, 223-236.

Henny, C. J., Blus, L. J., and Prouty, R. M.: 1982, 'Organochlorine residues and shell thinning in Oregon seabird eggs', *Murrelet* **63**, 15-21.

Henny, C. J., Blus, L. J., Thompson, S. P., and Wilson, U. W.: 1989, 'Environmental Contaminants, human disturbance and nesting of Double-crested cormorants in northwestern Washington', *Colonial Waterbirds* **12**, 198-206.

Hughes, K. D., Weseloh, D. V., and Braune, B. M.: 1998, 'DDE:PCB ratios in Great Lakes herring gull eggs and their use in interpreting contaminant data', *J. Great Lakes Res.* **24**, 12-31.

Kaiser, K. L. E.: 1978. 'The rise and fall of mirex', *Environ. Sci. Technol.* **12**, 520-528.

Kubiak, T., Harris, H. J., Smith, L. M., Schwartz, T. R., Stalling, T. R., Trick, J. A., Sileo, L., Docherty, K. E., and Erdman, T. C.: 1989, 'Micro-contaminants and reproductive impairment of the Forster's tern on Green Bay, Lake Michigan', *Arch. Environ. Contam. Toxicol.* **18**, 706-727.

Kury, C. R.: 1969, 'Pesticide residues in a marine population of double-crested cormorants', *J. Wildl. Manag.* **33**, 91-95.

Larson, J. M., Karasov, W. H., Sileo, L., Stromborg, K. L., Hanbidge, B. A., Giesy, G. P., Jones, P. D., Tillitt, D. E., and Verbrugge, D. A.: 1996, 'Reproductive success, developmental anomalies, and environmental contaminants in double-crested cormorants (*Phalacrocorax auritus*)', *Environ. Toxicol. Chem.* **15**, 553-559.

Ludwig, J. P.: 1984, 'Decline, resurgence and population dynamics of Michigan and Great Lakes double-crested cormorants', *Jack-Pine Warbler* **62**, 71-102.

Ludwig, J. P., Auman, H. J., Weseloh, D. V., Fox, G. A., Giesy, J. P., and Ludwig, M. E.: 1995, 'Evaluation of the effects of toxic chemicals in Great Lakes cormorants: Has causality been established', *Colonial Waterbirds* **18**(Special Publication 1), 60-69.

Ludwig, J. P., Kurita-Matsuba, H., Auman, H. J., Ludwig, M. E., Summer, C. L., Giesy, J. P., Tillitt, D. E., and Jones, P. D.: 1996, 'Deformities, PCBs and TCDD-Equivalents in double-crested cormorants (*Phalacrocorax auritus)* and Caspian terns (*Hydroprogne caspia)* of the upper Great Lakes 1986-1991: Testing a cause-effect hypothesis', *J. Great Lakes Res* **22**, 172-197.

Mineau, P., Fox, G. A., Norstrom, R. J., Weseloh, D. V., Hallett D.J., and Ellenton, J. A.: 1984, 'Using the herring gull to monitor levels and effects of organochlorine contamination in the Canadian Great Lakes', In *Toxic Contaminants in the Great Lakes*, pp. 425-452, eds. Nriagu, J. O. and Simmons, M. S., John Wiley and Sons Inc., New York.

Norstrom, R. J., and Simon, M.: 1991, 'Determination of specific polychlorinated dibenzo-*p*-dioxins and dibenzofurans in biological matrices by gel-permeation/carbon chromatography and gas chromatography-mass spectrometry', In *Environmental Carcinogens, Methods of Analysis and Exposure Measurement. Vol. 11. Polychlorinated Dibenzo-*p*-dioxins and Dibenzofurans*, pp. 281-297 (eds.) C. Rappe, H. R. Buser, B. Dodet and I. K. O'Neill,. World Health Organization, International Association for Research on Cancer, Lyon, France, Method 5.

Norstrom, R. J., Simon, M., and Muir, D. C. G.: 1990, 'Polychlorinated dibenzo-*p*-dioxins and dibenzofurans in marine mammals in the Canadian north', *Envron. Pollut.* **66**, 1-19.

Norstrom, R. J., Simon, M., Muir, D. C. G., and Schweinsburg, R.: 1988, 'Organochlorine contaminants in Arctic marine food chains: identification, geographical distribution and temporal trends in polar bears', *Environ. Sci. Technol.* **22**, 1063-1071.

Norstrom, R. J., and Won, T. J.: 1985, 'Long-term preservation of egg and tissue homogenates for retrospective determination of organochlorine compounds: freezing vs. freeze-drying', *J. Assoc. Offic. Anal. Chem.* **68**, 129-135.

Ohlendorf, H. M., Risebrough, R. W., and Vermeer, K.: 1978, 'Exposure of marine birds to enviromental pollutants', U.S. Fish and Wildlife Service Wildlife Research Report No. 9. 40 pp.

Oliver, B. G., and Bourbonniere, R. A.: 1985, 'Chlorinated contaminants in surficial sediments of Lakes Huron, St. Clair, and Erie: implications regarding sources along the St. Clair and Detroit rivers', *J. Great Lakes Res.* **11**, 366-372.

Palmer, R. S.: 1962, 'Handbook of North American Birds. Vol. 1", Yale University Press, New Haven, CT.

Pearce, P. A., Elliott, J. E., Peakall, D. B., and Norstrom, R. J.: 1989, 'Organochlorine contaminants in eggs of seabirds in the Northwest Atlantic, 1968-1984', *Environ. Pollut.* **56**, 217-235.

Pekarik, C., and Weseloh, D. V.: [in press], 'Organochlorine contaminants in herring gull eggs from the Great Lakes, 1974-1995: Change point analysis and short term regression', *Environ. Monitoring Assess.* **43**, 101-116.

Pettit, K. E., Bishop, C. A., Weseloh, D. V., and Norstrom, R. J.: 1994, 'An atlas of contaminants in eggs of fish-eating colonial birds of the Great Lakes (1989-92). Vol. 1 Accounts by species and locations', Technical Report Series No. 194, Canadian Wildlife Service, Ontario Region.

Postupalsky, S.: 1978, 'Toxic chemicals and cormorant populations in the Great Lakes', Canadian Wildlife Service, Wild. Toxicol. Div. Ms. Rep. No. 20, Ottawa. 70 pp.

Price, I. M., and Weseloh, D. V.: 1986, 'Increased numbers and productivity of double-crested cormorants, *Phalacrocorax auritus* on Lake Ontario', *Can. Field-Nat.* **100**, 474-482.

Ryan, T. A.: 1959, 'Multiple comparisons in psychological research', *Psychological Bulletin* **56**, 26-47.

Scharf, W. C., and Shugart, G. W.: 1981, 'Recent increases in of double-crested cormorants in the United States Great Lakes', *American Birds* **35**, 910-911.

Sokal, R. R., and F. J. Rolf.: 1981, 'Biometry, 2nd ed', W. H. Freeman and Co., New York.

Somers, J. D., Goski, B. C., and Barbeau, J. M.: 1993, 'Accumulation of organochlorine contaminants in double crested cormorants', *Environ. Pollut.* **80**, 17-23.

Suns, K., Curry, C., Rees, G. A., and Crawford, G.: 1981, 'Organochlorine contaminant declines and their present geographic distribution in Great Lakes spottail shiners (*Notropis hudsonius*)', Ontario Ministry of the Environment, Rexdale, Ontario.

Suns, K., Hitchin, G., and Toner, D.: 1991, 'Spatial and temporal trends of organochlorine contaminants in spottail shiners (*Notropis hudsonius*) from the Great Lakes and their connecting channels (1975-1988)', Report of Waters Resources Branch, Ontario Ministry of the Environment, Toronto. 97 pp.

Tillit, D. E. G., Ankley, D. T., Giesy, J. P., Ludwig, J. P., Kurita-Matsuba, Weseloh, D. V., Ross, P. S., Bishop, C. A., Sileo, K., Stromborg, K. L., Larson, J., and Kubiak, T. J.: 1992, 'Polychlorinated biphenyl residues and egg mortality in double-crested cormorants from the Great Lakes', *Environ. Toxicol. and Chem.* **11**, 1281-1288.

Turle, R., and Collins, B.: 1992, 'Validation of the use of pooled samples for monitoring of contaminants in wildlife', *Chemosphere* **25**, 463-469.

Tyson, L. A., Belant, J. L., Francesca, J. C., and Weseloh, D. V.: 1997, 'Nesting populations of double-crested cormorants in the United States and Canada', Proceedings of Symposium on Double-crested Cormorants: Population Status and Management Issues in the Midwest. 59th Midwest Fish and Wildlife Conference, Mil. Wisconsin, Univ. of Colorado Press, Denver.

Vermeer, K., and Peakall, D. B.: 1977, 'Toxic chemicals in Canadian fish-eating birds', *Mar. Pollut.* **8**, 205-210.

Vermeer, K., and Reynolds, L. M.: 1970, 'Organochlorine residues in aquatice birds in the Canadian Prairie Provinces', *Can. Field-Nat.* **84**, 117-129.

Weseloh, D. V., Ewins, P. J., Struger J., Mineau, P., Bishop, C. A., Postupalsky, S., and Ludwig, J. P.: 1995, 'Double-crested cormorants of the Great Lakes: changes in population size, breeding distribution and reproductive output between 1913 and 1991', *Colonial Waterbirds* **18** (Special Publication 1), 48-59.

Weseloh, D. V., Mineau, P., and Struger, J.: 1990, 'Geographical distribution of contaminants and productivity measures of herring gulls in the Great Lakes: Lake Erie and connecting channels 1978/79', *Sci. Total Environ.* **91**, 141-159.

Weseloh, D. V., Teeple, S. M., and Gilbertson M.: 1983, 'Double-crested cormorants of the Great Lakes: egg laying parameters, reproductive failure, and contaminant residues in eggs, Lake Huron 1972-73', *Can. J. Zool.* **61**, 427-436.

Yamashita, N., Tanabe, S., Ludwig, J. P., Kurita, H. , Ludwig, M. E., and Tatsukawa, R.: 1993, 'Embryonic abnormalities and organochlorine contamination in double-crested cormorants (*Phalacrocorax auritus*) and Caspian terns (*Sterna caspia*) from the upper Great Lakes in 1988', *Environ. Poll.* **79**, 163-173.

Zar, J. H.: 1984, 'Biostatistical Analysis, 2nd ed', Prentice-Hall, Englewood Cliffs, New Jersey.

Zitko, V.: 1976, 'Levels of chlorinated hydrocarbons in eggs of double-crested cormorants from 1971 to 1975', *Bull. Environ. Contam. Toxicol.* **16**, 399-405.

# TRENDS OF CONTAMINANTS AND EFFECTS IN BALD EAGLES OF THE GREAT LAKES BASIN

William W. Bowerman[1], David A. Best[2], Teryl G. Grubb[3], Gregory M. Zimmerman[4], and John P. Giesy[5]

*[1]Gale Gleason Environmental Institute, Lake Superior State University, Sault Sainte Marie, Michigan, USA, 49783, [2]U.S. Fish and Wildlife Service, East Lansing Field Office, 2651 Coolidge Road, East Lansing, Michigan, USA, 48823, [3]U.S. Forest Service, Rocky Mountain Forest and Range Experiment Station, 2500 S. Pine Knoll Drive, Flagstaff, Arizona, USA, 86001, [4]Department of Biology, Lake Superior State University, Sault Sainte Marie, Michigan, USA, 49783, [5]Department of Zoology, Pesticide Research Center, Institute for Environmental Toxicology, Michigan State University, East Lansing, Michigan, USA, 48824-1222*

**Abstract**. Bald eagle *(Haliaeetus leucocephalus)* numbers in North America have increased since the ban of DDT and other organochlorine compounds in the 1970s. The decrease in the environmental concentrations of *p,p'*-DDE has lead to the lessening of egg-shell thinning and has been a major reason for the current resurgence of bald eagle populations in temperate North America, however, this recovery has not been uniform. Eagles nesting along the shorelines and islands of the Great Lakes have continued to experience impaired productivity. In order to examine some of the reasons for the current recovery of bald eagles in the Great Lakes Basin and the potential use of eagles as a bioindicator species of Great Lakes water quality, we analyzed trends in reproductive activity, concentrations of PCBs and *p,p'*-DDE in unhatched eggs, and rates of developmental deformities. Numbers of occupied nests, fledged young, and yearly productivity rates have increased across the basin. No trends have been observed in changes in the concentrations of *p,p'*-DDE nor Total PCBs in unhatched eggs. An increasing rate in the incidence of developmental deformities in nestlings has been observed in Michigan. The recovery of the bald eagle population along the Great Lakes is most likely due to immigration of relatively uncontaminated adults from Interior regions.

## 1. Introduction

The number of breeding pairs of bald eagles *(Haliaeetus leucocephalus)* in North America has increased since the ban of DDT and other organochlorine compounds in the 1970s. The decrease in the environmental concentrations of *p,p'*-DDE has lead to the lessening of egg-shell thinning and has been a major reason for the current resurgence of bald eagle populations in temperate North America (Grier 1982, Postupalsky 1978, Grier 1982, Wiemeyer *et al.* 1984, Colborn 1991). However, the recovery has not been uniform and several regions where populations are not reproducing at a level considered to be healthy continue to exist (Colborn 1991, Bowerman 1993, Best *et al.* 1994, Bowerman et al. 1994). One of these areas is the Basin of the Laurentian Great Lakes, where *p,p'*-DDE and polychlorinated biphenyls (PCBs) have been linked to poor reproductive success (Kozie and Anderson 1991, Bowerman 1991, Bowerman 1993, Best et al., 1994). Other areas which have lower reproductive success associated with high levels of environmental contaminants include the lower Columbia River in Oregon and Washington, Maine, and the Channel Islands in California (Anthony *et al.* 1993, Welch 1994, Garcelon *et al.* 1989). Recent proposals to alter the status of the eagle under the Federal Endangered Species Act (Federal Register 1990) focused solely on the increasing numbers of breeding pairs and overall reproductive rate in the contiguous United States. It is equally important to understand and incorporate the dynamics of the population recovery and the role of PCBs and similar toxicants, as well as *p,p'*-DDE, in this decision.

Bald eagle productivity is correlated with some types of chlorinated hydrocarbon compounds especially polychlorinated diaromatic hydrocarbons (PCDH). Productivity is correlated with concentrations, determined in addled eggs, of some of these compounds, while exposure to other compounds has been suggested as causing teratogenic effects based on studies of surrogate species (Wiemeyer *et al.* 1984, Kubiak *et al.* 1989, Gilbertson *et al.* 1991, Wiemeyer *et al.* 1993, Bowerman *et al.* 1994a, Giesy *et al.* 1994, Bowerman *et al.* 1994b, Bowerman *et al.* 1995). Consequently, the bald eagle has been suggested as a biological indicator species of toxic effects of organochlorine compounds on piscivorous wildlife and the effects of bioaccumulation and biomagnification in the Great Lakes. Eagles forage primarily on fish and other vertebrates associated with the coastal zone of the Great Lakes, rivers, and interior aquatic systems. Concentrations of *p,p'*-DDE and PCBs in the plasma of nestlings reflect exposure to these compounds from the prey species within the breeding area (Frenzel 1985, Bowerman 1993). To examine some of the reasons for the current recovery of bald eagles in the Great Lakes Basin and the potential use of eagles as a bioindicator species of Great Lakes water quality, we analyzed trends in reproductive activity, concentrations of PCBs and *p,p'*-DDE in unhatched bald eagle eggs, and rates of developmental deformities in nestling eagles.

## 2. Trends in Regional Bald Eagle Reproduction

The bald eagle is currently distributed as a breeding species along the shorelines and islands of four of the five Great Lakes: Lake Superior; Lake Michigan; Lake Huron; and Lake Erie (Bowerman, 1993). Eagles nest along the Great Lakes shorelines of the states of Michigan, Minnesota, Ohio, Wisconsin, and the Province of Ontario. Surveys for bald eagle reproductive activity are conducted yearly for the entirety of Michigan and Ohio, and for smaller areas of Minnesota, Wisconsin, and Ontario. Two aerial surveys are conducted in Michigan, Minnesota, Wisconsin, and Ohio, while ground observations are utilized in Ontario. Timing of the aerial surveys coincide with nest initiation and post-hatch, and are dependent on local conditions and previous experience (Postupalsky 1974). Nestling eagles are banded at 5-9 weeks of age and these observations are utilized to update aerial survey data. In recent years, data for the states of Minnesota and Wisconsin have been incomplete for the second aerial survey which is used to determine the number of fledged young produced.

Regional bald eagle reproductive data were examined for the period 1973-1995 for the four states bordering the Laurentian Great Lakes that had eagles that nest along their shorelines or islands, Michigan, Minnesota, Ohio, and Wisconsin. Reproductive parameters examined included number of occupied nests and number of fledged young. Productivity was calculated by dividing the total number of fledged young by the number of occupied nests, using the method of Postupalsky (1974).

The number of occupied nests in all four states increased between 1973 and 1995 (Figure I). The greatest increases in numbers of nests occurred in Minnesota and Wisconsin, where the number of occupied nests increased from 115 to 618

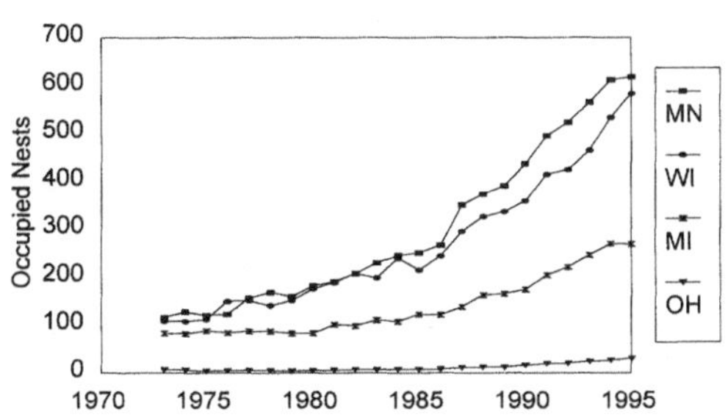


**Figure I.** Number of Occupied Bald Eagle Nests in Minnesota (MN), Wisconsin (WI), Michigan (MI), and Ohio (OH), 1973-1995. Source: U.S. Fish and Wildlife Service

(537%), and from 108 to 583 (540%), respectively. The increases of occupied nests in Michigan and Ohio were from 83 to 268 (323%), and from 7 to 30 (429%), respectively. In all four states, the number of occupied breeding areas have had the greatest increases since 1985.

The number of fledged young in all four states also increased between 1973 and 1995 (Figure II). The greatest increases in numbers of fledged young occurred in Minnesota and Wisconsin, where the number of fledged young increased from 113 to 797 (705%), and from 106 to 694 (655%), respectively. The increases of fledged

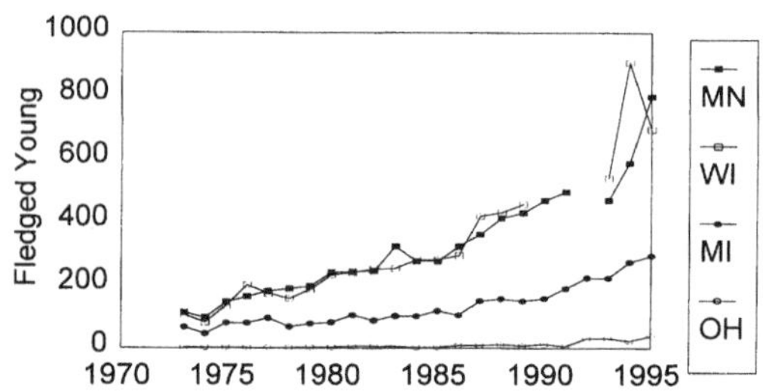


**Figure II.** Number of Young Fledged from Occupied Bald Eagle Nests in Minnesota (MN), Wisconsin (WI), Michigan (MI), and Ohio (OH), 1973-1995. Source: U.S. Fish and Wildlife Service

young in Michigan and Ohio were from 66 to 292 (442%), and from 2 to 38 (1900%), respectively.

The number of young per occupied nest, or annual productivity rate, increased in all four states between 1973 and 1995 (Figures III and IV). The greatest increases occurred in Ohio and Michigan, where annual productivity rates increased from 0.29 to 1.27, and from 0.80 to 1.09, respectively. The increases of annual productivity rates in Minnesota and Wisconsin were from 0.98 to 1.29, and from 0.98 to 1.19, respectively. Recent increases in Michigan and Ohio in excess of the 1.0 productivity rate associated with a healthy population are rare when one compares the rate over the entire period, while eagles in Wisconsin and Minnesota have consistently exceeded this rate.

The continuity of data collection from the annual surveys for determining whether there are continued effects of Great Lakes pollutants on bald eagle reproduction is threatened by budget constraints. Michigan and Ohio maintain statewide annual surveys. However, in Minnesota and Wisconsin, surveys of the

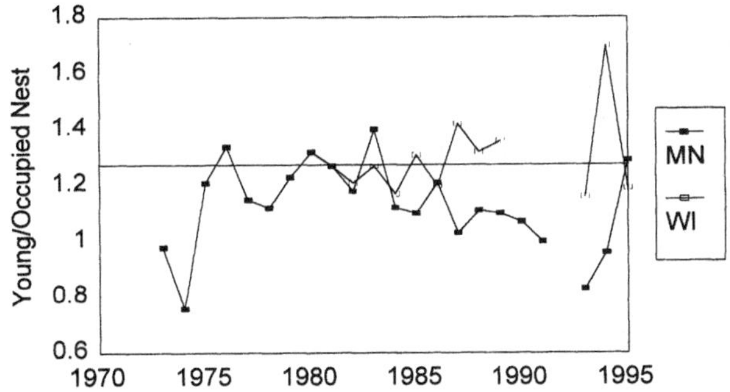


**Figure III.** Number of Young Fledged per Occupied Bald Eagle Nest in Minnesota (MN), and Wisconsin (WI), 1973-1995. The line at 1.0 young per occupied nest is the rate associated with healthy eagle populations (Sprunt *et al.*, 1973). Source: U.S. Fish and Wildlife Service.

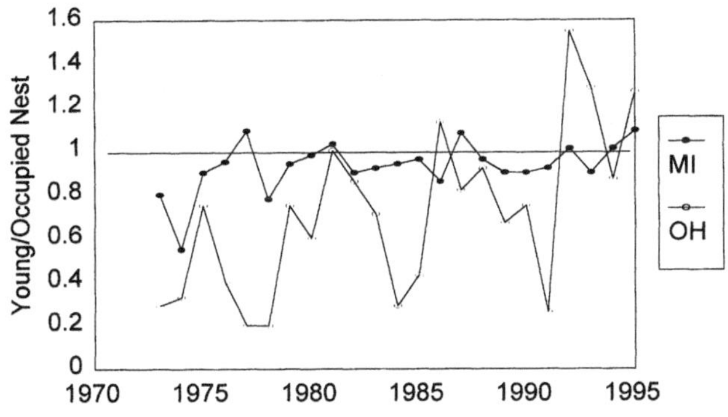


**Figure IV.** Number of Young Fledged per Occupied Bald Eagle Nest in Michigan (MI), and Ohio (OH), 1973-1995. The line at 1.0 young per occupied nest is the rate associated with healthy eagle populations (Sprunt *et al.*, 1973). Source: U.S. Fish and Wildlife Service

number of young produced have been conducted during the 1990s only as estimates based on smaller regions within each state. This has prevented utilization of these data in direct comparisons on a statewide basis.

## 3. Trends in Bald Eagle Reproduction in Michigan

The bald eagle is currently distributed along the Michigan shorelines and islands of Lake Superior, Lake Michigan, Lake Huron, and Lake Erie (Bowerman 1993). Surveys of bald eagle reproductive activity have been conducted annually for the entire State of Michigan since the early 1960s and include annual site visits and banding of young at over 85% of all occupied nests. We used Michigan since it had the only complete data set of sufficient sample size for comparing differences in bald eagle reproductive performance among regions near to and remote from the Great Lakes shoreline. While Ohio has a long term program for monitoring of their eagle population, the yearly sample size is less than 40 breeding pairs, with fewer than 10 of these pairs located in interior areas.

Within Michigan, regional bald eagle reproductive data were examined for the period 1961-1995 in 'Great Lakes' and 'Interior' regions. These are, respectively,

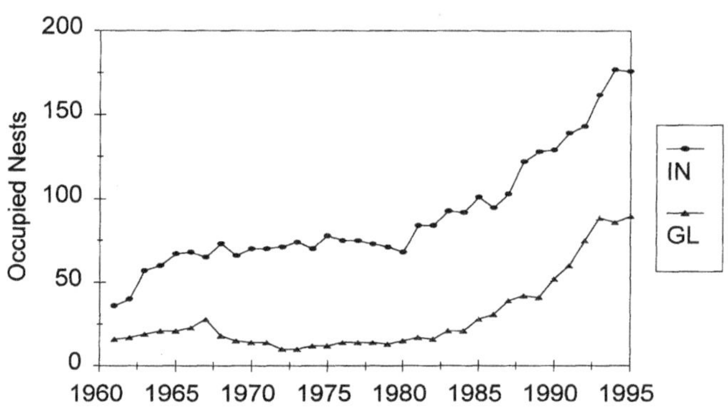


**Figure V.** Number of Occupied Bald Eagle Nests Between 1961 and 1995 in Two Regions of Michigan. Great Lakes (GL) includes those eagles nesting within 8.0 km of the shorelines of the Great Lakes and along rivers accessible to spawning runs of Great Lakes fish, and Interior (IN) includes those eagles nesting greater than 8.0 km from the Great Lakes and not along rivers accessible to spawning runs of Great Lakes fish. Source: U.S. Fish and Wildlife Service

bald eagles nesting within 8.0 km of the shorelines or on islands of the Great Lakes or along rivers open to spawning runs of Great Lakes fishes, and those nesting greater than 8.0 km from a Great Lake or nesting along a river that is not open to Great Lakes spawning runs of fishes.

The number of occupied nests in both regions increased between 1961 and 1995 (Figure V). The number of occupied nests increased from 16 to 90 (563%), and from 36 to 176 (489%), in Great Lakes and Interior regions, respectively.

The number of fledged young in both regions increased between 1961 and 1995 (Figure VI). The number of fledged young increased from 0 to 81, and from 34 to 207, in Great Lakes and Interior regions, respectively.

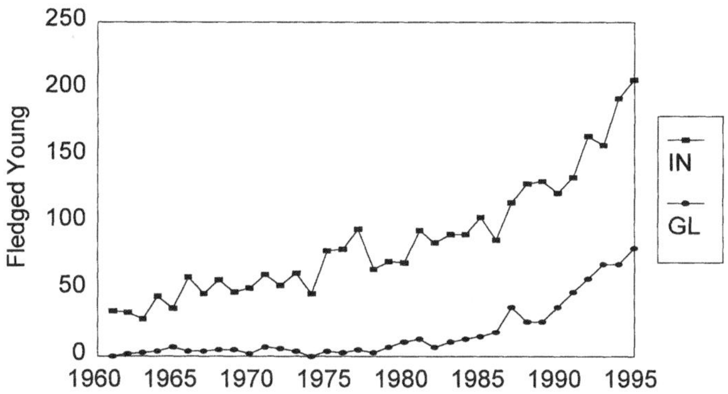


**Figure VI.** Number of Young Fledged from Occupied Bald Eagle Nests Between 1961 and 1995 in Two Regions of Michigan. Great Lakes (GL) includes those eagles nesting within 8.0 km of the shorelines of the Great Lakes and along rivers accessible to spawning runs of Great Lakes fish, and Interior (IN) includes those eagles nesting greater than 8.0 km from the Great Lakes and not along rivers accessible to spawning runs of Great Lakes fish. Source: U.S. Fish and Wildlife Service

The number of young per occupied nest, or annual productivity rate, also increased between 1961 and 1995 for both regions (Figure VII). The annual productivity rates increased from 0.00 to 0.90, and from 0.94 to 1.19, in Great Lakes and Interior regions, respectively.

The number of Interior occupied nests doubled from 1980 to 1995, while the Great Lakes nests doubled from 1989 to 1995. Increases in new occupied nests and

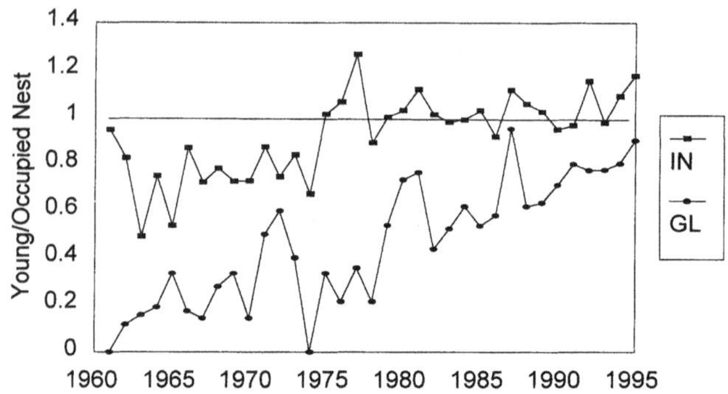


**Figure VII.** Number of Young Produced from Occupied Bald Eagle Nests Between 1961 and 1995 in Two Regions of Michigan. Great Lakes (GL) includes those eagles nesting within 8.0 km of the shorelines of the Great Lakes and along rivers accessible to spawning runs of Great Lakes fish, and Interior (IN) includes those eagles nesting greater than 8.0 km from the Great Lakes and not along rivers accessible to spawning runs of Great Lakes fish. The line at 1.0 young per occupied nest is the rate associated with healthy eagle populations (Sprunt *et al.*, 1973). Source: U.S. Fish and Wildlife Service

fledged young in the Great Lakes region correspond to an increase in the productivity rate for the entire State of Michigan. Based on the productivity rate for the Great Lakes region prior to 1989, this increase in new occupied nests can only be explained by immigration of adults from the Interior regions of Michigan, Minnesota, and Wisconsin (Colborn 1991, Bowerman 1993, Best *et al.* 1994, Bowerman *et al.*, 1995).

## 4. Trends in PCBs and DDE in Eggs

We examined the changes in Total PCBs and *p,p'*-DDE concentrations in 61 unhatched eggs collected from Michigan, Ohio, and Wisconsin, along the shorelines and islands of the Great Lakes since 1968 (Figures VIII and IX). Concentrations were reported on a fresh, wet weight concentration to account for moisture loss prior to collection (Bowerman *et al.* 1994c). The No Observable Adverse Effect Concentrations (NOAEC) based on a productivity rate of 1.0 were determined by weighted sigmoidal regression analysis to be 2.7 mg/kg for *p,p'*-DDE (Wiemeyer *et al.* 1993) and by regression analysis to be 4.0 mg/kg for Total PCBs (Wiemeyer *et.*

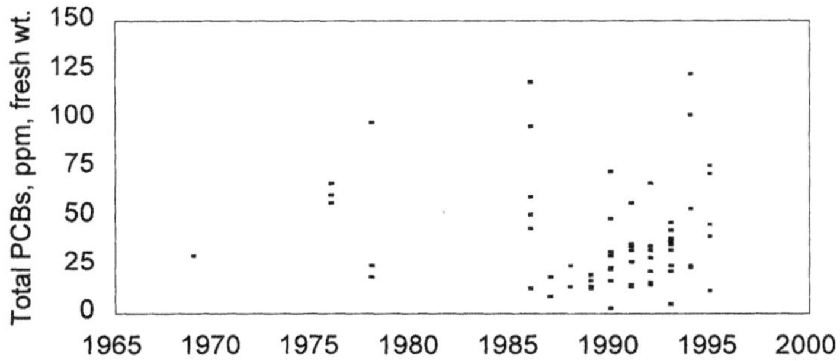


**Figure VIII.** Concentrations of Total PCBs, mg/kg, Fresh Wet Weight, in Unhatched Bald Eagle Eggs Collected from the Great Lakes, 1968-1995. Source: Wiemeyer *et al.*, 1984 and Best, unpubl. data.

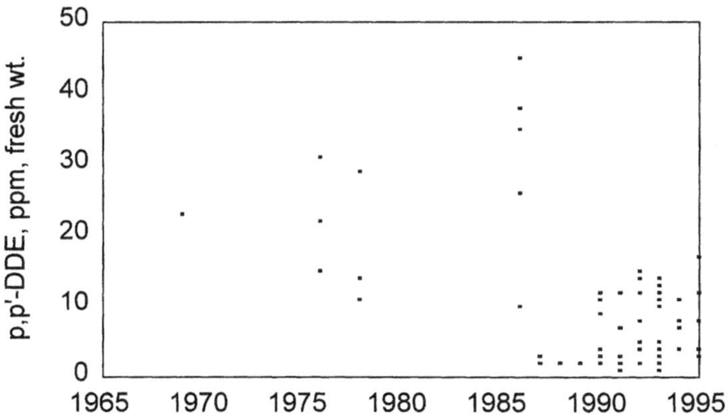


**Figure IX.** Concentrations of *p,p'*-DDE, mg/kg, Fresh Wet Weight, in Unhatched Bald Eagle Eggs Collected from the Great Lakes, 1968-1995. Source: Wiemeyer *et al.*, 1984 and Best, unpubl. data.

*al.* 1984). Concentrations associated with near total reproductive failure, i.e., a productivity rate of <0.10, were determined by weighted sigmoidal regression analysis to be 27 mg/kg for *p,p'*-DDE (Wiemeyer *et al.* 1993) and by regression analysis to be 35 mg/kg for Total PCBs (Wiemeyer *et al.* 1984).

There were no significant trends using autocorrelation and autocorrelation on first differences (Box and Jenkins, 1976) for the period 1968-1995 in the concentrations of either Total PCBs (Figure VIII) nor *p,p'*-DDE (Figure IX) in unhatched eggs of bald eagles from the Great Lakes region. Only one egg (n=60) collected since 1985 from the Great Lakes region has had Total PCB concentrations less than the calculated NOAEC, while 41.7% (n=25) have had concentrations greater than the concentration associated with nearly complete reproductive failure. Few eggs (n=16, 17.7%) collected since 1985 have had *p,p'*-DDE concentrations less than the calculated NOAEC, but only of these five eggs (8.3%) have had concentrations greater than the concentration associated with 15% egg shell thinning (16 mg/kg *p,p'*-DDE), a concentration associated with significant decreases in eagle productivity rates (Wiemeyer *et al.* 1984; Wiemeyer *et al.* 1993).

A problem encountered with interpretation of the effects of particular chemicals on bald eagle productivity rates lies in the intercorrelation among various organochlorine compounds (Wiemeyer *et al.* 1984, Wiemeyer *et al.* 1993, Best *et al.* 1994, Bowerman *et al.* 1994c, Bowerman *et al.* 1995). We have previously observed that eggs collected from the Great Lakes from 1986-1990 have shown a greater inverse correlation between Total PCBs ($r^2$= 0.80) and productivity than with *p,p'*-DDE concentrations ($r^2$= 0.63) (Best *et al.* 1994; Bowerman *et al.* 1994c).

## 5. Trends in Observed Abnormalities

The first nestling bald eagle with a crossed bill was reported by Grier (1968). More recently, our group has reported five additional observations: one near the Menominee River in the western Upper Peninsula of Michigan in 1968; one in the northern Lower Peninsula of Michigan in 1982; two in northern Wisconsin near the Menominee River in 1986 and 1987; and one in northern Minnesota in Voyageurs National Park in 1989 (Bowerman *et al.* 1994b).

Since 1989, six additional observations of deformities have been observed in bald eagles in Michigan. In 1992, near Hardwood Lake within the Saginaw Bay Watershed, a nestling with fused vertebrae and deformed legs which resulted in inability to stand upright, was observed. In 1993, four nestlings were observed with deformities, three with crossed bills and one without a crossed-bill, but with deformities in both feet. The nestlings with crossed bills occurred at nests located along the Raisin River and Wood Tick Peninsula near Lake Erie, and at Tomahawk Flooding in the northern Lower Peninsula of Michigan. The nestling with deformed feet occurred at a nest located at Black River Flooding in the eastern Upper Peninsula of Michigan. In 1995, a second nestling with a crossed bill was observed at the same nest located along the Raisin River. During the period between finding the first and second deformed nestling at the Raisin River site, at least one of the two adults in this breeding pair was replaced.

The incidence of deformed nestling eagles within Michigan for the period 1968-1995 was 26.6 per 10,000. The rate has changed from 12.5 per 10,000 for the period 1968-1989, to 42.3 per 10,000 for the period 1990-1995. Since the actual percentage of nestling eagles examined and banded in Michigan has not changed since 1964, this change in incidence appears to be real. We theorize that the increase in incidence is most likely attributed to a decrease in the egg shell thinning influence of *p,p'*-DDE as a factor in nesting success of bald eagles in the region which has allowed for a greater rate of survival and thus expression of deformities. Based on results of studies on other colonial waterbirds and poultry, these teratogenic effects in nestling eagles are likely due to dioxin-like congeners of PCBs (Bowerman *et al.* 1994b, Giesy *et al.* 1994, Ludwig *et al.* 1996).

## 6. The Bald Eagle as an Indicator Species of Water Quality

The bald eagle has been proposed as a basin-wide bioindicator of Great Lakes water quality by the International Joint Commission (IJC 1994). There are a number of advantages and of utilizing the bald eagle in this manner. The major strength of using the bald eagle as a bioindicator species lies in the fact that it is one of the only species where we can obtain a near total count of all breeding pairs and their reproductive outcome. We therefore, have a population "measure" rather than relying on an estimate or "index" of the actual population. These data exist for the entire area of Michigan and Ohio, and along the Lake Erie shoreline of Ontario, from the early 1960s. A number of other areas within the basin have data sets that also date back to the mid-1960s. Individual breeding areas can be followed from that time period to the present and compared to regional contaminant concentrations in bald eagles and other "surrogates" to determine relative contamination of the Great Lakes.

The eagle is primarily a year-around resident of the Great Lakes Basin, and in many areas is non-migratory. Unhatched eggs of eagles, therefore, are reflective of the local contamination of their breeding area. This is also true of egg concentrations of the non-migratory population of Great Lakes herring gulls *(Larus argentatus)*. While terns *(Sterna spp.)* and double-crested cormorants *(Phalacrocorax auritus)*, are migratory , they accumulate contaminant levels at rates that are reflective of the degree of pollution of their local nesting environments (Weseloh *et al.* 1983, Kubiak *et al.* 1989, Fox *et al.* 1991, Tillitt *et al.* 1992, Giesy *et al.* 1994).

Another advantage lies in the amount of information that is known about the life history and habitat requirements of bald eagles. Their prey preferences during the breeding season, their preferences in nest trees, and potential habitat within 1.6 km of the Great Lakes shoreline have been identified (Bowerman 1993). In addition, the effects of some of the organochlorine compounds are correlated to reproductive performance.

The large body size of bald eagles make it easy for the collection of large samples of blood and eggs of sufficient volume to conduct a number of chemical as well as genetic tests. Blood samples can be analyzed for organochlorine contaminants and associated changes in biochemical markers, and evaluated genetically to determine parentage of nestlings (Bowerman 1993). Since most nests are accessible by an experienced climber, samples of blood and feathers can be easily obtained each year from known breeding areas and changes in concentrations of environmental contaminants can be determined over time.

There is a number of disadvantages, however, in using the bald eagle as a bioindicator species. Since this species has been a "threatened" or "endangered" species throughout most of its range, and is highly prized by the public, it is not a suitable laboratory animal. Its legal status and importance to the public also makes it difficult to justify collection of fresh eggs. There are several areas along the shorelines of the Great Lakes and the connecting channels where the bald eagle has not reestablished territories either as the result of continuing severe environmental contamination or as a result of habitat destruction, urbanization and disturbance. For example, bald eagles do not currently nest along Lake Ontario shorelines after being extirpated in the 1950s.

A further problem with use of the bald eagle as a bioindicator species is the difficulty in collecting addled eggs from areas of interest, due to predation by scavengers. This problem results in low numbers of samples collected throughout the Great Lakes, with many samples being collected from accessible sites rather than at random sites.

A final problem with the eagle as an ecosystem monitor is the difficulty of separating the factor of prey availability from the effects of environmental contaminants on nest productivity. For example, there was no correlation between estimates of 5-year nest productivity at sites along the Wisconsin shoreline of Lake Superior and measures of either prey delivery rates or environmental contaminants measured indirectly as *p,p'*-DDE and Total PCBs in nestling plasma (Dykstra *et al.*

1998). It is suggested that a comparison of egg concentrations, as a direct measure of exposure, with threshold levels might be preferable in investigating the contribution of these two factors on reproductive success.

Some of the weaknesses of using the bald eagle as a bioindicator species can be overcome by combining and contrasting contaminant and reproductive data on various colonial waterbird species with data on bald eagles from the same areas. The development of these interspecies correlations might permit prediction of the contaminant status of areas where either eagles or colonial waterbirds do not currently nest. This could be used not only to expand upon the shoreline areas which can be assessed throughout the basin, but also to predict the further degree of remedial progress required for restoration of populations of these indicator species.

## 7. Conclusions

Bald eagle populations have generally increased throughout the Great Lakes Basin after the suspensions or cancellations of the registered uses of DDT and other organochlorine compounds in the early 1970s. The populations in the western portion of the basin, in Minnesota and Wisconsin, have had the largest increase in numbers of breeding eagles and fledglings produced and are the primary source of the recovery of the basin's eagle population. Eagles nesting along the shorelines and islands of the Great Lakes continue to experience impairment of reproduction and are likely a population "sink". The key to sustaining the current recovery of the bald eagle population within the Great Lakes Basin lies in maintaining sufficient reproductive rates within the relatively clean Interior population to compensate for the continued reproductive impairment of the Great Lakes eagle population (Bowerman 1993).

There are no trends in concentrations of PCBs or DDE in eggs of bald eagles nesting along the Great Lakes, although in a small number of eggs collected recently, concentrations are below thresholds associated with reproductive impairment.

We conclude that the bald eagle would make a good choice for adoption as a bioindicator of Great Lakes water quality, in combination with several colonial waterbird species. This would strengthen the use of avian species as bioindicators of Great Lakes health, as well as appeal to the public as suitable species for use in educational programs related to the effects of environmental pollutants in the Great Lakes Basin.

## References

Anthony, R.G., Garrett, M.J., and Schuler, C.A.: 1993, 'Environmental contaminants in bald eagles in the Columbia river estuary', *J. Wildl. Manage.* **57**, 10-19.

Best, D.A., Bowerman, W.W., Kubiak, T.J., Winterstein, S.R., Postupalsky, S., Shieldcastle, M.C., and Giesy, J.P.: 1994, 'Reproductive impairment of bald eagles along the Great Lakes shorelines of Michigan and Ohio', *Raptor Conservation Today,* Meyburg B.U. and Chancellor R.D. (eds.). World Working Group on Birds of Prey, The Pica Press, East Sussex, Great Britain, pp.697-702.

Bowerman, W.W.: 1991, *Factors influencing breeding success of bald eagles in upper Michigan.* M.A. thesis, Northern Michigan University, Marquette, Michigan.

Bowerman, W.W.: 1993, *Regulation of Bald Eagle (Haliaeetus leucocephalus) Productivity in the Great Lakes Basin: An Ecological and Toxicological Approach.* Ph.D. Dissertation, Michigan State University, East Lansing, Michigan.

Bowerman, W.W., Evans, E.D., Giesy, J.P., and Postupalsky, S.: 1994a, 'Using feathers to assess risk of mercury and selenium to bald eagle reproduction in the Great Lakes region', *Arch. Environ. Contam. Toxicol.* **27,** 294-298.

Bowerman, W.W., Kubiak, T.J., Holt, J.B., Evans, D.B., Eckstein, R.J., and Sindelar, C.R.: 1994b, 'Observed abnormalities in mandibles of nestling bald eagles', *Bull. Environ. Contam. Toxicol.* **53,** 450-457.

Bowerman, W.W., Best, D.A., Kubiak, T.J., Martell, M.S., Postupalsky, S., and Giesy, J.P.: 1994c, 'The influence of environmental contaminants on bald eagle populations in the Laurentian Great Lakes, North America' *Raptor Conservation Today.* Meyburg B.U. and Chancellor R.D. (eds.), World Working Group on Birds of Prey, The Pica Press, East Sussex, Great Britain, pp. 703-707.

Bowerman, W.W., Giesy, J.P., Best, D.A., and Kramer, V.J.: 1995, 'A review of factors affecting productivity of bald eagles in the Great Lakes region: Implications for recovery', *Environ. Health Perspect.* **103,** 51-59.

Box, G.E.P., and Jenkins, G.M.: 1976, *Time Series Analysis: Forecasting and Control.* Holden-Day, San Francisco.

Colborn, T.: 1991, 'Epidemiology of Great Lakes bald eagles', *J. Toxicol. Environ. Health* **33,** 395-453.

Dykstra, C.R., Meyer, M.W., Warnke, D.K., Karasov, W.H., Andersen, D.E., Bowerman, W.W., IV, and Giesy, J.P.: 1998, 'Low reproductive rates of Lake Superior bald eagles: Low food delivery rates or environmental contaminants?', *J. Great Lakes Res.* **24**,32-44.

Federal Register: 1990, '50 CFR Part 17, Endangered and Threatened Wildlife and Plants; Advance Notice of a Proposal to Reclassify or Delist the Bald Eagle (*Haliaeetus leucocephalus*)', *Federal Register* **55,** 4209-4212.

Frenzel, R.W., 1985, *Environmental contaminants and ecology of bald eagles in southcentral Oregon.* Ph.D. Thesis, Oregon State University, Corvallis, Oregon.

Fox, G.A., Weseloh, D.V., Kubiak, T.J., and Erdman, T.C.: 1991, 'Reproductive outcomes in colonial fish-eating birds: A biomarker for developmental toxicants in Great Lakes food chains', *J. Great Lakes Res.* **17,** 153-157.

Garcelon, D.K., Risebrough, R.W., Jarman, W.M., Chartrand, A.B., and Littrell, E.E.: 1989, 'Accumulation of DDE by bald eagles *Haliaeetus leucocephalus* reintroduced to Santa Catalina Island in southern California', *Raptors in the Modern World,* Meyburg B.U. and Chancellor R.D. (eds.). World Working Group on Birds of Prey, Berlin, Germany, pp.491-494.

Giesy, J.P., Ludwig, J.P., and Tillitt, D.E.: 1994, 'Embryolethality and deformities in colonial, fish-eating waterbirds of the Great Lakes region: Assigning causality', *Environ. Sci. Technol.* **28,** 128A-135A.

Gilbertson, M., Kubiak, T.J., Ludwig, J.P., and Fox, G.A.: 1991, 'Great Lakes embryo mortality, edema, and deformities syndrome (GLEMEDS) in colonial fish-eating birds: Similarity to chick-edema disease', *J. Toxicol. Environ. Health* **33,** 455-520.

Grier, J.W.: 1982, 'Ban of DDT and subsequent recovery of reproduction in bald eagles', *Science* **212,** 1232-1235.

International Joint Commission: 1974, *Bioindicators as a measure of success for virtual elimination of persistent toxic substances.* Fox, G.A.(ed.), International Joint Commission, Windsor, Ontario, Canada.

Kozie, K.D., and Anderson, R.K.: 1991, 'Productivity, diet, and environmental contaminants in bald eagles nesting near the Wisconsin Lake Superior shoreline', *Arch. Environ. Contam. Toxicol.* **20,** 41-48.

Kubiak, T.J., Harris, H.J., Smith, L.M., Schwartz, T., Stalling, D.L., Trick, J.A., Sileo, L., Docherty, D., and Erdman, T.C.: 1989, 'Microcontaminants and reproductive impairment of the Forster's tern in Green Bay, Lake Michigan - 1983', *Arch. Envir. Contam. Toxicol.* **18,** 706-727.

Ludwig, J.P., Kurita-Matsuba, H., Auman, H.J., Ludwig, M.E., Summer, C.L., Giesy, J.P., Tillitt, D.E., and Jones, P.D.: 1996, 'Deformities, PCBs, and TCDD-equivalents in double-crested cormorants *(Phalacrocorax auritus)* and Caspian terns *(Hydroprogne caspia)* of the upper Great Lakes 1986-1991: Testing a cause-effect hypothesis', *J. Great Lakes Res.* **22,** 172-197.

Postupalsky, S.: 1974, 'Raptor reproductive success: some problems with methods, criteria, and terminology', *Management of Raptors*, Hamerstrom, F.N. Jr., Harrell, B.E. and Ohlendorff, R.R.(eds.). Proc. Conf. Raptor Conserv. Tech., Raptor Res. Report No. 2, pp. 21-31.

Postupalsky, S.: 1978, 'The bald eagles return', *Natural History* **87,** 62-63.

Sprunt, A., IV, Robertson, W.B., Postupalsky, S., Hensel, R.J., Knoder, C.E., and Ligas, F.J., 1973, 'Comparative productivity of six bald eagle populations', *Trans. N. Am. Wildl. Nat. Resour. Conf.* **38,** 96-106.

Tillitt, D.E., Ankley, G.T., Giesy, J.P., Ludwig, J.P., Kurita-Matsuba, H., Weseloh, D.V., Ross, P.S., Bishop, C., Sileo, L., Stromberg, K.L., Larson, J., and Kubiak, T.J.: 1992, 'Polychlorinated biphenyls residues and egg mortality in double-crested cormorants from the Great Lakes', *Environ. Toxicol. Chem.* **11,** 1281-1288.

Welch, L.J.: 1994, *Contaminant burdens and reproductive rates of bald eagles breeding in Maine.* M.S. thesis, University of Maine, Old Town, Maine.

Weseloh, D.V., Teeple, S.M., and Gilbertson, M.: 1983, 'Double-crested cormorants of the Great Lakes: Egg laying parameters, reproductive failure and contaminant residues in egg, Lake Huron 1972-1973', *Can. J. Zool.* **61,** 427-436.

Weseloh, D.V., Custer, T.W., and Braune, B.M.: 1989, 'Organochlorine contaminants in eggs of common terns from the Canadian Great Lakes', *Environ. Pollut.* **59,** 141-160.

Wiemeyer, S.N., Lamont, T.J., Bunck, C.M., Sindelar, C.R., Gramlich, F.J., Fraser, J.D., and Byrd, M.A.: 1984, 'Organochlorine pesticide, polychlorobiphenyl, and mercury residues in bald eagle eggs -1969-1979- and their relationships to shell thinning and reproduction', *Arch. Environ. Contam. Toxicol.* **13,** 529-549.

Wiemeyer, S.N., Bunck, C.M., and Stafford, C.J.: 1993, 'Environmental contaminants in bald eagle eggs -1980-84- and further interpretations of relationships to productivity and shell thickness', *Archiv. Environ. Contam. Toxicol.*, **24,** 213-227.

# LONG TERM TRENDS IN LIVER NEOPLASM EPIZOOTICS OF BROWN BULLHEAD IN THE BLACK RIVER, OHIO

PAUL C. BAUMANN[1] and JOHN C. HARSHBARGER[2]

[1] *Biological Resources Division, USGS, The Ohio State University, School of Natural Resources, 2021 Coffey Road, Columbus, Ohio 43210;* [2] *Registry of Tumors in Lower Animals, George Washington University, Department of Pathology, 2300 Eye Street, NW, Washington, DC 20037.*

**ABSTRACT.** Since 1980, liver neoplasms in brown bullhead and polynuclear aromatic hydrocarbons (PAH) in sediment have been researched in a series of studies on the Black River in Lorain, Ohio. In the early 1980s the liver cancer prevalence in fish of age 3 and older was high, ranging from 22% to 39% of the adult population. These high cancer rates corresponded to high levels of PAHs in the sediment resulting from long-term releases by an upstream coking facility (USX). In 1983 this coking plant was closed, and by 1987 the PAH in sediment had declined by about two orders of magnitude. Coincidentally the tumor prevalence in 1987 was only about one-fourth of that in the early 1980s. In 1990, the most contaminated sediments were dredged. Neoplasm surveys in 1992 and 1993 found liver tumor frequencies in mature bullhead were as high as or higher than in the early 1980s. However liver tumor incidence declined in 1994, especially among age 3 fish, where neoplasm incidence was zero. These age 3 fish were the first group sampled that were not present during the 1990 dredging. These data are consistent with a hypothesis that the increase in tumor prevalence in 1992 and 1993 was caused by exposure to buried PAH-contaminated sediments released by the dredging. This research points out the insight provided by long term effects studies.

## 1. Introduction

Epizootics of neoplasms in benthic fish have been linked to carcinogens in sediment, particularly carcinogenic polynuclear aromatic hydrocarbons (PAH), in a variety of species from both freshwater and marine habitats (Harshbarger and Clark, 1990; Myers *et al.* 1990; Baumann 1992). Liver neoplasms, both hepatocellular and cholangiocellular, have been the most strongly correlated with environmental carcinogens (Harshbarger and Clark 1990). However only a few of these locations have had long term monitoring (ten or more years) with sufficient sampling to determine changes in neoplasm prevalence over time and to supply data on possible causes for changes noted.

One of these locations is the Black River, Ohio, where epizootiologic studies were first initiated in 1980 (Baumann *et al.* 1982). Research on this system has continued periodically through 1994. Evaluation of this data from the early 1980s through 1987 delineated a decline in liver tumors correlated with a decline in sediment PAH after a steel industry downturn in 1992 and a coke plant closure in

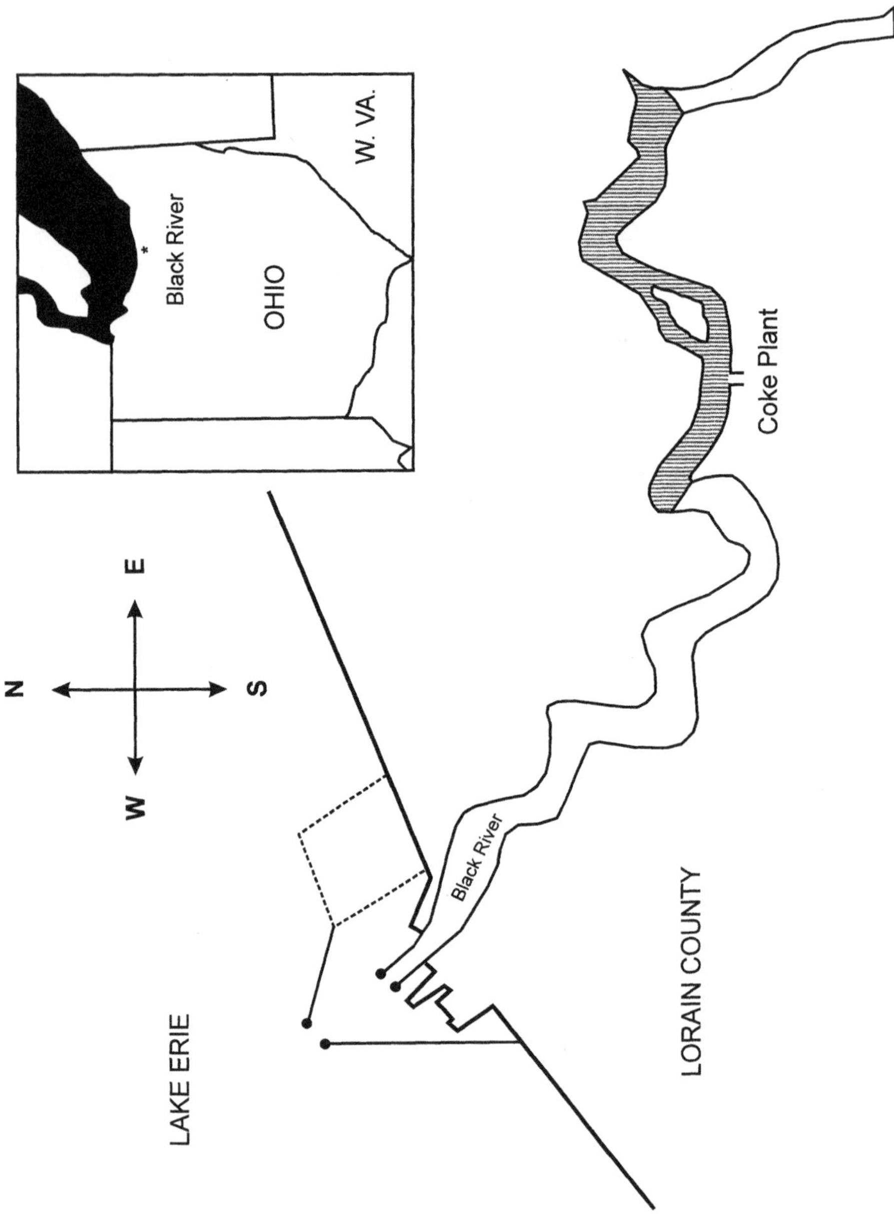


**Figure I** Location of the Black River, Ohio; Fish were Sampled from the Shaded Portion of the River.

1993 (Baumann and Harshbarger, 1995). In this paper we will continue to follow changes in tumor prevalence through 1994, a time period that included remedial dredging in 1990, and will hypothesize the causative factors for changes noted.

## 2. Methods

The Black River has its confluence with Lake Erie at Lorain, Ohio (Figure I). The principle industry on the lower river is the USS/KOBE Steel Company. A coking facility belonging to this company located about 5.5 km upstream from Lake Erie was closed in 1983. Sediment samples were taken with a small dredge in a series of studies from 1980 to 1994 (Baumann and Harshbarger 1995). Samples represent approximately the top two centimeters of sediment. The 1980 sediments were collected within 100 m of the coke plant outfall (Baumann *et al.* 1982). All other samples were collected from at least 5 nearshore depositional sites downstream and within 0.5 km of the coke plant outfall (Baumann and Harshbarger 1995). These locations would be expected to represent maximum PAH concentrations, and are within the area from which the brown bullhead were also sampled (Figure I). PAHs in sediment were identified and quantified by capillary column gas chromatography - mass spectrometry (Fabacher *et al.* 1988 and Smith *et al.* 1994).

Surveys in which adult brown bullhead *(Ameiurus nebulosus)* were randomly selected for necropsy and histopathology were carried out in 1982, 1987, 1992, 1993, and 1994. Surveys were also conducted in 1980, 1981, 1985 in which grossly visible liver tumors were subjected to histopathology but sectioning of livers without gross lesions did not occur. Data on gross lesions from these later years was used as a conservative estimate of cancer frequency. Statistical differences were determined only among years when surveys used random histopathology. Live fish were collected by overnight sets of fyke nets in all years. Net locations extended about 0.5 km above and below the coke plant outfall (Figure I). Only fish over 250 mm in length were necropsied in an attempt to ensure that all fish were sexually mature (age 3+ or greater). Age was confirmed by counting annular rings on pectoral spines as previously described (Baumann and Harshbarger 1995).

Necropsy of fish in all surveys in this manuscript was performed in the field under the direction of P. Baumann. Fish were anesthetized, sacrificed, and examined externally and internally for lesions. Livers were removed and examined separately. Livers with suspected lesions were always preserved, and during the years 1982, 1987, 1992, 1993, and 1994 all livers were preserved. Livers were preserved in 10% neutrally buffered formalin except for 1994 when NoTox was used as the preservative. Histopathology and diagnosis of all lesions in this manuscript were performed at the Registry of Tumors in Lower Animals under the direction of John

**Table I.** Black River Sediment Concentrations (μg/g Dry Sediment) of Various PAH Compounds. Data not presented in source material is designated 'not applicable'.

| COMPOUND | 1980[a] | 1982[b] | 1984[c] | 1987[d] | 1992[e] | 1994[f] |
|---|---|---|---|---|---|---|
| Acenaphthylene | 40.0 | 8.00 | 17.0 | 0.08 | 1.10 | 0.013 |
| Acenaphthene | 36.0 | 7.20 | 2.50 | 0.14 | NA | 0.055 |
| Phenanthrene | 390.0 | 100.0 | 52.0 | 0.73 | 1.75 | 1.86 |
| Fluoranthene | 220.0 | 76.0 | 33.0 | 0.79 | 2.30 | 3.46 |
| Pyrene | 140.0 | 56.0 | 24.0 | 0.93 | 2.95 | 2.86 |
| Benz(a)anthracene | 51.0 | 23.0 | 11.0 | 0.37 | 2.40 | 0.313 |
| Chrysene/triphenylene | 51.0 | 25.0 | 10.0 | 0.37 | 1.71 | NA |
| Benzo(a)pyrene | 43.0 | 21.0 | 8.80 | 0.24 | 2.65 | 0.237 |
| Benzofluoranthenes | 75.0 | 37.0 | 15.0 | 0.58 | NA | 0.838 |
| Benzo(ghi)perylene | 24.0 | 13.0 | 5.40 | 0.03 | NA | NA |
| Indeno(1,2,3-cd)pyrene | 26.0 | 15.0 | 6.40 | 0.01 | 1.75 | 0.195 |
| TOTAL[g] | 1096.0 | 381.2 | 185.1 | 4.27 | 16.6 | 9.826 |

[a] Baumann *et al.* (1982)
[b] West *et al.* (1985)
[c] Farbacher *et al.* (1988)
[d] Smith *et al.* (1994)
[e] Non-estimated mean values of River Mile 3.2-3.45. (Black River Remedial Action Plan 1994)
[f] Munkittrick, personal communication
[g] Total PAH content represents the sum of all values reported in the table.

Harshbarger. Tissue blocks were cut to include grossly visible lesions; other blocks were cut from four to seven cross sections as the size of the liver increased. Blocks were sectioned at 5 $\mu$m and were stained with Mayer's progressive hematoxylin and counter-stained with alcoholic eosin phloxine. Resulting slides were archived at the Registry of Tumors in Lower Animals.

Hepatocellular lesions were diagnosed to one of three categories by criteria previously described (Baumann *et al.* 1990). Those at a pre-neoplastic or incipient state were classified as hepatocellular alteration, those at a benign stage as hepatocellular adenoma, and those advanced to cancer as hepatocellular carcinoma. Biliary lesions were classified as benign stage (cholangioma) or cancer (cholangiocarcinoma). The category "cancer" in this article will refer to both hepatocellular carcinoma and cholangiocarcinoma. The term "total neoplasms" will refer to all lesions mentioned above except for hepatocellular alteration. Livers listed as normal were not affected by any lesions including hepatocellular alteration. Individual fish diagnosed with multiple lesions were scored for only the most advanced lesion of each cell type. Statistical differences were determined by contingency table analysis using chi-square values.

## 3. Results

Total PAH concentrations in sediment declined steeply from 1980 through 1987 (Table I) (Baumann and Harshbarger, 1995). Individual PAHs including carcinogens such as benzo(a)pyrene all display the same pattern. When analyzed in 1992, total PAH were about four times greater than 1987 levels. B(a)P concentrations were ten times greater in 1992 than in 1987. By 1994 PAH concentrations had again declined, being approximately midway between levels present in 1987 and 1992. Again this same trend was displayed by most of the individual PAH; B(a)P levels in 1994 were equivalent to those in 1987.

Liver cancer frequency was highest in the early 1980s, prior to the closure of the coking facility, and in 1992 and 1993 (Figure II). Cancer prevalences in 1987 and 1994 were significantly lower ($p<0.001$) than in 1982, 1992, and 1993. Total neoplasm frequencies mimic cancer rates to some extent, with 1982, 1992, and 1993 being significantly higher ($p<0.001$) than 1987 and 1994 (Figure III). However bullhead captured in 1987 had twice the incidence of neoplasms as did bullhead captured in 1994, a difference that was significant at the $p<0.05$ level. The ratio of normal livers, those lacking both neoplasms and foci of cellular alteration, to neoplastic livers was much greater in 1994 than in 1987 (Figure III). The percentage of normal livers in 1994 was significantly greater ($p<0.01$) than in any of the other years. Conversely the bullhead population in 1982 had a significantly ($p<0.05$) lower

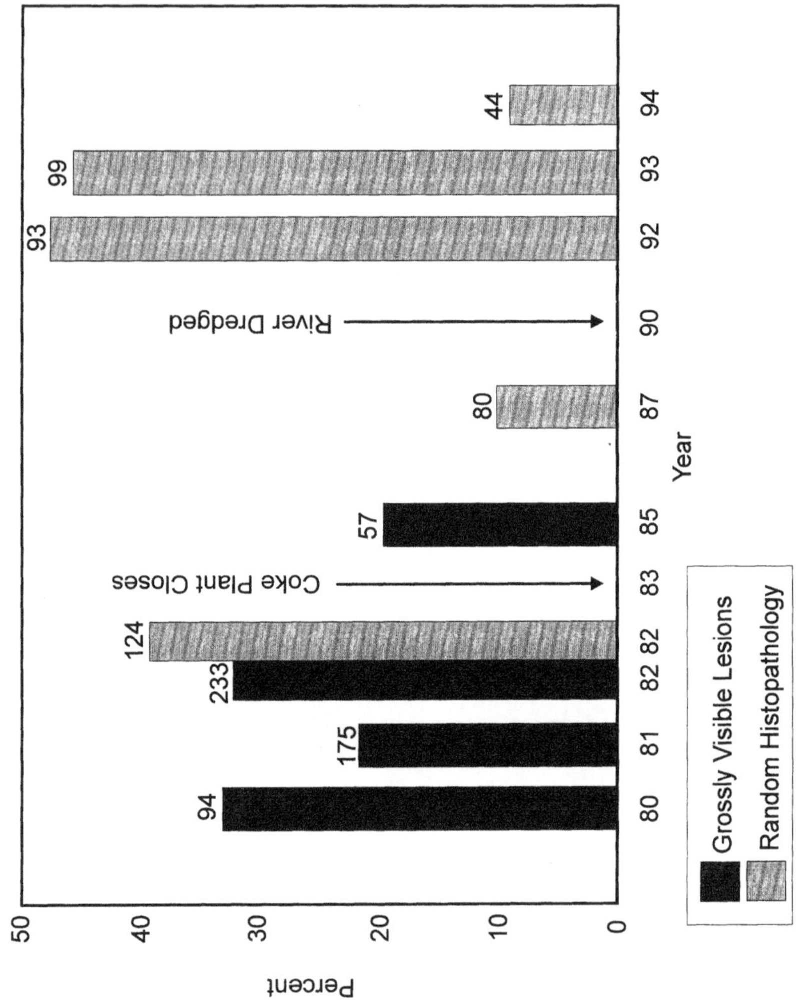


**Figure II** Cancer Prevalence in Brown Bullhead of Age 3 or Older from the Black River, Ohio. Sample Sizes are given at the Top of the Bars.

percentage of animals with normal livers than did any of the other years sampled except for 1993.

By comparing fish of a single age, bias caused by varying proportions of the age groups captured in different years can be eliminated. By focusing on fish of age 3, we can compare liver pathology in those exposed during coke plant operation (1982); those not hatched until after the coke plant ceased production, but prior to remedial dredging (1987); those exposed during the 1990 dredging either as age 1 (1992) or as young-of-the-year (1993); and those not hatched until the year after dredging was completed (1994) (Figure IV). Prevalence of liver pathology in the neoplastic process was greatest in 1982, but similar to the prevalence in 1993. Total pathology was somewhat reduced in 1987 and 1992, although actual neoplasms were significantly ($p<0.05$) more prevalent in 1992, with 1987 having a high incidence of hepatocellular alteration. Age 3 fish in 1994 had no neoplasms, and a significantly ($p<0.01$) higher prevalence of normal livers.

## 4. Discussion

The declining levels of PAH in the surficial sediment of the Black River in the mid 1980s were followed by a decline in liver cancer in age 3 and older brown bullhead as documented in the survey taken in 1987 (Baumann and Harshbarger, 1995). Therefore the increase in liver cancer noted in 1992 and 1993 should have been associated with an increase in PAH exposure, if the cause-effect hypothesis for PAH and wild fish neoplasms is correct. Unfortunately no PAH sediment values are available for 1988 through 1991. However PAH concentrations in 1992 were four-fold higher than those found in 1987. While these concentrations are still an order of magnitude less than those noted as recently as 1985, they indicate that an event resulting in input of PAH to the system may have happened between 1987 and 1992.

One event that might have had this result was the dredging of the area of the Black River near the coking facility in 1990 as part of a Consent Decree concluded between the USX steel company and the U.S. EPA (Figure I). The dredge used for this operation may have redistributed buried sediments with high PAH concentrations into the river. Although the mechanism was not investigated, PAH concentrations in sediment were higher after the dredging as were neoplasm prevalence in bullhead, presumably due to increased PAH exposure.

The histopathology data from 1992 through 1994 is consistent with such a scenario. Two and three years after the dredging the adult bullhead population had a liver cancer prevalence that was statistically indistinguishable from that documented in the early 1980s when the coking plant was operational. Perhaps most convincing is the absence of neoplasms, and the great increase in the percentage of completely

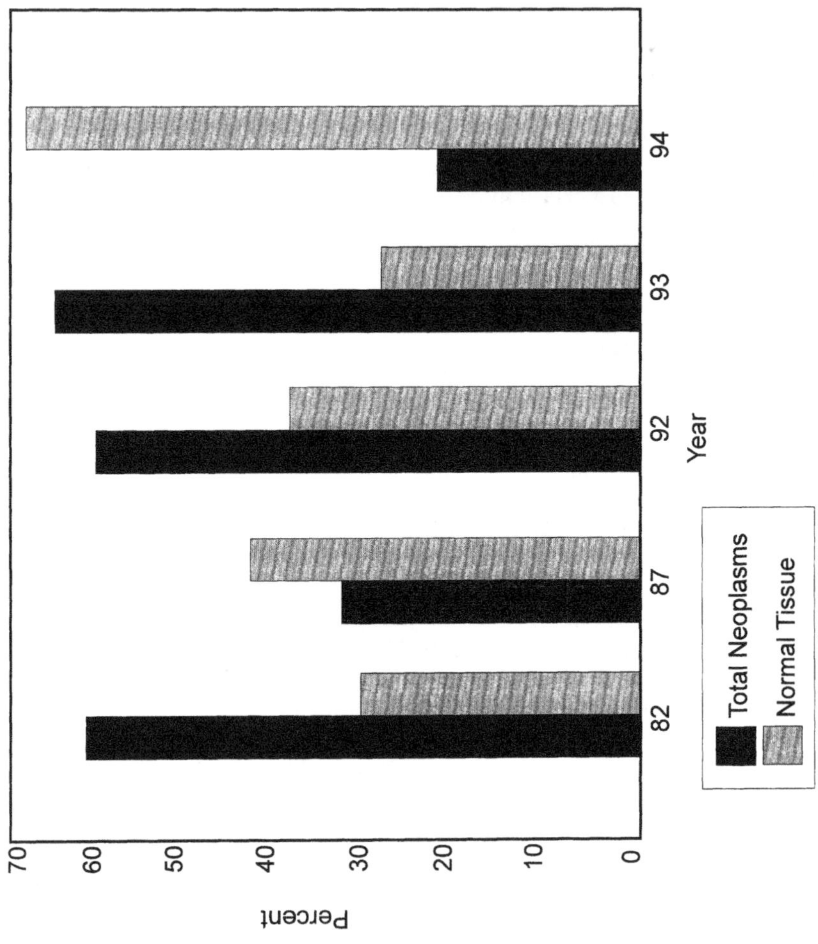


**Figure III Neoplasm Prevalence and Normal Liver Prevalence in Brown Bullhead of Age 3 or Older from the Black River, Ohio as Determined by Random Histopathology**

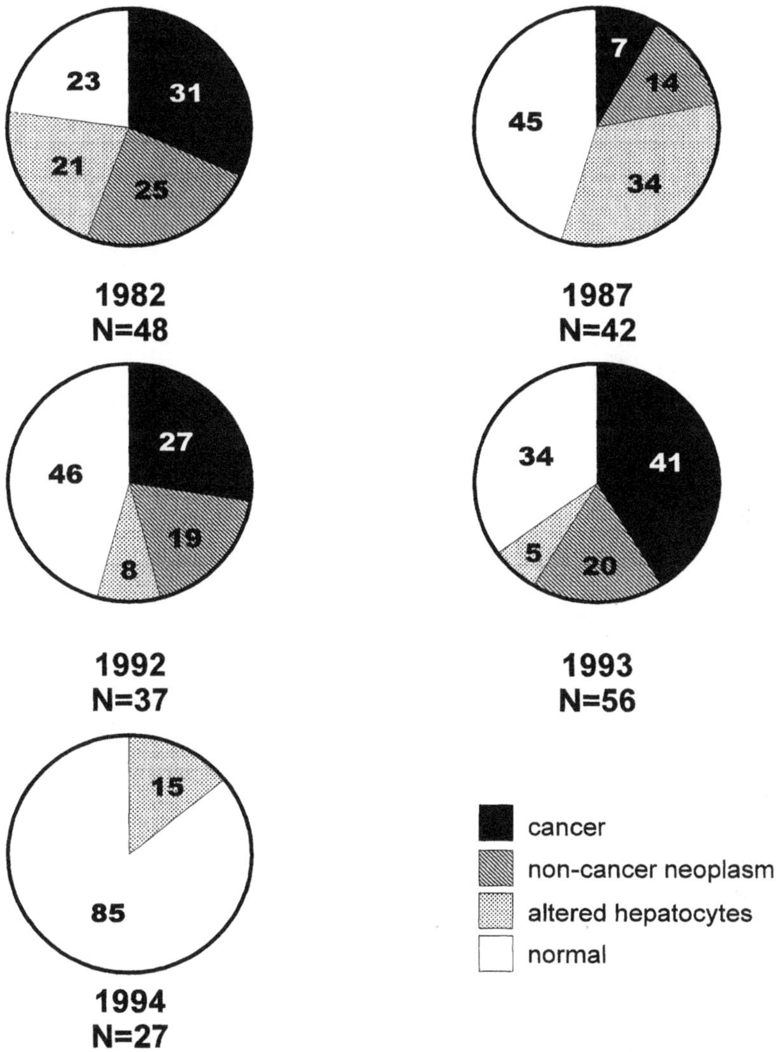


**Figure IV** Percentage of Age 3 Brown Bullhead from the Black River having Various Liver Lesions as Determined by Random Histopathology

normal livers, in age 3 fish taken in 1994. This was the first year when fish old enough to be included in the survey (age 3) would not have been present during the dredging in 1990. The percentage of fish with neoplasia as opposed to normal livers in this age group during 1982, 1992 and 1993 is in stark contrast to age 3 fish from 1994 (Figure IV). This difference indicates a marked change in carcinogen exposure for fish hatched in 1991.

A large number of studies have documented an increased prevalence of liver tumors in benthic fish from habitats containing elevated levels of PAH in sediment. These include marine fishes, such as English sole *(Pleuronectes vetulus)* in Puget Sound (Malins *et al.* 1984 and Myers *et al.* 1990); estuarine fishes such as mummichog *(Fundulus heteroclitus)* from the Elizabeth River, Virginia (Vogelbein *et al.* 1990); and freshwater fishes such as bullhead and white sucker *(Catostomus commersoni)* from the Great Lakes (Baumann *et al.* 1996). The series of studies undertaken over the last 15 years on the Black River have bolstered these findings by first demonstrating a decline in tumor incidence coincident with a decline in surficial PAH (Baumann and Harshbarger 1996). Now the cause-effect hypothesis of PAH and liver cancer is further supported by documenting an increase in liver cancer after the dredging of contaminated sediment followed by a subsequent decline in tumors as the year classes exposed passed through the system. These studies and the insight they allowed demonstrate the value of long term research on the effects of contaminants on wild populations of fish.

## References

Baumann, P.C.: 1992, 'The Use of Tumors in Wild Populations of Fish to Assess Ecosystem Health', *J. Aquat. Ecosystem Health* **1**,135-146.

Baumann, P.C., Harshbarger, J.C., and Hartman, K.L.: 1990, 'Relationship Between Liver Tumors and Age in Brown Bullhead Populations from Two Lake Erie Tributaries', *Sci. Total Environ.* **94**,71-87.

Baumann, P. C. and Harshbarger, J.C.: 1995, 'Decline in Liver Neoplasms in Wild Brown Bullhead Catfish After Coking Plant Closes and Environmental PAHs Plummet', *Environ. Health Perspect.* **2**,168-170.

Baumann P. C., Smith, I.R., and Metcalfe, C.D.: 1996, 'Linkages Between Chemical Contaminants and Tumors in Benthic Great Lakes Fish', *J. Great Lakes Res.* **22**,131-152.

Baumann, P. C., Smith W.D., and Ribick, M.: 1982, 'Hepatic Tumor Rates and Polynuclear Aromatic Hydrocarbon Levels in Two Populations of Brown Bullheads *(Ictalurus nebulosus)*', In: M.W. Cooke, A.J. Dennis & G.L. Fisher (eds), *Polynuclear Aromatic Hydrocarbons: Sixth International Symposium on*

*Physical and Biological Chemistry*, pp. 93-102. Battelle Press, Columbus, Ohio.

Black River Remedial Action Plan Coordinating Committee.: 1994, 'Black River Remedial Action Plan Stage One Report: Impairment of Beneficial Uses of Pollution in the Black River Area of Concern', , Elyria, OH.

Fabacher, D.L, Schmitt, C.J., Besser, J.M., and Mac, M.J.: 1988, 'Chemical Characterization and Mutagenic Properties of Polycyclic Aromatic Compounds in Sediment from Tributaries of the Great Lakes', *Environ. Toxicol. Chem.*, 7,529-543.

Harshbarger, J.C. and Clark, J.B.: 1990, 'Epizootiology of Neoplasms in Bony Fish of North America', *Sci. Total Environ.*, **94**,1-32.

Malins, D.C., McCain, B.B., Brown, D.W., Chan, S., Myers, M.S., Landahl, J.T., Prohaska, P.G., Friedman, A.J., Rhodes, L.D., Burrows, D.G., Gronlund W.D., and Hodgins, H.O.: 1984, 'Chemical Pollutants in Sediments and Diseases of Bottom-Dwelling Fish in Puget Sound, Washington', , **18**,75.

Myers, M.S., Landahl, J.T., Krahn, M.M., Johnson, L.L., and McCain, B.B.: 1990, 'Overview of Studies on Liver Carcinogenesis in English Sole from Puget Sound; Evidence for a Xenobiotic Chemical Etiology I: Pathology and Epizootiology', *Sci. Total Environ.* **94**, 33-50.

Smith, S.B., Blouin, M.A., and Mac, M.J.: 1994.: 'Ecological Comparisons of Lake Erie Tributaries with Elevated Incidence of Fish Tumors', *J. Great Lakes Res.* **20**,701-716.

Vogelbein, W.K., Fournie, J.W., Van Veld, P.A., and Huggett, R.J.: 1990, 'Hepatic Neoplasms in the Mummichog *Fundulus heteroclitus* from a Creosote-Contaminated Site', *Cancer Res.*, **50**,5978-5986.

West, W.R., Smith, P.A., Stoker, P.W., Booth, G.M., Smith Oliver, T., Butterworth, B.E., and Lee, M.L.: 1985, 'Analysis and Genotoxicity of a PAC-polluted River Sediment'. In: M. Cooke and J.D. Anthony, (eds), *Polynuclear Aromatic Hydrocarbons-Mechanisms: Mechanisms, Methods and Metabolism*, pp. 395-1411, Battelle Press, Columbus, OH,.

# REPTILES AND AMPHIBIANS: SHY AND SENSITIVE VERTEBRATES OF THE GREAT LAKES BASIN AND ST. LAWRENCE RIVER

CHRISTINE A. BISHOP[1*] and ANDRÉE D. GENDRON[2]

[1]*Canadian Wildlife Service, Environment Canada, 867 Lakeshore Road, Box 5050, Burlington, Ontario, L7R 4A6, Canada;* [2]*Department Biological Science, University of Québec at Montréal, Succ. Centre-ville, Montréal, Québec, H3C 3P8, Canada*

**Abstract.** Bioaccumulation of environmental contaminants has been documented in amphibians and reptiles inhabiting the Great Lakes, St. Lawrence River and elsewhere. The effects of pollutants on the physiology and reproduction of amphibians and reptiles has also been reported but this research has largely been restricted to laboratory studies. Much less work has been conducted to quantify the effects of toxic chemical exposures on these cryptic animals in the wild. In the Great Lakes basin and St. Lawrence River, this work has only been performed in detail since 1981 although some samples collected in 1974 indicated a high level of contamination. Results in the 1990s on aquatic salamanders, frogs and turtles indicate that adults and embryos are currently experiencing toxic effects and, in some species and locations, there are indications that population declines are influenced by environmental pollution exposure. In this review, we describe the existing literature on contaminant levels and effects in reptiles and amphibians of the St. Lawrence River and Great Lakes basin.

## 1. Introduction

The unique life histories of amphibians and reptiles make their role in food webs diverse and important. They are predators and prey of vertebrates and invertebrates and they are a large component of the faunal biomass in North American ecosystems (Burton and Likens, 1975). Despite their cryptic habits, approximately 6000 reptile species (Conant and Collins, 1991) and 4500 species of amphibians (Vial and Saylor, 1993) have been identified and are distributed from Nearctic to tropical regions of our planet.

* To whom correspondence should be addressed.

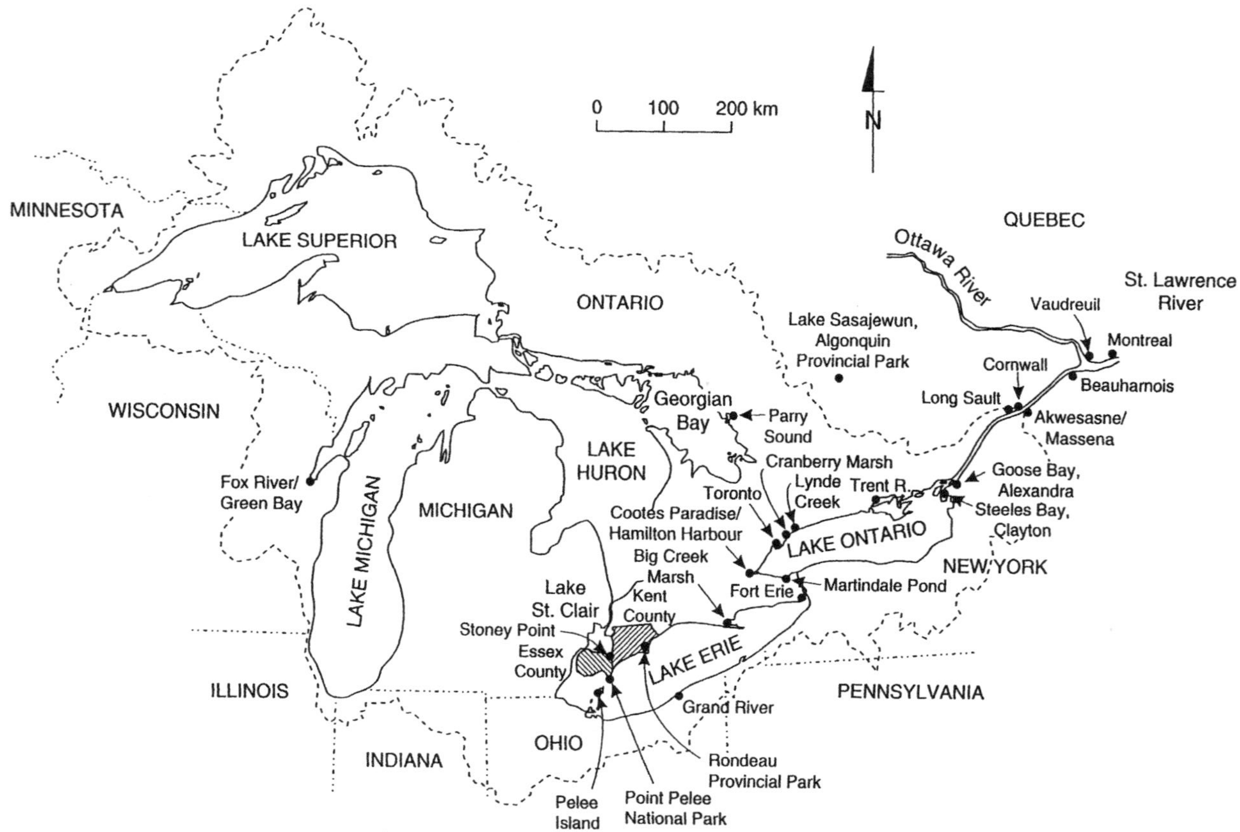


**Figure I.** Sites Noted in Text as Study Sites of the Levels and Effects of Environmental Contaminants on Reptiles and Amphibians.

As adults, many amphibians and reptiles are secondary and tertiary predators which makes them susceptible to bioaccumulation of persistent environmental contaminants. Many types of pollutants have been measured in a large variety of reptile and amphibian species. The value of these animals as indicators of local contamination in the environment has been recognized (Hall, 1980; Meyers-Shone and Walton, 1994; Harfenist *et al.*, 1989). However, no long term studies exist that examine trends and effects of contamination in amphibians and reptiles. Unexplained declines in amphibian populations and the disappearance of species in remote and urbanized areas of the world (Vial and Saylor, 1993) has lead to the suggestion that pesticides and other environmental contaminants have had a harmful effect on these populations (Bishop and Pettit, 1992; Hall and Henry, 1992; Carey, 1993). But the impact of pollutants on these animals has rarely been quantified and monitored in the wild. Only recently has contamination of amphibians and reptiles been documented in the Great Lakes and indications of their sensitivity to the effects of persistent chemicals has come to light. Research and monitoring has focussed on contamination in turtles and their eggs and an aquatic salamander, the mudpuppy (*Necturus maculosus*). Our purpose is to summarize the geographic and temporal trends in persistent contaminants and their effects on these animals in the Great Lakes and St. Lawrence River.

## 2. Reptiles

### 2.1 CONTAMINANT BIOACCUMULATION

The earliest collections of reptiles for contaminant analysis were made in 1974 by Craig Campbell under contract to the Canadian Wildlife Service (Campbell, 1974; Ontario Research Foundation, 1975). Eggs and tissues of Midland painted (*Chrysemys picta marginata*), map (*Graptemys geographica*), snapping (*Chelydra serpentina serpentina*), Blandings (*Emydoidea blandingi*) and eastern spiny softshell (*Apalone spiniferus*) turtles and one northern water snake (*Nerodia sipedon sipedon*) and three Lake Erie water snakes (*Nerodia sipedon insularum*) were collected at Parry Sound on Georgian Bay, and in southwestern Ontario including sites on the Great Lakes coastal marshes, Pelee Island, and inland at Essex and Kent Counties and the Thames River (Figure I; Table I). Among the turtles, the highest concentrations were in eastern spiny softshell eggs from Rondeau Provincial Park. (0.72 $\mu$g/g *p,p'*-DDE; 5.68$\mu$g/g Aroclor 1254:1260). The most contaminated sample collected was a Lake Erie water snake from Pelee Island, Lake Erie in which the liver contained 0.46 $\mu$g/g *p,p'*-DDE and 6.58 $\mu$g/g Aroclor 1254: 1260. Maximum mercury concentration among all samples was 0.39 $\mu$g/g which is actually lower than levels

**Table I.** Organochlorine Residues (μg/g Wet Weight) in Reptiles and Amphibians Collected in 1974 in the Canadian Great Lakes Basin (Campbell 1974; Ontario Research Foundation 1975).

| SAMPLE AREA | SPECIES | TISSUE | % LIPID | ORGANOCHLORINE PESTICIDES | | | PCB | MERCURY |
|---|---|---|---|---|---|---|---|---|
| | | | | *p,p'*- DDE | Dieldrin | Hexachloro-benzene | Aroclor 1254:1260 | |
| Parry Sound, Georgian Bay, Lake Huron | *Chrysemys picta marginata* | Hind leg muscle | 0.0 | 0.01 | ND | ND | 0.04 | 0.08 |
| | *C. p. marginata* | Egg | 5.2 | 0.24 | 0.01 | 0.06 | 0.54 | 0.01 |
| | | Egg | 5.5 | 0.09 | 0.01 | 0.02 | 0.26 | 0.03 |
| Rondeau Provincial Park, Lake Erie | *C. p. marginata* | Liver | 5.4 | 0.04 | 0.02 | <0.01 | 0.22 | 0.39 |
| | *Apalone spiniferus* | Egg | 8.0 | 0.72 | 0.05 | 0.01 | 5.68 | 0.06 |
| | *Graptemys geographica* | Egg | 3.2 | 0.06 | 0.01 | 0.01 | 0.26 | 0.01 |
| | *G. geographica* | Egg | 3.6 | 0.08 | 0.01 | 0.01 | 0.26 | 0.01 |
| | *Chelydra serpentina* | Egg | 4.4 | 0.15 | 0.02 | 0.01 | 1.39 | 0.09 |
| | *C. serpentina* | Egg | 5.2 | 0.21 | 0.03 | 0.01 | 1.92 | 0.08 |
| | *Bufo fowleri* | Whole body | 1.6 | 0.53 | 0.01 | 0.01 | 0.5 | NA |
| Long Point, Lake Erie | *C. p. marginata* | Egg | 5.9 | 0.05 | 0.01 | 0.02 | 0.12 | <0.01 |
| | *C. p. marginata* | Egg | 3.9 | 0.07 | 0.01 | 0.01 | 0.19 | <0.01 |
| | *Emydoidea blandingi* | Egg | 6.4 | 0.16 | 0.02 | 0.01 | 0.33 | 0.04 |
| | *E. blandingi* | Egg | 8.0 | 0.16 | 0.02 | 0.01 | 0.34 | 0.03 |
| | *C. serpentina* | Tail muscle | 4.1 | <0.01 | ND | ND | 0.09 | 0.39 |
| | *B. fowleri* | Whole body | 0.8 | 0.54 | 0.01 | 0.01 | 0.25 | NA |
| Pelee Island, Lake Erie | *C. p. marginata* | Egg | 3.6 | 0.28 | 0.02 | 0.01 | 2.52 | 0.05 |
| | *C. P. marginata* | Egg | 5.6 | 0.44 | 0.03 | 0.01 | 3.51 | 0.06 |
| | *Nerodia sipedon insularum* | Whole body | 0.4 | 0.52 | 0.04 | <0.01 | 5.7 | NA |
| | *N. s. insularum* | Whole body | 4.0 | 0.41 | 0.03 | 0.01 | 1.93 | NA |
| | *N. s. insularum* | Liver | 4.0 | 0.46 | 0.16 | 0.01 | 6.58 | NA |
| | *Acris crepitans* | Pool of four whole bodies | 1.2 | 0.17 | 0.02 | <0.01 | 0.1 | NA |
| | *A. crepitans* | Pool of two whole bodies | 0.0 | 0.12 | 0.02 | 0.01 | 0.16 | NA |
| Bradley's Marsh, Kent County, | *G. geographica* | Egg | 8.7 | 0.59 | 0.05 | 0.04 | 3.1 | 0.07 |
| | *N. s. sipedon* | Liver | 2.4 | 0.12 | 0.05 | 0.01 | 0.44 | NA |
| Paincourt, Essex County | *G. geographica* | Egg | 5.2 | 0.37 | 0.03 | 0.02 | 2.01 | 0.09 |
| Shrewbury, Essex County | *G. geographica* | Liver | 9.1 | 0.23 | 0.60 | 0.01 | 0.55 | 0.21 |

found in snapping turtle eggs collected in 1990 in the Cornwall area of the St. Lawrence River (Bonin *et al.* 1995). To date, the concentrations of *p,p'*-DDE in the eastern spiny softshell eggs from Rondeau Provincial Park of Lake Erie are the highest ever found in any reptile egg (see Meyers-Shone and Walton, 1994). The PCB[1] concentrations in the eastern spiny softshell and northern water snake are the third highest concentrations found in reptile tissues from the Canadian Great Lakes. The only higher concentrations were found in snapping turtle eggs from more industrialized areas such as Hamilton Harbour and the Bay of Quinte in 1984 (Struger *et al.* 1993).

The study of trends in contaminants and their effects in reptiles in the Great Lakes and St. Lawrence River has been mainly confined to the examination of levels and effects in the common snapping turtle. The common snapping turtle is an opportunistic omnivore for which fish comprise about one-third of its diet (Alexander, 1943). This species is a common inhabitant of wetlands throughout eastern North America including the Great Lakes basin (Weller and Oldham, 1988).

The first account of contaminated snapping turtle tissue from the Great Lakes was published by Olaffson *et al.* (1983). They analyzed one animal for total PCBs, mirex and *p,p'*-DDE. The animal was collected from Irondequoit Bay at Rochester, New York (Figure I). Olaffson *et al.* (1983) found 633.2 $\mu$g/g total polychlorinated biphenyl (PCB), 87.1 $\mu$g/g *p,p'*-DDE, and 28.3 $\mu$g/g mirex in that snapping turtle's fat. Ryan *et al.* (1986) then collected three snapping turtles in 1984, one specimen from each of Steele's Bay, Clayton, New York; Goose Bay east, Alexandria, NY; and East Massena/ Akwesasne, NY along the St. Lawrence River (Figure I). They found 2,3,7,8-tetrachlorodibenzo-*p*-dioxin (TCDD) concentrations ranged from 232 to 474 pg/g in their fat with lower levels of 32 to 107 pg/g in liver. Ryan *et al.* (1986) also collected one Blanding's turtle (*Emydoidea blandingii*) from the Ottawa River, Ontario in 1980 and did not find detactable 2,3,7,8-TCDD in liver, the only tissue analyzed. Such concentrations are high considering that eggs of fish-eating predators such as herring gulls (*Larus argentatus*) from eastern Lake Ontario in 1987 contained 80 pg/g 2,3,7,8-TCDD and 7.8 $\mu$g/g total PCB (Bishop *et al.*, 1992a and b). Grey seal (*Halichoerus grypus*) fat from the Gulf of Bothnia, Sweden in 1981 contained only 100 $\mu$g/g total PCB (Rappe *et al.*, 1981).

Stone *et al.* (1980) measured PCB contamination in snapping turtle tissues

---

[1] Total PCB concentrations have been analytically and arithmetically determined using different methods in each paper cited except that methods in the following papers/abstracts are the same and comparable: Hebert *et al.*, 1993; Struger *et al.*, 1994; Gendron *et al.*, 1994; 1995; Bonin *et al.*, 1995; Bishop *et al.*, 1992; . Bishop *et al.*, 1996 . Consult individual papers/authors for details of methodologies.

collected during 1976-1978 from a highly contaminated area near the Great Lakes, the Hudson River in New York state. In adult turtles they found total PCB concentrations up to 3560 $\mu$g/g (lipid weight basis) in fat as well as PCBs in eggs, and muscle. PCB concentrations were as high as 230 $\mu$g/g in soup made from those turtles. Since snapping turtles are trapped extensively in North America for food, for example, 30,000- 50,000 snapping turtles per year were harvested from Ontario in the early 1980s (Lovisek, 1982), a collection of adult turtles from the Great Lakes and analysis of their muscle was conducted in 1988/89 by the Ontario Ministry of Natural Resources (OMNR). The OMNR found up to 2.12 $\mu$g/g total PCBs (Cornwall Island, St. Lawrence River); 1.1 $\mu$g/g *p,p'*-DDE (Martindale Pond, Lake Ontario); 0.0093 $\mu$g/g mirex (Cornwall Island); and 2.4 $\mu$g/g octachlorostyrene (Stoney Point, Lake St. Clair) in muscle of adult snapping turtles in Ontario (Hebert *et al.,* 1993b) (Figure I). Due to high contaminant concentrations, the Ontario Ministry of Natural Resources severely restricted the trapping of turtles in Ontario in 1990. The catch limit is now two turtles per day with a possession limit of five turtles and a fishing licence is required. This change was welcomed by environmental groups and biologists. Prior to that time there were no restrictions on the season or number of animals that could be trapped. Also in 1990 the New York State Dept. of Health advised that all children under 15 years and women of childbearing age should not consume snapping turtle tissues.

## 2.2 Geographic Trends in Organochlorine Concentrations in Common Snapping Turtle Eggs

Common snapping turtles sexually mature at approximately seven years in the lower Great Lakes regions and can take up to 18 years to mature in the most northern areas of its range in Ontario. As a result, the adult population is extremely sensitive to the collection and killing of adult turtles to measure contaminant concentrations in their tissues (Brooks *et al.*, 1988; Galbraith and Brooks, 1987). In contrast, natural rates of predation on snapping turtle eggs are extremely high, therefore Struger *et al.* (1994) chose to monitor lipophilic contaminant levels in eggs from the Great Lakes. Also, contaminant concentrations in turtle fat and liver increase with age and size of the turtle. This makes it difficult to assess annual changes in contamination in body burdens (Bishop, 1990; Hebert *et al.*, 1993a). Female snapping turtles lay a single clutch of eggs each year, and contamination in their eggs reflect annual changes in chlorinated hydrocarbon exposure in the female turtle (Bishop *et al.*, 1994). This enhances the use of the eggs as annual biomonitoring tools. To date, twenty organochlorine pesticides, total mercury, 48 polychlorinated biphenyl (PCBs) congeners including six non-*ortho* PCBs, 8 polychlorinated dibenzodioxins (PCDDs), and 14 polychlorinated dibenzofurans (PCDFs) have been measured in snapping turtle eggs from the Great Lakes-St. Lawrence River collected during 1981-1991.

Struger *et al.* (1994) sampled snapping turtle eggs from nine locations in the Great Lakes as well as a remote reference site, Lake Sasajewun, in 1981 and/or 1984. Total PCBs ranged from 0.057 to 4.76 mg/kg (wet wt.) among the Great Lakes-St. Lawrence River samples. Mean total PCB concentration at Lake Sasajewun, Algonquin Provinical Park was 0.187 mg/kg (Figure I). Eggs from Hamilton Harbour, Port Franks, Bay of Quinte/Murray Canal and Lake St. Clair (Figure I) were significantly more contaminated among the ten sample locations (Struger *et al.* 1994).

A pool of snapping turtle eggs collected from Hamilton Harbour in 1984 contained 67 ng/kg of 2,3,7,8-tetrachlorodibenzo-*p*-dioxin and 14.0 ng/kg of 2,3,4,7,8-pentachlorodibenzofuran (Struger *et al.*, 1994) (Figures I and II). Some dioxin congeners were present in turtle eggs at concentrations higher or equal to those in herring gull eggs from Hamilton Harbour (Bishop *et al.*, 1992a; 1992b). Comprehensive GC/MS analysis of the Hamilton Harbour eggs revealed trace amounts of *o,p*-dicofol, octachlorostyrene and toxaphene.

During 1989-1991, collections of turtle eggs were made at least once at five sites studied earlier by Struger *et al.* (1994). The geographic trends detected for PCBs and organochlorine pesticides in eggs during 1981-1984 were consistent with trends in 1988-1991(Figure II). For polychlorinated dioxins and furans (PCDD/Fs), eggs from Cootes Paradise/Hamilton Harbour and Lynde Creek on Lake Ontario contained the highest PCDD concentrations (e.g. 2,3,7,8-TCDD= 25.6 pg/g at Hamilton and 55 pg/g at Lynde Creek) and number of detectable PCDF congeners among all sites (Figure I). Twenty PCDD/F congeners were detected at Lynde Creek and Hamilton while eggs from other sites only contained six congeners. Eggs from Cranberry Marsh on Lake Ontario contained organochlorine pesticide and PCB concentrations similar to those from Big Creek Marsh and Rondeau Provincial Park on Lake Erie, although PCDD and PCDF concentrations were higher and more congeners were detectable in Cranberry Marsh eggs (Figures I and II).

Snapping turtle eggs were also collected at ten sites on the St. Lawrence River and its tributaries in 1989 and 1990 ( Bonin *et al.* 1995). Most of these sites had not been sampled previously. Eggs from Massena/Akwesasne area (Figure I) were highly contaminated with mean total PCBs of 5.09 $\mu$g/g and a maximum concentration in one clutch of 11.0 $\mu$g/g. These are the highest PCB concentrations reported in snapping turtle eggs from any location. At three sites in the Cornwall area, located upstream of Akwesasne, total PCB concentrations were much lower at concentrations of 0.7 to 2.8 $\mu$g/g.

Also in 1989/1990, organochlorine pesticides were highest in eggs from the Cornwall area. Eggs collected from Akwesasne and from Long Sault (Figure I), near

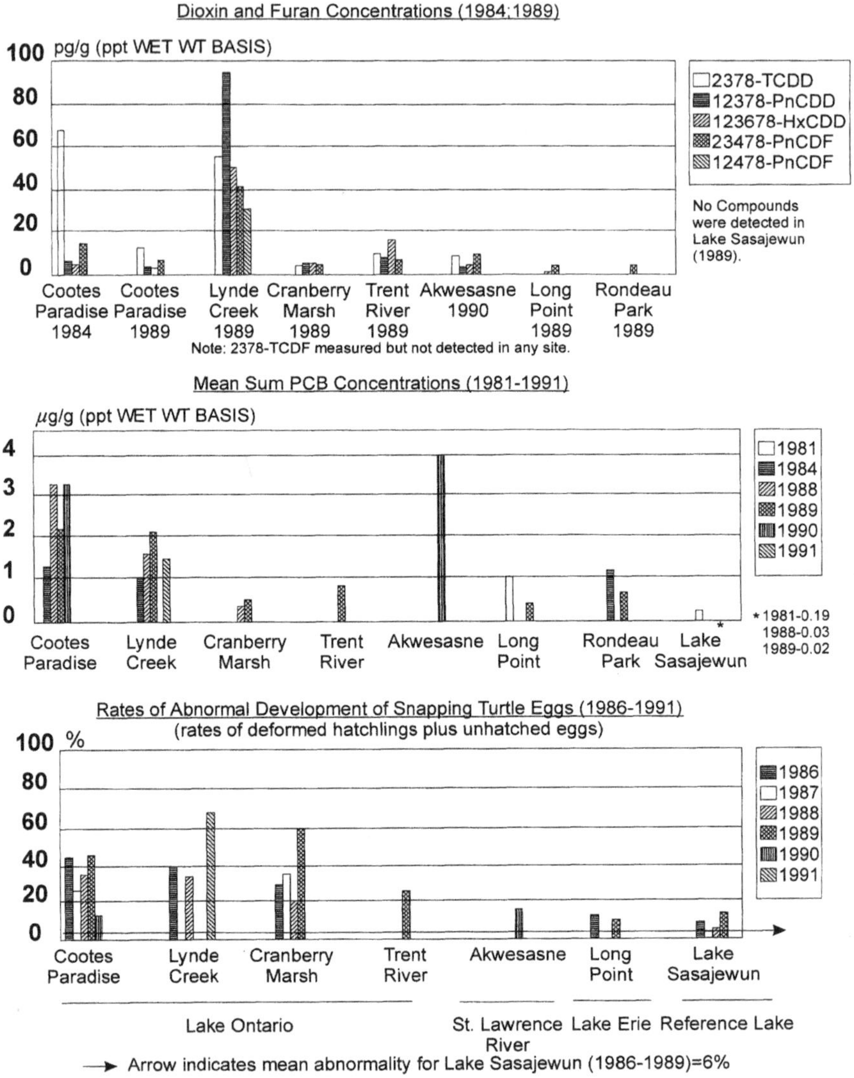


**Figure II.** Polychlorinated Hydrocarbon Concentrations and Rates of Abnormal Development in Common Snapping Turtle Eggs

Cornwall, upstream of Akwesasne, contained organochlorine pesticide concentrations that were two to three times higher than those from the other sites studied on the St. Lawrence River (Bonin *et al.* 1995). The concentrations of pesticides in eggs from the Cornwall area were similar to those found in eggs sampled in 1984 at that general area by Struger *et al.* (1994). They were also similar to concentrations in eggs from Hamilton sampled during 1984-1991 (Struger *et al.*, 1984; Bishop *et al.*, 1996).

## 2.3 Temporal Trends in Organochlorine Concentrations in Eggs

From 1984 to 1991, significant increases occurred in several organochlorine chemicals in snapping turtle eggs from Lake Ontario whereas decreasing or stable concentrations were noted at Lake Erie sites and the inland reference site, Lake Sasajewun (Figure II). Significant decreases of hexachlorobenzene, mirex and PCBs were found in turtle eggs collected in 1981/84 relative to 1989 at Big Creek Marsh and Rondeau Provincial Park in Lake Erie (Figures I and II). There were no significant changes in concentrations of *p,p'*-DDE and dieldrin. In Lake Ontario eggs, concentrations of PCBs, *p,p'*-DDE and dieldrin increased significantly between 1984 and 1991 at Cootes Paradise and/or Lynde Creek (Figures I and II). Notably, 2,3,7,8-TCDD and other PCDD/F concentrations in Hamilton eggs from Lake Ontario had declined markedly since 1984 (Figures I and II). At the remote reference location, Lake Sasajewun, concentrations of all organochlorines except dieldrin had significantly decreased between 1981 and 1991 (Figures I and II).

## 2.4 Effects in Reptiles

Recently, effects such as altered sex steroid concentrations and abnormal gonadal morphology were found in juvenile American alligators (*Alligator mississippiensis*) in Florida exposed to concentrations of 5.8 $\mu$g/g *p,p'*-DDE in eggs (Guillette *et al.*, 1994; 1996). Although contamination can be high in adult turtles collected in the Great Lakes and elsewhere, biological effects in adult or juvenile turtles have never been studied in detail. Stone *et al.* (1980) and Ryan *et al.* (1986) both reported that the adult snapping turtles they collected 'appeared to be normal'. Bishop (1990) reported an apparently high rate of 21 dead turtles during an 8 month period of 1986-1987 in Cootes Paradise / Hamilton Harbour. One of those dead turtles contained 1.3 $\mu$g/g total PCB; 1.8 $\mu$g/g *p,p'*-DDE and 0.081-0.66 $\mu$g/g of heptachlor epoxide, total chlordane and mirex in fat (Dept. Pathology, University of Guelph, 1985). Other snapping turtles (N=11) from Cootes Paradise/ Hamilton Harbour were analyzed in the same years and PCB[1] concentrations were as high as 72.3 $\mu$g/g in fat (Bishop, 1990). The turtle containing 72.3 $\mu$g/g PCB was collected from Windermere Basin in Hamilton Harbour and PCB concentrations appeared to reflect the high degree of local contamination at that site (Envirosearch, 1988). Research to examine the effects of such high exposures to potential endocrine disrupting chemicals (Colborn *et al.*, 1993) in snapping turtles is certainly needed.

Although adult turtles have been overlooked, there has been an extensive study on the hatching success and deformity rates of hatchling snapping turtles. Snapping turtle eggs were artificially incubated in 1986-1988 from four sites in the Great Lakes and a reference site. The eggs with the highest chlorinated hydrocarbon concentrations also had the poorest developmental success (Bishop *et al.*, 1991). The frequency of unhatched eggs, deformed hatchlings and the severity of deformities was highest in eggs from Lake Ontario at Cootes Paradise/Hamilton (43.1% ), Lynde Creek (37.5%), and Cranberry Marsh (33.7%), followed by Lake Erie at Big Creek Marsh (11.0%) and the reference site, Lake Sasajewun (7.4%) in 1986-1988 (Bishop *et al.* 1991) (Figures I and II).

During 1989-1991, developmental abnormalities were again assessed in embryos and hatchlings from the same sites studied in 1986-1988 as well as two more sites in Ontario, Canada (Trent River; Rondeau Provincial Park) and Akwesasne-Massena, New York, USA (Figures I and II). The mean summed rate of unhatched eggs and deformed hatchlings was 12.1% at Lake Sasajewun, Algonquin Provincial Park in 1989 whereas the rates ranged from 8.0 to 67.5% in the Great Lakes and St. Lawrence River. These rates were not significantly different than those found in 1986-1988 at the same sites. Relative to the reference site, the highest percentage increase in risk of abnormalities occurring in eggs was 1351% in Lynde Creek eggs. There was significantly higher risk of abnormalities occurring in eggs contaminated with seven individual PCDDs, 11 PCDFs, and 10 PCB congeners including 2,3,4,3',4'-pentaCB (PCB #105) and 2,4,5,3',4'-pentaCB (PCB #118) (Bishop *et al.* in press). This was consistent with the findings in the 1986-1988 study (Bishop *et al.*, 1991). In contrast, toxic equivalent concentrations (Walker and Peterson, 1991; Safe, 1994; Ahlborg *et al.*, 1994; Kennedy *et al.*, 1996) in eggs collected in 1989-1991 were not correlated with developmental abnormalities. This suggests that the concentrations of individual PCDD/Fs, and the combined effect of chemicals may be more important in embryotoxicity than overall PCB contamination. It may also mean that 'Toxic Equivalency Factors' developed for fish, birds and human risk assessment using rodents are not predictive of toxicity in reptiles.

Three biochemical indicators of chlorinated hydrocarbon exposure and *Ah*-receptor mediated response (Safe, 1994) were also measured in snapping turtles from the Great Lakes. In livers of hatchling turtles from Lynde Creek and Lake Sasajewun, activity of 7-ethoxyresorufin O-deethylase and cytochrome P4501A concentration were both significantly higher in turtles from Lynde Creek. In the same livers, there were no differences in concentrations of uro-, hexa-, and hepta-carboxyl porphyrins between sites (Bishop *et al.*, in press).

## 3. Amphibians

### 3.1 CONTAMINANT BIOACCUMULATION

Campbell (1975) also collected a few amphibians from Lake Erie in 1974 (Table I). Notably, the Canadian population and geographic range of the two species collected, Fowler's toad (*Bufo fowleri*), and Blanchard's cricket frog (*Acris crepitans*), have declined severely since 1974. Fowler's toad no longer occurs at Fort Erie and Blanchard's cricket frog has disappeared from Pelee Island, two sites where Campbell's collections were made (Weller and Oldham, 1988; Table I). Concentrations of contaminants were relatively elevated in the Fort Erie samples with 0.72 $\mu$g/g *p,p'*-DDE and 1.34 $\mu$g/g Aroclor 1254: 1260 in a whole body sample of Fowler's toad (Table I).

At some of the same locations where snapping turtle eggs were collected on the St. Lawrence River (Bonin *et al.* 1995), PCBs and organochlorine pesticides were measured in mudpuppies (Bonin *et al.,* 1995) in 1988 and 1990. The mudpuppy can live up to 25 years (Pfingsten and White, 1989) and is a common but rarely seen resident of riverine systems in north-eastern United States and south-eastern Canada (Conant and Collins, 1991). They are carnivorous, feeding mainly on crayfish and other benthic animals (Pfingsten and White, 1989). While PCB congener patterns in snapping turtles are most similar to those in fish-eating birds (Bonin *et al.*, 1995; Bishop *et al*, 1996) in which lower chlorinated PCBs appear to be metabolized and higher chlorinated congeners persist, PCB patterns in mudpuppies are similar to fish (Bonin *et al.*, 1995) with lower chlorinated PCBs being predominant. This may be due to the benthic organisms in the mudpuppy diet and/or to the hepatic enzymes in amphibians which appear to be less responsive to exposure to PCBs (Ronis and Walker, 1985).

As in snapping turtles, PCB contamination in mudpuppies was highest in samples near or downstream of Akwesasne. Concentrations of PCBs, total mercury and organochlorine pesticides were elevated in mudpuppies from Vaudreuil, at the junction of the Ottawa and St. Lawrence Rivers and the Beauharnois area downstream of Akwesasne and near Montreal (Figure I). Total PCBs were as high as 1834 ng/g and *p,p'*-DDE was 195.6 $\mu$g/g in gonads from mudpuppies from Beauharnois, *p,p'*-DDE was 330.2 ng/g in liver from Vaudreuil (Figure I) (Bonin *et al.*, 1995). Maximum total mercury (0.44 $\mu$g/g) was found at Beauharnois in a carcass analyzed without gonads.

Organochlorine compounds have also been measured in frogs in the Great Lakes but only from a single location. In 1993, spring peepers (*Pseudacris crucifer*) from Point Pelee National Park (Figure I) contained 1.0 μg/g *p,p'*-DDE and lower concentrations of seven other pesticides (Russell *et al.*, 1995). This is the consequence of intensive DDT use for mosquito control in wetlands of the park until 1967 (Park records in Russell *et al.*, 1995). Three species of amphibians have been extirpated from Point Pelee since 1972. Given the sensitivity of the European ranid, the common frog (*Rana temporaria*), to *p,p'*-DDT in laboratory studies (Cooke, 1970; 1973), there is concern that the spraying program contributed to species losses in the park however no direct evidence of this exists.

### 3.2 Effects in Amphibians

Gendron *et al.* (1994) examined the rates of digital deformities and corticosteroid responses in mudpuppies from nine sites on the St. Lawrence/ Ottawa Rivers as well as chlorinated hydrocarbon concentrations in female gonads during 1992-1993. High rates of skeletal deformities (50%; N=50) were found in samples from the most polluted site (total PCBs= 58.2 μg/g; Gendron, unpublished), Akwesasne, where mudpuppy adults were significantly more at risk to display limb defects than the reference site (8.7%; N=69). The frequency of terata such as oligodactyly and polydactyly significantly increased with higher exposure to chlorinated hydrocarbons. This suggested that pollutants may interfere with the developmental or regenerational limb processes, and contribute with other environmental factors to an increase in background prevalence of skeletal defects. There was also a shift toward older age structure in the Akwesasne site while there were few differences in age structure among other sites. This may reflect a reduction in recruitment at the most polluted site. In contrast, fecundity, gonado-somatic index and circulating levels of 17 ß-estradiol and testosterone in females with vitellogenic eggs were not correlated with the degree of contaminant bioaccumulation occurring in animals from different sites.

In a more detailed examination of hormone function in mudpuppies, Gendron *et al.* (1997) studied circulating levels of corticosterone (an interrenal secretory product) following standardized capture and handling stress in order to identify potential interference of contaminants with the normal activation of the corticosterone-producing endocrine axis in response to stress. Among nine populations in the Ottawa and St. Lawrence Rivers, there was a weak trend for lower plasma corticosterone in females with higher organochlorine pesticides and PCBs in their tissues. In a follow up study, responsiveness to corticotrophin (ACTH) challenge, which is normally secreted by the pituitary to stimulate the release of corticosterone, was compared between a contaminated and reference population.

Exogenous ACTH administered after a 24 hour recovery in a field adapted flow through system elicited a significantly lower surge in plasma corticosterone in the more contaminated group from Des Prairies River compared to Quesnel Bay (Figure I). Overall, these results suggest that contamination by chlorinated hydrocarbons might reduce to some degree the ability of mudpuppies to display a healthy corticosterone response to stress.

The sensitivity of mudpuppies, frog tadpoles and adult frogs to 3-trifluoromethyl-4-nitrophenol (TFM) use in the Great Lakes has been noted on many occasions. TFM has been used annually since 1958 for sea lamprey (*Petromyzon marinus*) control throughout the Great Lakes. Amphibians have been regularly found dead in creeks immediately after TFM treatment (Gilderhus and Johnson, 1980; Matson, 1990). Laboratory tests have confirmed that species native to the Great Lakes basin such as gray tree frog (*Hyla versicolor*), leopard frog (*Rana pipiens*), bullfrog (*Rana catesbeiana*) are sensitive to field applied rates of TFM (Chandler and Marking, 1975). In Grand River, Ohio (Figure I), Matson (1990) found that mudpuppy population size decreased by a minimum of 29% after a spray event.

There are also other indications that persistent contaminants can contribute to degradation of amphibian populations. Species diversity and density has been assessed in the Fox River, which is highly contaminated with PCBs, and in wetlands in Green Bay, which are cleaner and less disturbed than the Fox River (Figure I). Fewer species and lower numbers of amphibians, and reduced hatching success and survival to metamorphosis were found in the Fox River during 1993-1995 (Jung *et al*, 1996).

The Blanchard's cricket frog (*Acris crepitans blanchardii*) is among the species that have disappeared from Point Pelee since the 1970s but a population persisted on Pelee Island, Lake Erie (Figure I), until recently. Twelve to fifteen breeding sites were known for this species on Pelee Island during the 1970s but by the 1980s only one confirmed site was used for breeding (Oldham, 1992). In the 1990s the Blanchard's cricket frog is thought to have been extirpated from Pelee Island and Canada. Although fertilizers and pesticides are used on the island there are a variety of other possible agents which alone or together may have contributed to the loss of this species; factors such as habitat loss and degradation, fluctuating water levels that may have introduced predatory fish or caused drought in some sites, and the introduction of bullfrogs, and bait fishing may all have been important (Oldham, 1992).

Since the 1960s, the northern cricket frog (*Acris crepitans*) has become rare in Wisconsin (Vogt, 1981; Mossman and Hine, 1985), Iowa, north and central Illinois, and northern Indiana (Beasley *et al.*, 1995) although reasons for this decline are not understood. Beasley *et al.* (1995) surveyed southern to northern Illinois and found the sites with the lowest index of summer reproduction of northern cricket frogs had the most elevated concentrations of atrazine and metolachlor in water and sediment and extremely low oxygen saturation. Several cricket frogs with intersex condition were also encountered during the study.

Most recently, media reports in 1997 described deformed frogs at up to 100 locations in the state of Minnesota, however, to date, there are no scientific studies or publications on these occurrences. Leopard frogs with malformed or more than two hind legs have been most commonly reported by the general public and biologists. Further research and monitoring into this phenomenon is warranted to determine if pollutants are linked to these deformities. It is known that a trematode parasite isolated in wetlands in California can induce malformed and supernumerary limbs in frogs and salamanders (Sessions and Ruth, 1990). Detailed and statistically defensible methods are still required to account for all the possible factors contributing to these occurrences in Minnesota.

## 4. Conclusions

Snapping turtle eggs and mudpuppies sampled in the late 1980s and early 1990s contained contaminant levels which were generally similar to or lower than levels in herring gull and other colonial waterbird eggs in Lake Ontario and the St. Lawrence (Pettit *et al.*, 1994). These levels of exposure in reptiles and amphibians were correlated with teratogenic effects whereas colonial waterbirds did not display significant rates of deformities in these areas at this time (Fox *et al.*, 1991). Despite this, there is a lack of consistent monitoring of contamination and its effects on amphibians and reptiles. Research is relatively recent and sparse compared to the extensive studies devoted to both levels and effects in fish and birds in the Great Lakes and elsewhere (Hebert *et al.*, 1993b). Furthermore, almost all reptile and amphibian studies suggesting that contaminant effects are occurring are based on correlative evidence derived from field research. Mechanistic studies of the physiological responses to contaminant exposure are lacking. Preliminary evidence suggests that amphibians do not exhibit typical *Ah*-receptor biochemical responses and sensitivities (Jung and Walker., 1997; Gendron, unpublished data) and this needs to be examined more thoroughly.

## Acknowledgements

Research on environmental contaminants in snapping turtles and mudpuppies conducted by or in cooperation with the Canadian Wildlife Service- Environment Canada has been funded by the Great Lakes Action Plan and grants of the Natural Sciences and Engineering Research Council to Ronald Brooks, University of Guelph, Ontario. We would like to thank the National Water Research Institute's Graphic Arts Unit and Glenn Barrett for technical support with the production of this paper.


## References


Ahlborg, U.G., Becking, G.C., Birnbaum, L.S., Brouwer, A., Derks, H.J.G.M., Feeley, M., Golog, G., Hanberg, A., Larsen, J.C., Liem, A.K.D., Safe, S.H., Schlatter, C., Waern, F., Younes, M., Yrjanheikki, E.: 1994, 'Toxic Equivalency Factors for Dioxin-like PCBs', *Chemosphere* **28**, 1049-1067.

Alexander, M.: 1943, 'Food Habits of the Snapping Turtle in Connecticut', *J. Wildl. Mgmt.* **7**, 278-282.

Beasley, V., Faeh, S., Wikoff, B., Eisold, J., Nichols, D., Brown, L., and Greenwell, M.: 1995, 'Evidence for a Potential Role of Pesticides in Population Declines of the Cricket Frog (*Acris crepitans*)', Annual Meeting of the Society of Environmental Toxicologists and Chemists, Vancouver, British Columbia, 5-9 Nov 1995.

Bishop, C.A.: 1990, '*The Snapping Turtle* (Chelydra s. serpentina) *as an Indicator of Organochlorine Contamination in Wetlands*', MSc thesis, York University, Toronto, Ontario, 135 pp.

Bishop, C.A., Brooks, R.J., Carey, J.H., Ng, P., Norstrom, R.J., and Lean D.R.S.: 1991, 'The Case for a Cause-Effect Linkage Between Environmental Contamination and Development in Eggs of the Common Snapping Turtle (*Chelydra serpentina*) from Ontario, Canada', *J. Toxicol. Environ. Health* **33**, 512-547.

Bishop, C.A. and Pettit, K.E.: 1992, '*Declines in Canadian Amphibian Populations: Designing a National Monitoring Strategy*', Occasional Paper No. 76, Canadian Wildlife Service, Environment Canada, Ottawa, Ontario, 120 pp.

Bishop, C.A., Weseloh, D.V., Burgess, N.B., Norstrom, R.J., Turle, R., and Logan K.A.: 1992, '*An Atlas of Contaminants in Eggs of Colonial Fish-Eating Birds of the Great Lakes (1970-1988): Volume I: Accounts by Location*', Technical Report No. 152, Canadian Wildlife Service Ontario Region, Burlington, Ontario, 540 pp.

Bishop, C.A., Weseloh, D.V., Burgess, N.M., Struger, J., Norstrom, R.J., Turle, R., and Logan, K.A.: 1992a, '*An Atlas of Contaminants in Eggs of Fish-Eating Colonial Birds of the Great Lakes (1970-1988): Volume I*', Technical Report Series No. 152, Canadian Wildlife Service Ontario Region, Burlington, Ontario, 279 pp.

Bishop, C.A., Weseloh, D.V., Burgess, N.M., Struger, J., Norstrom, R.J., Turle, R., and Logan, K.A.: 1992b, '*An Atlas of Contaminants in Eggs of Fish-Eating Colonial Birds of the Great Lakes (1970-1988): Volume II*', Technical Report Series No. 153, Canadian Wildlife Service Ontario Region, Burlington, Ontario, 315 pp.

Bishop, C.A., Brown, G.P., Brooks, R.J., Lean, D.R.S., and Carey, J.H.: 1994, 'Organochlorine Contaminant Concentrations in Eggs and their Relationship to Body Size, and Clutch Characteristics of the Female Common Snapping Turtle (*Chelydra serpentina serpentina*) in Lake Ontario, Canada', *Arch. Environ. Contam. Toxicol.* **27**, 82-87.

Bishop, C.A., Ng, P., Norstrom, R.J., Brooks, R.J., and Pettit, K.E.: 1996, 'Temporal and Geographic Variation of Organochlorine Residues in Eggs of the Common Snapping Turtle (*Chelydra serpentina serpentina*) (1981-1991) and Comparisons to Trends in the Herring Gull (*Larus argentatus*) in the Great Lakes Basin in Ontario, Canada', *Archiv. Environ. Contam. Toxicol.* **31**, 512-524.

Bishop, C.A., Ng, P., Brooks, R.J., Kennedy, S., Stegeman, J.J., and Norstrom, R.J.: 1998, 'Environmental contamination and developmental abnormalities in eggs and hatchlings of the common snapping turtle *(Chelydra serpentina serpentina)* from the Great Lakes-St. Lawrence River basin (1989-91)', *Environ. Poll.* **99**,1-14.

Bonin, J., DesGranges, J.-L., Bishop, C.A., Rodrigue, J., Gendron, A., and Elliott, J.E.: 1995, 'Comparative Study of Contaminants in the Mudpuppy (Amphibia) and Common Snapping Turtle (Reptilia), St. Lawrence River, Canada', *Archiv. Environ. Contam. Toxicol.* **28**, 184-194.

Brooks, R.J., Galbraith, D.A., Nancekivell, E.G., and Bishop, C.A.: 1988, 'Developing Management Guidelines for Snapping Turtles', In: Szaro, R.C., Severson, K.E., and Patton, D.R. (eds) *Management of Amphibians, Reptiles, and Small Mammals in North America*', USDA Forest Service Technical Report RM-166, pp 174-179.

Burton, T.M. and Likens, G.E.: 1975, 'Salamander Populations and Biomass in the Hubbard Brook Experimental Forest', *Copeia* **1975**, 541-546.

Campbell, C. 1974. '*Reproduction and Ecology of Turtles and other Reptiles and Amphibians of Lakes Erie and St. Clair in Relation to Toxic Chemicals*'. Report to Canadian Wildlife Service, Contract 7475/022.

Carey, C.: 1993, 'Hypothesis Concerning the Disappearance of Boreal Toads from the Mountains of Colorado', *Cons. Biol.* **7**, 355-362.

Chandler, J.H. and Marking, L.L.: 1975, 'Toxicity of the Lampricide 3-Trifluoromethyl-4-nitrophenol (TFM) to Selected Aquatic Invertebrates and Frog Larvae', *Invest. Fish Control* **62**, 1-7.

Colborn, R, vom Saal, F.S., and Soto, A.M. 1993.: 'Developmental Effects of Endocrine-Disrupting Chemicals in Wildlife and Humans'. *Environ. Health Perspect.* **101**, 378-384.

Conant, R. and Collins, J.T.: 1991, '*A Field Guide to Reptiles and Amphibians*', 3rd edition, Houghton-Mifflin, Boston, Massachusetts, 450 pp.

Cooke, A.S.: 1970, 'The Effects of *p,p'*-DDT on Tadpoles of the Common Frog (*Rana temporaria*)', *Environ. Pollut.* **1**, 57-71.

Cooke, A.S.: 1973, 'Responses of *Rana temporaria* Tadpoles to Chronic Doses of *p,p'*-DDT', *Copeia.* **1973**, 647-652.

Dept. Pathology.: 1985, '*Necropsy No. W 253-85*', University of Guelph, Guelph, Ontario.

Envirosearch.: 1988, 'Windermere Basin Rehabilitation Project: Initial Assessment for the Partial Dredging and Filling of Windermere Basin,' p.12, 21.

Fox, G.A., Collins, B., Hayadawa, E., Weseloh, D.V., Ludwig, J.P., Kubiak, T.J., Erdman, T.C.: 1991, 'Reproductive outcomes in colonial fish-eating birds: a biomarker for developmental toxicants in Great Lakes food chains. II. Spatial variation in the occurrence and prevalence of bill defects in young double-crested cormorants in the Great Lakes, 1979-1987', *J. Great Lakes Res.* **17**, 158-167.

Galbraith, D.A., Brooks, R.J.: 1987, 'Survivorship of Adult Females in a Northern Population of Snapping Turtles (*Chelydra serpentina*)', *Can. J. Zool.* **65**, 1581-1586.

Gendron, A.D., Bishop, C.A., Van Der Kraak, G., Fortin, R., and Hontela, A.: 1994, 'Multi-Level Detection of Toxic Stress in the Mudpuppy (Amphibian, Salamander)', Annual Meeting of the Society of Environmental Toxicologists and Chemists, Denver, Colorado, 30 Oct- 4 Nov 1994.

Gendron, A.D., Bishop, C.A., Fortin, R., and Hontela, A.: 1997, 'In vivo testing of the functional integrity of the corticosterone-producing axis in mudpuppy (Amphibia) exposed to chlorinated hydrocarbons in the St. Lawrence and Ottawa Rivers', *Environ. Toxicol. Chem.* **16**, 1694-1706.

Gilderhus, P.A. and Johnson, B.G.H.: 1980, 'Effects of Sea Lamprey (*Petromyzon marinus*) Control in the Great Lakes on Aquatic Plants, Invertebrates, and Amphibians', *Can. J. Fish. Aquat. Sci.* **37**, 1895-1905.

Guillette, L.J., Gross, T.S., Masson, G.R., Matter, J.M., Percival, H.F., and Woodward, A.R.: 1994, 'Developmental Abnormalities of the Gonad and Abnormal Sex Hormone Concentrations in Juvenile Alligators from Contaminated and Control Lakes in Florida', *Env. Health Persp.* **102**, 680-688.

Guillette, L.J., Pickford, D.B., Crain, D.A., Rooney, A.A., Percival, H.F.: 1996, 'Reduction in Penis Size and Plasma Testosterone Concentrations in Juvenile Alligators Living in a Contaminated Environment', *Gen. Comp. Endocrin.* **101**, 33-42.

Hall, R.J.: 1980, '*Effects of Environmental Contaminants on Reptiles: a Review*', U.S. Fish and Wildlife Service Special Scientific Report - Wildlife No. 228, United States Dept. of the Interior, Washington, D.C.

Hall, R.J., and Henry, P.F.P.: 1992, 'Review: Assessing Effects of Pesticides on Amphibians and Reptiles: Status and Needs', *Herpetological J.* **2**, 65-71.

Harfenist, A., Power, T., Clark, K.L. and Peakall, D.B.: 1989, '*A Review and Evaluation of the Amphibian Toxicological Literature*', Technical Report Series No. 61, Canadian Wildlife Service Ontario Region, Ottawa, Ontario, 222 pp.

Hebert, C.E., Bishop, C.A., and Weseloh, D.V.: 1993a, '*Evaluation of Wetland Biomonitors for the Great Lakes: a Review of Contaminant Levels and Effects in Five Vertebrate Classes*', Technical Report Series No. 182, Canadian Wildlife Service Ontario Region, Burlington, Ontario, 103 pp.

Hebert, C.E., Glooshenko, V., Haffner, G.D., and Lazar, R.: 1993b, 'Organic Contaminants in Snapping Turtle (*Chelydra serpentina*) Populations from Southern Ontario, Canada', *Arch. Environ. Contam. Toxicol.* **24**, 35-43.

Jung, R.E., Karasov, W.H., Huang, Y.-W., and Kersten, S.M.: 1996, 'Amphibian Ecotoxicology in Green Bay, Wisconsin: Are Toxicants Influencing Amphibian Distributions?', Abstracts of Great Lakes Declining Amphibian Conference. Milwaukee, Wisconsin, 30 March 1996.

Jung, R.E., and Walker, M.K.: 1997, 'Effects of 2,3,7,8-tetrachlorodibenzo-*p*-dioxin (TCDD) on Development of Anuran Amphibians'. *Environ. Chem. Toxicol.* **16**, 230-240.

Kennedy, S.W., Lorenzen, A., Norstorm, R.J.: 1996, 'Chicken Embryo Heptacyte Bioassay for Measuring Cytochrome P4501A-based 2,3,7,8-Tetrachlorodibenzo-*p*-dioxin Equivalent Concentrations in Environmental Samples'. *Environ. Sci. Tech.* **30**, 706-715.

Lovisek, J.: 1982, '*Survey of Harvesting of Turtles in Ontario*', Wildlife Branch, Ontario Ministry of Natural Resources, Queen's Park, Toronto, Ontario.

Matson, T.O.: 1990, 'Estimation of Numbers for a Riverine *Necturus* Population Before and After TFM Lampricide Exposure', *Kirtlandia* **45**, 33-38.

Meyers-Schöne L. and Walton, B.T.: 1994, 'Turtles as Monitors of Chemical Contaminants in the Environment', *Rev. Environ. Contam. Toxicol.* **135**, 93-153.

Mossman, M.J. and Hine, R.L.: 1985, '*Wisconsin's Frog and Toad Survey, 1984*', Wisconsin Endangered Resources Report 16, Wisconsin Department of Natural Resources, Bureau of Endangered Resources, Madison, Wisconsin.

Olafsson, P.G., Bryan, A.M., Bush, B., and Stone, W.: 1983, 'Snapping Turtles - a Biological Screen for PCBs', *Chemosphere* **12**, 1525-1532.

Oldham, M.J.: 1992, 'Declines in Blanchard's Cricket Frog in Ontario', In: Bishop, C.A. and Pettit, K.E. (eds) *Declines in Canadian Amphibian Populations: Designing a National Monitoring Strategy*, Occasional Paper No. 76, Canadian Wildlife Service, Environment Canada, Ottawa, pp 30-31.

Ontario Research Foundation. 1975. Laboratory Report 75-4. Missisauga, Ontario.

Pettit, K.E., Bishop, C.A., Weseloh, D.V., and Norstrom R.J.: 1994, '*An Atlas of Contaminants in Eggs of Colonial Fish-Eating Birds of the Great Lakes (1989-1992): Volume I: Accounts by Location*', Technical Report No. 193, Canadian Wildlife Service Ontario Region, Burlington, Ontario, 400 pp.

Pfingsten, R.A. and White, A.M.: 1989, '*Necturus maculosus* (Rafinesque), Mudpuppy', In: Pfingsten, R.A. and Downs, F.L. (eds) *Salamanders of Ohio*, Bulletin of the Ohio Biological Survey, Columbus, pp. 72-78.

Rappe, C., Buser, H.R., Stalling, D.L., Smith, L.M., and Dougherty, R.C.: 1981, 'Identification of Polychlorinated Dibenzofurans in Environmental Samples', *Nature*, **292**, 524-526.

Ronis, M.J.J. and Walker, C.H.: 1985, 'Species Variations in the Metabolism of Liposoluble Organochlorine Compounds by Hepatic Microsomal Monooxygenase: Comparative Kinetics in Four Vertebrate Species', *Comp. Biochem. Physiol.* **82**, 445-449.

Russell, R.W., Hecnar, S.J., and G.D. Haffner.: 1995, 'Organochlorine Pesticide Residues in Southern Ontario Spring Peepers', *Environ. Toxicol. Chem.* **14**, 815-817.

Ryan, J.J., Lau, B.P.-Y., Hardy, J.A., Stone, W.B., O'Keefe, P., and Gierthy, J.F.: 1986, '2,3,7,8-Tetrachlorodibenzo-*p*-dioxin and Related Dioxins and Furans in Snapping Turtle (*Chelydra serpentina*) Tissues from the Upper St. Lawrence River', *Chemosphere* **15**, 537-548.

Safe, S.H.: 1994, 'Polychlorinated Biphenyls (PCBs): Environmental Impact, Biochemical and Toxic Responses, and Implications for Risk Assessment', *Crit. Rev. Toxicol.* **24**, 87-149.

Sessions, S.K. and Ruth, S.B.: 1990. 'Explanation for Naturally Occurring Supernumerary Limbs in Amphibians'. *J. Exp. Zool.* **254**, 38-47.

Stone, W.B., Kiviat, E., and Butkas, S.A.: 1980, 'Toxicants in Snapping Turtles', *New York Fish and Game J.* **27**, 39-50.

Struger, J., Elliott, J.E., Bishop, C.A., Obbard, M.E., Norstrom, R.J., Weseloh, D.V., Simon, M., and Ng, P.: 1993, 'Environmental Contaminants in Eggs of the Common Snapping Turtle (*Chelydra serpentina serpentina*) from the Great Lakes-St. Lawrence River Basin in Ontario, Canada (1981, 1984)', *J. Great Lakes Res.* **19**, 681-694.

Vial, J.L., and Saylor, L.: 1993, '*The Status of Amphibian Populations: A Compilation and Analysis*', Working Document No. 1, Declining Amphibian Task Force, IUCN/SSC, 56 pp.

Vogt, R.C.: 1981, '*Natural History of Amphibians and Reptiles in Wisconsin*', The Milwaukee Public Museum, Milwaukee, Wisconsin.

Walker, M.K., and Peterson, R.E. 1991. 'Potencies of Polychlorinated Dibenzo-*p*-dioxin, Dibenzofuran and Biphenyl Congeners, Relative to 2,3,7,8-Tetrachlorodibenzo-*p*-dioxin for Producing Early Life Stage Mortality in Rainbow Trout *(Oncorynchus mykiss)*', *Aquatic Toxicol.* **21**: 219-238.

Weller, W.F. and Oldham, M.S.: 1988, '*Ontario Herpetofaunal Summary*', Publ. Ont. Field Herpetologists, Ontario, Canada, 221 pp.

# HUMAN HEALTH EFFECTS OF ENVIRONMENTAL POLLUTANTS: NEW INSIGHTS

D.O. CARPENTER
*School of Public Health, University at Albany, SUNY, One University Place, Rensselaer, New York 12144-3456*

**Abstract.** While a variety of effects of toxic chemicals are known in animals exposed both in the laboratory and *in situ*, it has proven more difficult to obtain definitive information relating harm to humans resulting from environmental contamination. Until quite recently it has been generally assumed that cancer was the human disease of greatest importance. In fact, the majority of regulations of exposure to toxic chemicals by most governments are designed on the basis of presumed cancer risk. The evidence that hazardous chemicals can cause cancer is strong, and concern of cancer risks is appropriate. However, recent evidence has triggered a reevaluation of the assumption that cancer is the sole disease of concern. New evidence has emerged suggesting that exposure to hazardous chemicals may lead to a variety of non-cancer endpoints, and that these effects may occur at low concentrations. Of particular concern is evidence for irreversible effects on the embryo and very young children which influence intelligence, attention span, sexual development and immune function. Some of these actions appear to be direct effects on the brain and other organ systems while others are mediated via disruption of endocrine systems. While these effects are subtle and difficult to quantify, the aggregated evidence is sufficiently compelling as to necessitate reevaluation of those health outcomes upon which regulations are based.

## 1. Introduction

Contamination of the environment with either man-made or natural hazardous substances has the possibility of affecting the ecosystem, animal life and humans in a great variety of ways, since there are an enormous variety of contaminants and each may have different effects. The purpose of this article is to briefly review what is known about environmental hazards to human health, and to discuss the changes and evolution in our perceptions of which effects on human health may be of greatest importance and occur at the lowest exposure levels. While it is not possible to present a comprehensive review of all studies of environmental effects on human health, the goal here is to identify new areas of concern and to discuss how this new information can be incorporated in regulatory actions concerning hazardous exposures.

## 2. Health Effects of Pollutants

### 2.1 CANCER

If one were to ask the average person in the street what disease is of greatest concern from environmental contamination, they would undoubtedly say "cancer". The public is particularly concerned about cancer from organic contaminants, but a number of metals are also carcinogenic (Ennever, 1994). Cancer has also been the primary, but not exclusive, focus of governmental regulations and research efforts to determine the degree of hazard from specific substances. Cancer is easily quantified, even though the cause of origin of a specific cancer may not be so easy. There has been a widely held view that any agent which causes cancer in more than one person in a million poses an unacceptable risk, and this value has become the basis of much of the policy of the US and other governments. Another indication of the attention to cancer is the distribution of resources from the National Toxicology Program, which in 1992 devoted 83% of its' total effort to carcinogenesis investigations (Silbergeld and Tonat, 1994).

Cancer is certainly not unimportant. Table I shows particular kinds of cancers that are increasing in incidence frequency in recent years. There is reason to believe that several of these are related to environmental factors. The increase in melanomas probably reflects the depletion of the ozone layer (Jones, 1987), and lung cancer is due primarily to smoking (Bartecchi *et al.*, 1994). The increase in the endocrine cancers, testis, prostrate and breast, may reflect effects of pollutants which alter sex steroid hormone systems as discussed below. Wolff *et al.* (1993) have shown that the risk of breast cancer increases with serum levels of DDE (the principal metabolite of DDT) and PCBs, and that at least for DDE there is a clear dose-response relationship. Finally, cancers such as brain, leukemia, lymphoma, gastrointestinal and kidney have also been linked to environmental pollutants (Gold *et al.*, 1979; Preston-Martin *et al.*, 1982; Sandler and Collman, 1987; Linet *et al.*, 1993; Aschengrau *et al.*, 1993; Koivusalo *et al.*, 1994; Doyle *et al.*, 1997; Rothman *et al.*, 1997). It is difficult to accurately determine the degree to which environmental contamination contributes to incidence of these and other cancers, and genetic susceptibility is certainly an important variable. But the ubiquitous exposure of human populations to many of these agents makes it possible that pollutants are more significant a factor than at present can be considered proven.

**Table I.** Increases in Certain Cancers: 1973 - 1987

| Cancer Type | % Increase |
|---|---|
| Melanoma | 83.3 |
| Non-Hodgkins Lymphoma | 50.9 |
| Prostate | 45.9 |
| Testis | 39.0 |
| Lung | 31.5 |
| Kidney | 27.0 |
| Breast | 24.2 |
| Brain/CNS | 23.0 |

From Henderson *et al.* 1991

## 2.2 BIRTH DEFECTS

It was Love Canal that brought the issues of hazardous wastes to the public's attention as never before, and raised public awareness of health effects from chemicals in addition to cancer. While much of the initial concern focused on the possibility of an increase in cancer incidence, none was ever detected (Janerich *et al.*, 1981). What was found was a clear increase in low birth weight (Vianna *et al.*, 1984; Goldman *et al.*, 1985) and suggestive incidence of short stature, learning problems, hyperactivity, headaches, skin rashes, abdominal pain and incontinence in children (Paigen *et al.*, 1985; 1987) .

Some of the best evidence that hazardous waste can cause birth defects comes from a recent paper which used the congenital malformations registry in New York to access the incidence of birth defects in relation to residential proximity to 590 hazardous waste sites (Geschwind *et al.*, 1992). Small, but statistically significant, elevations in incidence of birth defects were documented with exposure when considering all malformations and also with specific defects of the nervous system, musculoskeletal system and skin.

## 2.3 EFFECTS ON OTHER ORGAN SYSTEMS

While Love Canal raised concern about birth defects and environmental contamination, it is now clear that environmental contaminant agents, whether in water, air, soils or foodstuff, can affect multiple organ systems, and that each individual chemical can exert different effects (see Buffler *et al.*, 1985; Carpenter, 1994). Furthermore, there are different effects of specific agents as a function of developmental age, and at least some actions cause subtle, but long lasting, alteration of normal physiology. As a result of animal studies, there has been an enormous increase in our understanding of the variety of ways contaminants affect people, even though the definitive documentation of human health effects secondary to environmental agents is at present lacking in many cases. A significant factor in this new understanding relates to the interactions among the various body systems outlined in Figure I. While the functions of the nervous system, the immune system, the endocrine system and reproduction are all determined by genetic factors, these organ systems also are extensively interconnected and use common messengers (hormones, transmitters, cytokines). Thus environmentally-induced direct alterations on one organ system may result in indirect changes in other functions.

**Figure I.** Interactions Among Organ Systems

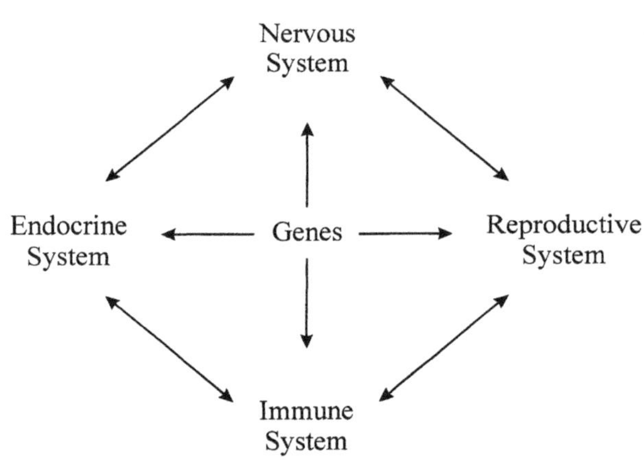

### *2.3.1 Immunotoxicity*

That the immune system is altered by pollutants has been known for many years (Dean *et al.*, 1994). In a healthy person the immune system is in balance - not so active as to promote excessive rashes and autoimmunity, but still responsive to appropriate foreign agents. Different pollutants affect the immune system differently. Many metals, alcohols and solvents promote a hyperimmune state, leading to skin rashes, asthma, autoimmune diseases and sometimes to anaphylaxis (Lawrence, 1985; Sosroseno, 1995). On the other hand, many pesticides, chlorinated hydrocarbons and polyaromatic hydrocarbons (PAHs), cause immunosuppression, which can lead to increased susceptibility to infection and possibly can contribute to an increased incidence of cancer (Silkworth *et al.*, 1986; Davila *et al.*, 1995). It is very difficult to attribute disorders of the immune system to a specific exposure, but it appears likely that hazardous substances contribute to immune disorders more commonly than generally appreciated.

### *2.3.2 Endocrine Disruption: Hypothyroidism*

Thyroid hormone regulates cellular maturation and development, and is essential for normal growth and especially for brain development (Porterfield and Hendrich, 1993). Lack of normal thyroid function in the extreme results in cretinism, with severe retardation. However, even less severe forms of hypothyroidism result in deficits, and these are often permanent if the hypothyroidism occurs during early development (Table II), even if the problem is treated at birth (Porterfield, 1994; van Wassenaer *et al.*, 1997).

Unfortunately there is structural similarity between thyroid hormone and the persistent organics like some PCB congeners and dioxin. As a result these substances interfere with thyroid function by binding to the thyroid transport proteins, reducing production or increasing metabolism of T4 or reducing conversion of T4 to T3 (McKinney *et al.*, 1985; Morse *et al.*, 1993). The result is a functional hypothyroidism, most likely due to the displacement of the endogenous hormone from the binding proteins which promotes excretion of the thyroid hormone. Given the clear evidence that hypothyroidism can cause a reduction in mental development, it is certainly possible that some of the reduced IQ and behavioral deficits resulting from exposure to these persistent organics is secondary to the hypothyroidism (cf. Porterfield, 1994; Hauser *et al.*, 1997).

**Table II.** Maternal Hypothyroidism and/or Treated Congenital Hypothyroidism

| |
|---|
| Reduced IQ |
| Minimal Brain Dysfunction |
| Difficulty Reasoning |
| Reduced Language Comprehension |
| Defective Fine Motor Skills |
| Behavioral Problems |
| Decreased Learning and Memory Skills |

From Porterfield, 1994; van Wassenaer *et al.*, 1997

*2.3.3 Endocrine Disruption: Estrogen Mimicry and Anti-Estrogenic Actions*

While it has been known for some time that a number of xenobiotics and other chemicals exhibit weak estrogenic activity (xenoestrogens), only relatively recently has considerable attention been given to the possibility that this activity might be important in animals and humans (Table III). It is now clear that estrogenic activity is characteristic of a great variety of substances, including many pesticides, mycotoxins, some PCBs and chemicals such as alkyl phenols, some phthalates and bisphenol A (Birnbaum, 1994; Jobling *et al.*, 1995; Danzo, 1997; vom Saal *et al.*, 1998). While the concentration of these substances in animals or man is usually relatively low, and their estrogenic potency is also low relative to estradiol, the fact that these substances are persistent suggests that they might have significant effects. In fact, there is strong evidence from wildlife studies that male reproductive capacity in such species as the Florida panther, alligator, mink and wood duck has been compromised, associated with and possibly as a result of elevated concentrations of estrogenic chemicals (see Colborn *et al.*, 1993; Kavlock *et al.*, 1996).

**Table III.** Possible Effects of Xenoestrogens in Humans

| |
|---|
| Decreased Sperm Count |
| Increased Breast Cancer |
| Increased Testicular Cancer |
| Increased Cryptorchidism |
| Increased Hypospadias |
| Altered Sexual Preference |

From Colborn *et al.*, 1993; Kavlock *et al.*, 1996

To add to the complication of possible sex hormone effects of xenobiotics, is the clear evidence that some substances are anti-estrogenic in action. The best documented is dioxin (Safe *et al.*, 1991), but antiestrogenic actions of other substances are also known. Furthermore, small changes in xenobiotic structure can shift the sex hormone action from antiestrogenic to estrogenic. For example, some PCB congeners are antiestrogenic, but their hydroxylated breakdown products are estrogenic (Gierthy *et al.*, 1997). At least for PCBs the hydroxylated derivatives tend to have much more estrogenic activity than the parent compounds (Gierthy *et al.*, 1997).

There was one large human experiment with administration of estrogen, which was widely given in the form of diethylstilbestrol (DES) for the prevention of spontaneous abortion. After decades of use it became apparent that a certain percentage of young women who had been exposed to this synthetic estrogen *in utero* developed clear cell carcinoma of the vagina (Herbst *et al.*, 1971). Later it became clear that there were other effects of DES exposure, including increased deformities of the reproductive tract in both men and women, increased incidence of autoimmune disease, altered sexual orientation in women, and possibly increased incidence of psychiatric disorders in both men and women (Colborn and Clement, 1992).

Unlike the situation with DES, there is at present little definitive evidence in humans that estrogenic chemicals in the environment alter physiologic function, but the DES experience in humans, together with the wildlife and experimental animal studies, provide strong support for this hypothesis (Newbold, 1995). The animal studies indicate that the developing male is particularly vulnerable to effects of estrogenic substances. In addition, there is some information on alterations of human reproductive function consistent with the hypothesis that these changes are caused by estrogenic contaminants. Some of these concerns are as follows:

#### *2.3.3.1 Hormone-Dependent Cancers in Women*

One of the clearest risk factors for breast cancer is estrogen, and therefore it is reasonable to suspect that xenoestrogens might increase the risk of breast cancer (Davis *et al.*, 1993). The study of Wolff *et al.* (1993) reported that there was a concentration-dependent relationship between serum DDE and increase in risk of breast cancer, and a less convincing relationship with PCB concentration. A relationship between ovarian cancer and herbicides has also been reported (Donna *et al.*, 1984). Not all studies have confirmed the suggestion that the concentration of persistent chlorinated organics is elevated in cancerous breast tissues (Hunter *et al.*, 1997), but this may be due to a variety of factors. For example, most studies of PCBs have measured only total PCB concentrations, but some congeners are estrogenic while others are antiestrogenic (Gierthy *et al.*, 1997). Therefore the total PCB concentration would not be expected to necessarily be correlated with cancer incidence, although elevation of specific congeners may well be.

*2.3.3.2 Sperm Count*

There are numerous reports that the human sperm count and semen volume has declined significantly over the past fifty years (Auger *et al.*, 1995; Carlsen *et al.*, 1992; Topari *et al.*, 1996; Swan *et al.*, 1997), and most of the investigators attribute this decline to xenoestrogens and their effects on the developing male. More direct evidence that persistent organochlorine compounds affect sperm function comes from the study of Bush *et al.* (1986) who showed that there was a correlation between sperm motility and sperm count and semen concentrations of three PCB congeners. However, agents with xenoestrogenic activity may have other actions. For example, some PCB congeners are estrogenic, but also cause hypothyroidism, and Cooke *et al.* (1996) have shown that hypothyroidism alone will cause enlarged testis and increased sperm production in rats. Thus these compounds may have a direct action which tends to reduce sperm production, but an indirect action that causes it to increase.

*2.3.3.3 Deformities and Cancer of the Male Urogenital System*

There has been a dramatic increase in the incidence of testicular cancer in developed countries over the past several decades, along with an increase in undescended testis and hypospadias (see review by Topari *et al.*, 1996). While the latter two are clearly birth defects, there is a possibility that all are defects resulting from xenoestrogen actions during development.

*2.3.3.4 Endometriosis*

Endometriosis is a painful disease in women characterized by the growth and proliferation of endometrial cells at sites outside the uterus. The etiology of the disease is uncertain and many possible factors have been considered. The possibility that dioxins might be involved in the etiology of the disease was first posed by Rier *et al*, (1993) when an increased incidence and severity of endometriosis was documented in monkeys that had been exposed to 2,3,7,8-TCDD for a period of 4 years. Research is being undertaken to investigate possible mechanisms of action and whether there is a relationship between the incidence of endometriosis in women and exposure to dioxin.

*2.3.3.5 Disturbance of Sexual Preference*

The brains of male homosexuals are different in several of the sexually dimorphic regions from those of heterosexuals and from females (LeVay, 1991; Allen and

Gorski, 1992; Friedman and Downey, 1993). As mentioned above, sexual preference of DES-exposed females appears to be different from unexposed controls. At present there is little direct evidence to support the obvious hypothesis that xenoestrogens and xenoantiestrogens can influence the developing male and female brain in such a way as to influence sexual preference. At least in human populations, the hypothesis will be very difficult to prove since the presumed brain alteration occurs in early fetal development.

*2.3.4 Effects on the Nervous System and Behaviour*

Reduced IQ and learning deficits: In the children born to mothers who were highly contaminated with PCBs at Yu-Cheng, many of the adverse symptoms related to the brain and development (Table IV). The adverse effects on IQ and attention span of exposure to lead in the developing child has been well known and accepted since the work of Needleman *et al.* (1979). It is also clear that mercury can cause a decrement is short term memory (Smith *et al.*, 1983; Myers *et al.*, 1997). What is less well known is accumulating evidence for the same effects of some organics, especially PCBs. Human studies of children exposed to elevated PCBs have suggested permanent or at least transient deficits in IQ and rate of acquisition of new skills (Jacobson *et al.*, 1990; Gladen *et al.*, 1988; Chen *et al.*, 1992). Animals exposed to PCB mixtures and single congeners have shown impaired learning in water maze tests (Shiota, 1976), poorer visual discrimination (Holene *et al.*, 1995) and impaired performance in an operant conditioning fixed interval schedule test (Pantaleoni *et al.*, 1988; Lilienthal *et al.*, 1990). The recent study of humans by Jacobson and Jacobson (1996) has shown that prenatal exposure has the greatest effect, and they document a decrement of 6.2 IQ points in children with the greatest exposure to PCBs as a result of maternal fish consumption. It should be noted that this decrement in IQ is of a similar magnitude to that resulting from exposure to lead (Needleman *et al.*, 1979).

**Table IV.** Adverse Symptoms inYu-Cheng Children

| |
|---|
| Developmental Abnormalities |
| Lower IQ |
| Behavioral Problems |
| Higher Activity Levels |
| Lower Body Weight |
| Shortened Attention Span |

From Rogan *et al.*, 1988

*2.3.4.1 Attention Deficit Disorder (ADD)*

ADD is at present an "in" disease in suburbia, but it is a real disorder that occurs frequently, even though at present it may often be over diagnosed. ADD is associated with motor restlessness, impulsiveness, inattention and increased distractibility, and individuals with ADD have a higher risk of antisocial personality disorder and depression (Biederman *et al.*, 1990). Hyperactivity and ADD are more common in patients with hypothyroidism (Hauser *et al.*, 1993). In animal models, exposure to PCBs during development has demonstrated increased activity level during adolescence in monkeys (Bowman *et al.*, 1978; 1981) rats (Lilienthal *et al.*, 1990) and mice (Storm *et al.*, 1981). There is at present a lack of any direct evidence implicating environmental contaminants in the apparent rise in incidence of ADD, with the possible exception of the demonstration that children with lead intoxication have a reduced attention span (Bellinger *et al.*, 1994) and that children with ADD have higher average lead levels than controls (Gittelman and Eskenazi, 1983). However, it is a reasonable hypothesis that the syndrome of ADD may arise as a result of exposure of the developing brain to environmental contaminants which exert either direct damage or indirect damage via alteration of other regulatory mechanisms such as thyroid function. Furthermore, the detrimental effects of lead and PCBs may well be secondary to an altered attention span, since it is clearly impossible to learn as well if you cannot pay attention.

## 3. Conclusion

Environmental contaminants affect many aspects of human physiology and behavior. At present, there is definitive evidence for an association with certain forms of cancer with specific pollutants, and this evidence is the basis for much of our regulation of pollutant levels. But with the increasing knowledge of the effects of environmental contaminants on immune, reproductive and endocrine systems and direct and indirect actions on the nervous system, non-cancer endpoints assume an increasingly important consideration in protection of the health of the public, even though these parameters are very difficult to quantify. While it is easy to count cases of cancer and birth defects, it is much more difficult to determine the degree to which environmental contamination leads to subtle changes in immune or endocrine status, intelligence and behavior in the population. Because these are fundamental characteristics of the human existence, we must find more precise ways of determining the magnitude of the threat posed by pollutants on these parameters and utilize this information in environmental policy development.

## Acknowledgments

The author's research is supported by NIEHS P42 ES04913

## References

Allen, L.S. and Gorski, R.A.: 1992, *Proc. Natl. Acad. Sci. USA* **89**, 7199-7202.
Aschengrau., A., Ozonoff, D., Faulu, C., Coogan, P., Vezina, R, Heeren, T. and Zhang, Y.: 1993, *Arch. Environ. Health* **48**, 284-292.
Auger, J., Kunstmann, J.M., Czyglik, F. and Jouannet P.: 1995, *N. Engl. J. Med.* **332**, 281-285.
Bartecchi, C.E., MacKenzie, T.D. and Schrier, R.W.: 1994, *N. Engl. J. Med.* **330**, 907-912.
Bellinger, D., Leviton, A., Allred, E. and Rabinowitz, M.: 1994, *Environ. Res.* **66**, 12-30.
Biederman, J., Faraone, S.V., Keenan, K., Knee, K. and Tsuang, M.T.: 1990, *J. Am. Acad. Child Adolesc. Psychiatry* **29**, 526-533.
Birnbaum, L.S.: 1994, *Environ. Health Perspect.* **102**, 676-679.
Bowman, R.E., Heironimus, M.P. and Allen, J.R.: 1978, *Fundam. Appl. Toxicol.* **15**, 457-467.
Bowman, R.E., Heironimus, M.P. and Barsotti, D.A.: 1981, *NeuroToxicology* **2**, 251-268.
Buffler, P.A., Crane, M. and Key, M.M.: 1985, *Environ. Health Perspect.* **62**, 423-456.
Bush, B., Bennett, A.H. and Snow, J.T.: 1986, *Arch. Environ. Contam. Toxicol.* **15**: 333-341.
Carlsen, E., Giwercman, A., Keiding, N. and Skakkebaek, N.E.: 1992, *Br. Med. J.* **305**, 609-613.
Carpenter, D.O.: 1994, *Cent. Eur. J. Publ. Health Suppl.* **2**, 6-9.
Chen, Y.-C., Guo, Y.-L., Hsu, C.-C. and Rogan, W.: 1992, *J. Am. Med. Assoc.* **268**, 3213-3218.
Colborn, T. and Clement, C.: 1992, 'Chemically induced alterations in sexual and functional development: The wildlife/human connection', Colborn, T. and Clement, C. eds., Princeton Scientific Publishing Co., 403 pp.
Colborn, T., vom Saal, F.S. and Soto, A.M.: 1993, *Environ. Health Perspect.* **101**, 378-384.

Cooke, P.S., Zhao, Y.D. and Hansen, L.G.: 1996, *Toxicol. Appl. Pharmacol.* **136**, 112-117.

Danzo, B.J.: 1997, *Environ. Health Perspect.* **105**, 294-301.

Davila, D.R., Davis D.P., Campbell, K., Cambier, J.C., Zigmond, L.A. and Burchiel, S.W.: 1995, *J. Toxicol. Environ. Health* **45**, 101-126.

Davis, D.L., Bradlow, H.L., Wolff, M., Woodruff, T., Hoel, D.G. and Anton-Culver, H.: 1993, *Environ. Health Perspect.* **101**: 372-377.

Dean, J.H., Cornacoff, J.B., Rosenthal, G.J. and Luster, M.I.: 1994, 'Immune System, evaluation of injury', in *Principals and Methods of Toxicology*, 3rd Edition, Hayes, A.W., ed., Raven Press, New York, pp. 105-190.

Donna, A., Betta, P.G., Robutti, F., Crosignani, P., Berrion, F. and Bellinger, D.: 1984, *Carcinogenesis* **5**, 941-942.

Doyle, T.J., Zheng, W., Cerhan, J.R., Hong, C.-P., Sellers, T.A., Kushi, L.H. and Folsom, A.R.: 1997, *AJPH* **87**, 1168-1176.

Ennever, F.K.: 1994, 'Metals', in *Principals and Methods of Toxicology*, 3rd Edition, Hayes, A.W., ed., Raven Press, New York, pp. 417-466.

Friedman, R.C. and Downey, J.: 1993, *J. Neuropsychiatry Clin. Neurosci.* **5**, 131-153.

Geschwind, S.A., Stolwijk, J.A.J., Bracken, M., Fitzgerald, E., Stark, A., Olsen, C. and Melius, J.: 1992, *Am. J. Epidemiol.* **135**, 1197-1207.

Gierthy, J.F., Arcaro, K.F. and Floyd, M.: 1997, *Chemosphere* 34: 1495-1505.

Gittelman, R., and Eskenazi, B.: 1983, *Arch. Gen. Psychiatr.* **40**, 827-833.

Gladen, B.C., Rogan, W.J., Hardy, P., Thullen, J., Tingelstad, J. and Tully, M.: 1988, *J. Pediatr.* **113**, 991-995.

Gold, E., Gordis, L., Tonascia, J. and Szklo, M.: 1979, *Am. J. Epidemiol.* **109**, 309-319.

Goldman, L.R., Paigen, B., Magnant, M.M. and Highland, J.H.: 1985, *Hazard. Waste Hazard. Mat.* **2**, 209-224.

Hauser, P., Soler, R., Brucker-Davis, J. and Weintraub, B.D.: 1997, *Psychoneuroendocrinology* **22**, 107-114.

Hauser, P., Zametkin, A.J., Martinez, P., Vitello, B., Matachik, J.A., Mixson, A.J. and Weintraub, B.D.: 1993, *N. Engl. J. Med.* **328**, 997-1001.

Henderson, B.E., Ross, R.K. and Pike, M.C.: 1991, *Science* **254**, 1131-1138.

Herbst, A., Ulfelder, H. and Posskanzer, D.: 1971, *N. Engl. J. Med.* **284**, 878-881.

Holene, E., Nafstad, I., Skaare, J.U., Bernhoft, A., Engen, P. and Sagnolden, T.: 1995, *Environ. Toxicol. Chem.* **14**, 967-976.

Hunter, D.J., Hankinson, S.E., Laden, F., Colditz, G.A., Manson, J.E., Willett, W.C., Speizer, F.E. and Wolff, M.S.: 1997, *N. Engl. J. Med.* **337**, 1253-1258.

Jacobson, J.L. and Jacobson, S.W.: 1996, *N. Engl. J. Med.* **335**, 783-789.

Jacobson, J.L., Jacobson, S.W. and Humphrey, H.E.B.: 1990, *J. Pediatr.* **116**, 38-45.

Janerich, D.T., Burnett, W.S., Feck, G., Hoff, M., Nasca, P., Polednak, A.P., Greenwald, P. and Vianna, N.: 1981, *Science* **212**, 1404-1407.

Jobling, S., Reynolds, T., White, R., Parker, M.G. and Sumpter, J.P.: 1995, *Environ. Health Perspect.* **103**, 582-587.

Jones, R.R.: 1987, *Lancet* **2**, 443-446.

Kavlock, R.J., Daston, G.P., *et al.*: 1996, *Environ. Health Persp.* **104** (Suppl 4), 715-740.

Koivusalo, M., Jaakkola, J.J.K., Vartiainen, T., Hakulinen, T., Karjalainen, S., Pukkala, E. and Tuomisto, J.: 1994, *AJPH* **84**, 1223-1228.

Lawrence, D.A.:1985, 'Immunotoxicity of heavy metals', in *Immunotoxicology and Immunopharmacology*, Dean, J. *et al.* ed., Raven Press, New York, pp. 341-353.

LeVay, S.A.: 1991, *Science* **253**, 1034-1037.

Lilienthal H., Neuf, M,. Munoz, C. and Winneke, G.: 1990, *Fundam. Appl. Toxicol.* **15**, 457-467.

Linet, M.S., Malker, H.S.R., McLaughlin, J.K., Weiner, J.A., Blot, W.J, Ericsson , J.L.E. and Fraumeni, J.F.: 1993, *Br. J. Indust. Med.* **50**, 79-84.

McKinney, J.D., Fawkes, J., Jordan, S., Chae, K., Oatley, S., Coleman, R.E. and Briner, W.: 1985, *Environ. Health Perspect.* **61**, 41-53.

Morse, D.C., Groen, D., Veerman, M., Van Amerongen, C.J., Koeter, H.B.W.M., Smits Van Prooije, A.E., Visser, T.J., Koeman, J.H. and Brouwer, A.: 1993, *Toxicol. Appl. Pharmacol.* **122**, 27-33.

Myers, G.J., Davidson, P.W., Weitzman, M. and Lanphear, B.P.: 1997, *MRDD Research Reviews* **3**, 239-245.

Needleman, H.L., Gunne, C., Leviton, A., Reed, R., Peresie, H., Maher, C. and Barrett, P.: 1979, *N. Engl. J. Med.* **300**, 689-695.

Newbold, R.: 1995, *Environ. Health Perspect.* **103**, 83-87.

Paigen, B., Goldman, L.R., Highland, J.H., Magnanat, M.M. and Steegman, A.T.: 1985, *Hazard. Waste Hazard. Mat.* **2**, 23-43.

Paigen, B., Goldman, L.R., Magnant, M., Highland, J.H. and Steegman, A.T.: 1987, *Hum. Biol.* **59**, 489-508.

Pantaleoni, G., Fanini, D., Spnta, A.M., Palumbo, G., Giorgi, R. and Adams, P.M.: 1988, *Fundam. Appl. Toxicol.* **11**, 440-449.

Porterfield, S.P.: 1994, *Environ. Health Persp.* **102**, Suppl. 2, 125-130.

Porterfield, S.P. and Hendrich, C.E.: 1993, *Endocrine Rev.* **14**, 344-356.

Preston-Martin, S., Yu, M.C., Benton, B. and Henderson B.E., 1982, *Cancer Res.* **42**, 5240-5245.

Rier, S.E., Martin, D.C., Bowman, R.E., Dnowski, W.P. and Becker, J.L.: 1993, *Fund. Appl. Toxicol.* **21,** 433-441.

Rogan, W.J., Gladen, B.C., Hung, K.-L., Koong, S.-L., Shih, L.-Y., Taylor, J.S., Wu, Y.-C., Yang, D., Ragan, N.B. and Hsu, C.-C.: 1988, *Science* **241**, 334-336.

Rothman, N., Cantor, K.P., Blair, A., Bush, D., Brock, J.W., Helzisouer, K., Zahn, S.H., Needham, L.L., Pearson, G.R., Hoover, R.N., Comstock, G.W. and Strickland, P.T.: 1997, *Lancet* **350**, 240-244.

Safe, S., Astroff, B., Harris, M., Zacharewski, T., Dickerson, R., Romkes, M. and Biegel, L.: 1991, *Pharmacol. Toxicol.* **69**, 400-409.

Sandler, D.P. and Collman, G.W.: 1987, *Am. J. Epidemiol.* **126**, 1017-1032.

Shiota, K.: 1976, *Okajimas Fol. Anat. Jpn.* **53**, 105-114.

Silbergeld, E. and Tonat, K.: 1994, *Toxicol. Indust. Health* **10**, 675-682.

Silkworth, J.B., Antrim, L. and Sack, G.: 1986, *Toxicol. Appl. Pharmacol.* **86**, 380-390.

Smith, P.J., Langolf, G.D. and Goldberg, J.: 1983, *Brit. J. Indust. Med.* **40**, 413-419.

Sosroseno, W.: 1995, *Asian Pacific J. Allergy Immunol.* **13**, 173-181.

Storm, J.E., Hart, J.L. and Smith R.F.: 1981, *Neurobehav. Toxicol. Teratol.* **3**, 5-9.

Swan, S.H., Elkin, E.P., and Fenster, L.: 1997, *Environm. Health Perspect.* **105**,1228-1232.

Toppari, J., Larsen, J.C., *et al.*: 1996, *Environ. Health Perspect.* **104** (Suppl. 4), 741-806.

van Wassenaer, A.G., Kok, J.H., De Vijlder, J.J.M., Briët, J.M., Smit, B.J., Tamminga, P., van Baar, A., Dekker, F.W. and Vulsma, T.: 1997, *N. Engl. J. Med.* **336**, 21-26.

Vianna, N. and Polan, A.K.: 1984, *Science* **226**, 1217-1219.

Vom Saal, F.S., Cooke, P.S., Buchanan, D.L., Palanza, P., Thayer, K.A., Nagel, S.C., Parmigiani, S. and Welshons, W.V.: 1998, *Toxicol. Indust. Health* **14**, 239-260.

Wolff, M.S., Paolo, G., Toniolo, P., Lee Eric, W., Rivera, M. and Dubin, N.: 1993, *J. Nat. Cancer Inst.* **85**, 648-652.

**SPRINGER NATURE**

# GPSR Compliance

*The European Union's (EU) General Product Safety Regulation (GPSR) is a set of rules that requires consumer products to be safe and our obligations to ensure this.*

*If you have any concerns about our products, you can contact us on ProductSafety@springernature.com*

In case Publisher is established outside the EU, the EU authorized representative is:

Springer Nature Customer Service Center GmbH
Europaplatz 3
69115 Heidelberg, Germany

FSC
www.fsc.org
MIX
Papier | Fördert
gute Waldnutzung
FSC® C083411

Kolibri 360 GmbH
Ferdinand-Jühlke-Straße 7
99095 Erfurt, Deutschland
produktsicherheit@kolibri360.de